CAUSATION AND CREATION IN LATE ANTIQUITY

Written by a group of leading scholars, this unique collection of essays investigates the views of both pagan and Christian philosophers on causation and the creation of the cosmos. Structured in two parts, the volume first looks at divine agency and how late antique thinkers, including the Stoics, Plotinus, Porphyry, Simplicius, Philoponus and Gregory of Nyssa, tackled questions such as: is the cosmos eternal? Did it come from nothing or from something pre-existing? How was it caused to come into existence? Is it material or immaterial? The second part looks at questions concerning human agency and responsibility, including the problem of evil and the nature of will, considering thinkers such as Plotinus, Porphyry, Proclus and Augustine. Highlighting some of the most important and interesting aspects of these philosophical debates, the volume will be of great interest to upper-level students and scholars of philosophy, classics, theology and ancient history.

ANNA MARMODORO is an Official Fellow in Philosophy at Corpus Christi College, University of Oxford. She is the author of *Aristotle on Perceiving Objects* (2014), and editor of a number of volumes including *The Author's Voice in Classical and Late Antiquity* (co-edited with Jonathan Hill, 2013) and *The Metaphysics of Powers: Their Grounding and their Manifestations* (2010).

BRIAN D. PRINCE is a Postdoctoral Researcher at the University of Oxford. He has published articles in journals including *Apeiron: A Journal for Ancient Philosophy* and *Science and Plato: The Internet Journal of the International Plato Society*.

CAUSATION AND CREATION IN LATE ANTIQUITY

EDITED BY

ANNA MARMODORO

and

BRIAN D. PRINCE

CAMBRIDGE
UNIVERSITY PRESS

CAMBRIDGE
UNIVERSITY PRESS

University Printing House, Cambridge CB2 8BS, United Kingdom

One Liberty Plaza, 20th Floor, New York, NY 10006, USA

477 Williamstown Road, Port Melbourne, VIC 3207, Australia

314-321, 3rd Floor, Plot 3, Splendor Forum, Jasola District Centre, New Delhi-110025, India

79 Anson Road, #06-04/06, Singapore 079906

Cambridge University Press is part of the University of Cambridge.

It furthers the University's mission by disseminating knowledge in the pursuit of education, learning and research at the highest international levels of excellence.

www.cambridge.org
Information on this title: www.cambridge.org/9781107695320

First published 2015
First paperback edition 2018

A catalogue record for this publication is available from the British Library

Library of Congress Cataloging in Publication data
Causation and creation in late antiquity / edited by Anna Marmodoro and Brian D. Prince.
pages cm
Includes bibliographical references and index.
ISBN 978-1-107-06153-8 (Hardback)
1. Causation. 2. Cosmogony. 3. Creation. 4. Cosmology.
I. Marmodoro, Anna, 1975– editor.
BD591.C38 2014
122.09–dc23 2014026342

ISBN 978-1-107-06153-8 Hardback
ISBN 978-1-107-69532-0 Paperback

Contents

PART II: THE ORIGINS OF HUMAN AGENCY

Contributors

HAN BALTUSSEN is the Hughes Professor of Classics at the University of Adelaide and a Fellow of the Australian Academy of the Humanities. He has published on a wide range of topics in the history of philosophy. He is the author, editor or translator of five books, including *Theophrastus Against the Presocratics and Plato* (2000), *Philosophy, Science and Exegesis in Greek, Latin and Arabic Commentaries* (with P. Adamson and M. Stone, 2004), *Philosophy and Exegesis in Simplicius: The Methodology of a Commentator* (2008), *Simplicius' Commentary on Aristotle's* Physics *1.5–9* (with M. Share, M. Atkinson and I. Mueller, 2012) and *Greek and Roman Consolations: Eight Studies of a Tradition and its Afterlife* (2013).

RICCARDO CHIARADONNA is Associate Professor of Ancient Philosophy at Roma Tre University. His publications include *Sostanza Movimento Analogia: Plotino Critico di Aristotele* (2002), *Plotino* (2009), *Physics and Philosophy of Nature in Greek Neoplatonism* (with F. Trabattoni, 2009), *Universals in Ancient Philosophy* (with G. Galluzzo, 2013), as well as numerous articles on Plotinus, Neoplatonism, the Ancient Commentators and Galen.

GILLIAN CLARK FBA is Professor Emerita of Ancient History at the University of Bristol. She works on the social and intellectual history of late antiquity. Her publications include *Porphyry: On Abstinence* (2000), *Christianity and Roman Society* (2004), *Late Antiquity: A Very Short Introduction* (2011), *Body and Gender, Soul and Reason in Late Antiquity* (2011), and many papers on Augustine. Her main work in progress is a commentary project on *City of God* (www.epiphanius.org). She also co-edits *Oxford Early Christian Studies* and *Translated Texts for Historians 300–800*.

KEVIN CORRIGAN is Samuel Candler Dobbs Professor of Interdisciplinary Humanities and Director of the Graduate Institute of the Liberal

Arts at Emory University, Atlanta, Georgia. His recent works include *Evagrius and Gregory: Mind, Soul and Body in the 4th Century* (2009), *Religion and Philosophy in the Platonic and Neoplatonic Traditions: From Antiquity to the Early Medieval Period* (edited with J. D. Turner and P. Wakefield, 2012) and *Reason, Faith and Otherness in Neoplatonic and Early Christian Thought* (2013).

MARK EDWARDS is Tutor in Theology at Christ Church, Oxford, and Lecturer in Patristic Theology in the Faculty of Theology and Religion at the University of Oxford. He is the author of *Neoplatonic Saints* (2000), *Origen against Plato* (2002), *John through the Centuries* (2003), *Constantine and Christendom* (2004), *Culture and Philosophy in the Age of Plotinus* (2006), *Catholicity and Heresy in the Early Church* (2009) and *Image, Word and God in the Early Christian Centuries* (2012).

MICHAEL GRIFFIN is Assistant Professor in Classics and Philosophy at the University of British Columbia. He has published on Neoplatonism and the Aristotelian tradition in antiquity, and is the author of *Aristotle's* Categories *in the Early Roman Empire* (2014). He is co-editor of the Ancient Commentators on Aristotle series. The first volume of his translation of Olympiodorus of Alexandria's commentary on the *Alcibiades I* is due to be published with the same series in 2014.

D. M. HUTCHINSON is Assistant Professor of Philosophy at St Olaf College. He specialises in Plotinus and late ancient philosophy, and has published on Plotinus. He is currently writing a book on Plotinus' theory of consciousness.

AARON P. JOHNSON teaches classics at Lee University and works on Greek literature in late antiquity. He has authored *Ethnicity and Argument in Eusebius'* Praeparatio Evangelica (2006), *Religion and Identity in Porphyry of Tyre* (2013) and *Eusebius* (2014), as well as several discussions of various aspects of intellectual culture in the third and fourth centuries.

ANNA MARMODORO is an Official Fellow in Philosophy at Corpus Christi College, University of Oxford. She specialises in ancient philosophy and contemporary metaphysics, with research interests also in medieval philosophy, the philosophy of mind and philosophy of religion. Her publications span all these areas. Recent work in press include a monograph on *Aristotle on Perceiving Objects* (2014), a number of edited volumes, among them *The Metaphysics of Relations* (with David

Yates, 2015) and *The Metaphysics of Powers: Their Grounding and their Manifestations* (2010), and numerous articles in peer-reviewed journals and edited collections. Dr Marmodoro directs a large research group investigating Power Structuralism in Ancient Ontologies, based in the Faculty of Philosophy of the University Oxford.

CHRISTOPHER ISAAC NOBLE is a Post-doctoral Fellow at the Ludwig-Maximilians-Universität München. He works primarily on Plotinus, and has published articles on various aspects of ancient Greek psychology, physics and cosmology.

NATHAN M. POWERS is an Associate Professor of Philosophy at the State University of New York at Albany. He works primarily on Plato and on Hellenistic philosophy, and has published articles on various aspects of ancient Greek natural philosophy, epistemology and philosophical theology.

RICARDO SALLES is Reseacher at the Institute for Philosophical Research of the Universidad Nacional Autónoma de México, a former Fellow of the Center of Hellenic Studies in Washington DC (2003–4) and a former Member of the Institute for Advanced Study in Princeton (2010–11). His work centres on ancient ethics, metaphysics and science. He is the author of *The Stoics on Determinism and Compatibilism* (2005), the co-author of *Alejandro de Afrodisia, Sobre el Destino: Introducción, Traducción y Notas* (2009), the editor of *God and Cosmos in Stoicism* (2009), and co-author of *Los Filósofos Estoicos: Ontología, Lógica, Física y Ética* (2014).

RICHARD SORABJI is Honorary Fellow of Wolfson College, Oxford. He is the author of 16 books, editor of 100 volumes of translation of later Greek philosophy, and editor or co-editor of another 10 volumes. These books include *Necessity, Cause and Blame* (1980), *Time, Creation and the Continuum* (1983), *Matter, Space and Motion* (1988), *Animal Minds and Human Morals* (1993), *Emotion and Peace of Mind: From Stoic Agitation to Christian Temptation* (2000), *The Philosophy of the Commentators, 200–600 AD: A Sourcebook* (3 vols., 2004), *Self: Ancient and Modern Insights about Individuality, Life and Death* (2006), *Opening Doors: The Untold Story of Cornelia Sorabji, Reformer, Lawyer and Champion of Women's Rights in India* (2010), *Gandhi and the Stoics: Modern Experiments on Ancient Values* (2012) and *Moral Conscience through the Ages* (2014).

JAMES WILBERDING is Professor of Ancient and Medieval Philosophy at the Ruhr University, Bochum. He is the author of numerous articles on ancient moral psychology and the intersection of philosophy and ancient science, especially medicine and embryology, with a particular focus on Plato, the Platonic tradition, and the ancient commentators on Aristotle. His publications include *Plotinus' Cosmology: A Study of Ennead ii.1* (2006), *Philoponus: Against Proclus on the Everlastingness of the World 12–18* (2006), *Porphyry: To Gaurus on How Embryos are Ensouled and What is in our Power* (2011), *Neoplatonism and the Philosophy of Nature* (with C. Horn, 2012) and *Philosophical Themes in Galen* (with P. Adamson and R. Hansberger, 2014).

Acknowledgements

This book project began with a seminar series co-organised by the editors and Dr Neil McLynn in Corpus Christi College, Oxford, to which the three of them are affiliated. The seminar series, as well the subsequent work on the volume itself, were part of Anna Marmodoro's ongoing project, Power Structuralism in Ancient Ontologies, supported by a starting investigator award (number 263484) from the European Research Council. The editors are grateful for the generous support of the European Research Council which made this work possible. They also want to express gratitude to their colleagues and seminar participants for the intellectual stimulation that engendered this project; and to the Cambridge University Press anonymous readers for many insightful and helpful suggestions on an earlier draft of the manuscript.

Introduction

Anna Marmodoro and Brian D. Prince

How did the cosmos come into existence? Is it eternal? Was it created *ex nihilo*? If so, how? Or was it created from something that already existed? But then what was it created from, and by whom or what? Further, was this creation an instance of the same kind of causation we observe in the world we inhabit? Or was it another, special type of causation instead? Throughout the late antique period, from roughly the third through the seventh centuries CE, philosophers of all schools had something to say on these questions.

Friction between Christian and pagan philosophers produced many of the new ideas of this period, some of which disappeared from history after a short time, while others enjoyed centuries of currency. The philosophical community on which this book focuses is a loosely knit group of thinkers in mutual conversations, overlapping both geographically and temporally, from Gaza in the east, to North Africa in the south, to Italy in the north and west. Christian philosophers faced a particular difficulty with questions about creation, as their commitments bade them reconcile the emerging doctrine of creation *ex nihilo* in the Church with the strenuous denial of this – even as a metaphysical possibility – by the Greek philosophical heritage that they also wanted to accept. Since these thinkers had been educated in the tradition of Greek philosophy as well as in their (relatively) new religion, the issue was one of pressing urgency, and they had no choice but to try out innovative solutions, since there were simply no predecessors similarly situated to whom they could look.

The picture that emerges is of two partly overlapping communities of thinkers grappling with many of the same issues, drawing on the same philosophical heritage but applying different further assumptions. Readers will find some well-known debates and views in this volume, but also, we trust, many that are less familiar because they were dropped rather than developed in the history of Western thought, and which therefore come to us as new thoughts after centuries of dormancy.

We can think of the choices involved in reaching a philosophical view in this area as beginning with the choice between a universe that has always existed and one that has been created. For those choosing creation, the next step is to opt for creation from nothing or from something, and then to describe the process or mechanism by which creation was accomplished. For all Christian and at least some Neoplatonic philosophers, the cosmos was brought into existence by a divine being's creative act. The Neoplatonic thinkers especially drew on Plato's views in the *Timaeus*, which can be read with equal plausibility as endorsing either creation from pre-existing stuff or a non-creation account. The Christian thinkers, of course, do not take Plato as an authority on the question of how the universe was created, but nevertheless respond to him indirectly when engaging with Neoplatonists who do.

Viewed in one way, causation is a genus including creation as a special case; on this view, causation is the more basic concept, because more general. But conversely, creation may be taken as the more basic notion, either because creation happens first temporally or because it is prior in a metaphysical sense. From this perspective, understanding creation is the more basic task, on which explanations of *intra-cosmic* causation will be built afterward. From either perspective, though, causation and creation are closely linked. For example, if God creates the cosmos in a certain way, certain implications will follow about the nature of that cosmos (see Chapter 5), and thus for the behaviour of its inhabitants (see Chapters 10 and 11), including both (merely) physical objects (see Chapter 1) and human (and other) agents (see Chapters 12 and 13). Thinkers in both traditions, Neoplatonist and Christian, were concerned to understand the human situation in a certain light, and to produce corresponding prescriptions that would be grounded in their picture of the cosmos. We thus also reach questions about human agency and willing, as species of causation, familiar to other philosophical ages. Where does the human soul come from as an embryo/foetus grows and is born (see Chapter 9)? What is the (eventually) resulting human agent (see Chapter 11)? What can we learn about human agency by comparing it with the agency of God and that of angels? What allows humans to be morally responsible for their actions? How is evil possible, compatible with the assumption that God is the cause of everything in creation, without attributing evil to God? As with creation, we see both pagan and Christian thinkers grappling with similar issues, and offering a variety of paths into and out of them.

This volume divides its contributions into two groups: as focusing more on the creation of the cosmos, or more on issues of human agency and responsibility, although the close connections between these issues lie on the surface of many chapters. The volume does not aim to offer an exhaustive treatment of any of the subjects addressed, but rather to demonstrate that many important questions from this period remain unexplored. Our larger goal is to excite further interest, both in the particular thinkers discussed here and in their peers who cannot be included. Perhaps the most glaring omission is the third-century Christian philosopher Origen, whose views on the creation of the universe and on the causation involved in moral agency would be prominent in any complete treatment of these issues. These topics are too vast for a single volume to deal with adequately; the richness of material included here should not be taken to suggest that we have told the whole story.

Six chapters in Part I of this book examine questions about *the origin of the cosmos.* Chapter 1 looks at early Stoic accounts of how the cosmos begins, focusing on the genesis of the four basic elements. Chapters 2 and 3 discuss different aspects of Plotinus' startling view that the demiurge creates the cosmos without any conscious planning. Chapter 4 looks at arguments formulated by three Christian philosophers to the effect that the world has been created, drawing from their work many insights into their society, education and relations to other thinkers. Chapter 5 takes up a problem examined by the Christian Gregory of Nyssa: if God is immaterial, how does He create a material world, given that causation seems to involve the widely held principle that 'like causes like'? Finally, Chapter 6 shows how Simplicius used his predecessors from the whole earlier philosophical tradition to arrive at his own views about creation. Chapters 1, 2, 4 and 6 examine their subjects more by looking at historical developments, while Chapters 3 and 5 focus on giving philosophical analyses.

In Chapter 1, 'Two early Stoic theories of cosmogony', Ricardo Salles looks at the Stoic view that the cosmos is periodically destroyed by fire, and then recreated. Salles focuses on the creation of the basic elements (earth, water, air and fire). Two competing accounts arose, Zeno and Chrysippus teaching that the fire which engulfs a cosmos at the end of its life is completely extinguished before the next cosmos comes to be, but Cleanthes arguing that the fire is never completely put out, so that at least some portion of fire exists continuously from one cosmos to the next. Salles traces these differences to differing views about the four elements. Finally, he examines how the two accounts explained the creation of the sun.

In this chapter we have a link between Greek philosophy of the classical period – since the Stoics built their theory of creation by developing ideas from Plato's *Timaeus* – and the late antique period – since Stoic views remained a strong influence throughout this period. Philoponus, for example, has sometimes been characterised as a Christian Stoic (Sorabji 1983: 34). In addition, Stoicism took a different path from other schools in its choice to advocate creation, but a creation from pre-existing stuff in which both the 'before' and 'after' stages are material. This can be profitably set beside both the Neoplatonic and Christian accounts of creation. As the following chapters show, Plotinus faces the difficult issue of how to describe the creator's intentionality, while Christian accounts were sometimes felt to run into a problem because God seems to be creating something – matter – which is completely unlike Himself. The Stoic account described in Chapter 1 holds out an alternative on which *these* problems are minimised, although others may be rendered more grave. In spite of the temporal distance from the thinkers of other chapters, Salles' new account of classical Stoic ideas about creation forms an important counterpoint for the rest of the volume.

Plotinus' thought permeates the late antique period because of its originality and depth. His influence is also due to the manifold ways he could be, and was, read. For example, one might justifiably read Plotinus as holding that creation was a sort of 'automatic' process, devoid of intentionality yet still teleological. Riccardo Chiaradonna discusses this view of Plotinus in Chapter 2, 'Plotinus' account of demiurgic causation and its philosophical background', looking back at the debates to which Plotinus was responding as he formulated his views about the creation of our world. Did Plotinus adopt this view mainly because he wanted to reject Gnostic views, according to which the world was created by a god, but one who was either incompetent or malevolent? Chiaradonna argues that this is not Plotinus' principal motivation (but cf. Chapter 7), and that he was instead responding to debates between Platonists and Aristotelians from the second century.

In Chapter 3, 'Creation and divine providence in Plotinus', Christopher Isaac Noble and Nathan M. Powers analyse the premises that Plotinus relies on in order to clarify his innovative and influential account of creation without deliberation or planning – an account that represents a remarkably austere reinterpretation of the 'providential agency' of Plato's divine craftsman.

In Chapter 4, 'Waiting for Philoponus', Richard Sorabji expounds and examines the arguments produced by three mostly forgotten Christian philosophers from Gaza between 485 and 529 CE. Aeneas, Zacharias and Procopius respond to Proclus' arguments for the eternity of the world, trying out responses that Philoponus would refine and use to better effect. The exchange is illuminating partly for the ways in which Philoponus' precursors fail to find the more penetrating arguments he would employ, but also yields a wealth of detailed observations about the philosophical training and religious surroundings of the Christian thinkers. Because these thinkers knew and borrowed from one another, we can also, to some extent, watch their arguments grow from initial ideas to more refined and effective presentations, as each learns from his predecessors and takes some steps beyond what they had accomplished.

Meanwhile, in the late fourth century the Christian philosopher Gregory of Nyssa was running into a similar puzzle in his thinking about creation. If an immaterial God creates a material cosmos, then it looks as if either the causal principle that 'like causes like' isn't true, or else the intuitive and plausible view that the world is material turns out to be false. Since the principle that like causes like enjoyed a broad consensus in antiquity, neither horn of this dilemma was attractive. Gregory shows us one way of denying both theses, as we see in Anna Marmodoro's Chapter 5, 'Gregory of Nyssa on the creation of the world'. Gregory's discussion of the intelligibility of the qualities of material objects addresses the same thorny problem that Plato's puzzle of the self-predication of the Forms had first raised in philosophy, albeit in a different context. Gregory's contribution reveals intuitions about the individuation of abstract entities that have only recently resurfaced in the metaphysics of abstraction. Gregory's philosophical work does not tower over later centuries in the way that Plotinus does, but he draws our attention in this volume for equally good reasons. He exercised significant influence on later thinkers, even enjoying some continuing influence in the Latin west via John Scotus Eriugena in the ninth century. Gregory's philosophical ability and innovations have been too often overlooked or dismissed: his thought contains views that have not so far been sufficiently appreciated, and contribute interesting new possibilities to debates about creation. Other Christian thinkers of late antiquity are capable of furnishing further examples of overlooked philosophical views; here we offer Gregory as one case among others worthy of re-examination.

Part I closes with 'Simplicius on elements and causes in Greek philosophy: critical appraisal or philosophical synthesis?' by Han Baltussen (Chapter 6). Baltussen shows how Simplicius' astoundingly ambitious project in authoring commentaries led to the development of his own views about creation and causal principles: Simplicius wanted to produce nothing less than a synthesis of *all* previous Greek thinking, and not just a synthesis, but one that would show how all previous thinkers had been in harmony with one another. The result is a version of Aristotle's views, and yet also belongs distinctively to the sixth century CE.

Part II turns the focus to human agency in seven chapters examining how humans – and other agents – are able to bring about various causal results, including their own actions, their status as unified agents, and evil. Chapter 7 returns to Plotinus to examine three kinds of causation: creative contemplation, non-deliberative demiurgic production (for which cf. Chapters 2 and 3) and the relation of agency to divine self-causality. Chapter 8 is also concerned with Plotinus' view of agency; it argues that Plotinus introduces an unprecedented degree of inwardness as he claims that the principles of reality are located *in us*. Chapter 9 examines Neoplatonist theories of embryology, showing a fascinating parallel in explanations of plant life and of the growth of unborn animals. In Chapter 10 we look at fragments from an otherwise lost work of Porphyry of Tyre; unlike some portraits of Porphyry, which attribute views to him that seem to block freedom of action for humans, this work shows him taking human freedom seriously. Chapter 11 shows how Proclus' version of Neoplatonism furnishes him with a ready solution to a problem about ethical self-constitution raised recently by Christine Korsgaard. Chapter 12 looks at Augustine's account of the creation of angels, in particular what about that creation made evil possible. Finally, Chapter 13 examines Augustine's views about the connections between sin, guilt and desire.

Kevin Corrigan offers a subtle and wide-ranging study of Plotinus' developing views about creative agency in Chapter 7, 'Divine and human freedom: Plotinus' new understanding of creative agency'. Corrigan locates the heart of Plotinus' concerns in the relation between creation and production, arguing that in response to this basic preoccupation Plotinus developed three new theoretical models: first, a theory of creative contemplation or insight; second, a model of non-deliberative demiurgic production; and third, a view about agency and divine self-causality. Corrigan not only analyses Plotinus' development of these models but at the same time draws many illuminating comparisons with Plotinus' predecessors and successors.

In Chapter 8, 'Consciousness and agency in Plotinus', D. M. Hutchinson argues that Plotinus' views license and require a greater degree of inwardness than those of any of his predecessors. Hutchinson shows how Plotinus' account of consciousness produces the need to 'turn inwards and ascend upwards' in order to achieve full agency: awareness is a *sine qua non* for initiating actions and unifying our bodies. But of course Plotinus does not mean awareness of physical surroundings: instead of this, we must become aware of the higher principles in us and in the universe. Only this consciousness can form the basis for truly free actions. Hutchinson concludes that this is a greater demand for, and theory of, inwardness than was available in earlier thinkers.

James Wilberding examines the fascinating connections between theories of planthood and of developing embryos in Neoplatonic thought in Chapter 9, 'Neoplatonists on the causes of vegetative life'. Many philosophers had drawn this analogy previously, but the Neoplatonists take it in a new and surprising direction. Since the embryo is dependent on its mother not only for nourishment but also in order to remain alive, the Neoplatonists concluded that the vegetative soul of a plant exists not in the plant itself, but in the earth.

Chapter 10, 'Astrology and the will in Porphyry of Tyre', by Aaron P. Johnson, aims to balance a certain picture of Porphyry as granting too much credence to the 'sciences' of his day such as astrology. Johnson argues that, on the contrary, fragments of one of Porphyry's works show him firmly committed to enough freedom of action for humans to render us moral agents. Moreover, we can discern Porphyry's attempt to given an account consistent with Plato's Myth of Er, that is, with the idea that we arrive in our present lives by making a choice in the interim following a past life – a choice that is not perfectly free or informed, but nevertheless gives us enough freedom that our current lives are our own responsibility.

In Chapter 11, 'Proclus on the ethics of self-constitution', Michael Griffin looks at how we become moral agents from a different perspective. He connects this problem with recent discussions by Christine Korsgaard and others of a seeming paradox: in order to become moral agents we must achieve a kind of unity in our agency. But such unity can only be produced by an agent who is sufficiently unified to act in a unified way. Since we do not begin life as agents meeting this standard, it is puzzling how we are ever able to achieve it. Griffin, however, shows how Proclus' version of Neoplatonism has resources to solve this puzzle by appealing to the real self that exists already, even while one's phenomenal self is not yet unified.

The final two chapters turn to the other giant of late antique thought, Augustine. Augustine commands the attention of later thinkers in a way similar to Plotinus. His thinking about creation prioritises the origin of evils, and consequently also the nature of the will. How can God's creation of everything in the universe be consistent with the existence of evils within that universe? What sort of thing is a will, such that it can both be created by God, who creates only good things, and yet also be capable of turning away from God to produce evil?

Gillian Clark is the author of Chapter 12, 'Deficient causes: Augustine on creation and angels'. Here she explicates Augustine's answer to the question of the origin of evil. Evil is able to exist, Augustine decides, because the angels (as well as the rest of creation) were created *from nothing*. Thus the doctrine of creation *ex nihilo* returns here from Part I (especially in Chapter 5), this time explaining the origin and possibility of evil.

Mark Edwards brings this volume to a close with Chapter 13, 'Willed causes and causal willing in Augustine'. Edwards gives a nuanced examination of three areas of Augustine's thought, in each comparing the saint with the sources on which he drew and was most familiar. First, Augustine is a subtle philosopher of the will but relies in the first instance on theological, not philosophical, principles. Second, while Roman history in particular offered Augustine the example of a people with a habit of attributing bad acts to their own ancestors, Augustine's careful treatment of sin and evil is nevertheless unprecedented. Finally, earlier ideas about the generation of embryos had sometimes included claims that physical or psychic characteristics could be inherited; Augustine contributes a new kind of argument, however, to justify his view that the very *will* to sin is heritable.

PART I

The Origin of the Cosmos

CHAPTER I

Two early Stoic theories of cosmogony

Ricardo Salles

Stoic thinking on the cosmogony is firmly grounded on the Platonic idea that the cosmos was created by a divine demiurge from pre-existing matter. In Stoic ontology this demiurge and this matter, called 'god' (θεός) and 'matter' (ὕλη), are in fact the two fundamental principles of corporeal reality. They are, unlike everything else, absolutely primitive entities in the sense that all else is ultimately produced by the action of the former on the latter. The Stoics devoted much attention to how particular kinds of bodies were created by the action of god on matter, and their theories of the cosmogony envisage a sequence of three different stages. The first is one at which (a) the simplest homogeneous bodies – the four elements: fire, air, water and earth – come into being by the contraction and expansion of matter by god. The second is where (b) the complex homogeneous bodies – gold, flesh, wood and the like – are created out of the simple ones by mixture; and the third is the one at which (c) the heterogeneous bodies – e.g. animals and plants – are created by the assemblage of complex homogeneous bodies. There is a basic account of these three stages that was shared by nearly all major early Stoics.[1] However, as I shall argue, there was also a great deal of polemic, especially in connection with stage (a). According to Stoic cosmology the present cosmos and its cosmogony were preceded by a conflagration that destroyed a previous cosmos. But if so, when exactly did stage (a) begin? Did it begin once the fire of the conflagration was extinguished? Or was it rather just

Most of the material contained in this chapter was presented at the Centre Léon Robin, Paris, in May 2013, at the Universidad de Rosario, Argentina, in August 2013 and at the Institute of Philosophical Research of the UNAM, Mexico, in March and November 2013. I am grateful to these audiences for their questions and, especially, to Stephen Menn and André Laks for their written comments. The editors of the present volume suggested several editorial changes that greatly improved the original version of this chapter. Research supported by PAPIIT 400914 and CONACYT 221268.

[1] The agreement includes at least the doctrine of the two principles in (a), for which cf. especially *SVF* 2.526, 527, 528, 555, 580, 634 and 642, and the account of complex homogeneous bodies in terms of mixture in (b), for which cf. especially *SVF* 2.465, 470, 473, 479, 480, 555.

before that? In other words, is there any overlap between the end of the conflagration and the beginning of the cosmogony? In what follows, I argue that these questions divided the early Stoics into two parties.

In the first section I present one of the two parties in the dispute: that formed by Zeno and Chrysippus. According to Zeno, later followed by Chrysippus, the cosmogony begins after the fire of the conflagration is totally extinguished and, in particular, when the heat left by the conflagration cools down and becomes water; part of this water is transformed into air and part of this air into fire. In the second section I deal with the second party, formed by Cleanthes, the successor of Zeno as head of the Stoa. According to Cleanthes, the cosmogony begins before the fire of the conflagration is totally extinguished. In fact, one portion of this fire is never put out, and the current celestial fire is directly made of this original fire. This will lead us in the third section to the question of why Zeno argued in the first place that the fire of the conflagration must be totally extinct at some point (why this fire cannot burn forever), and to the question of how Cleanthes addressed this argument. As will be seen, the root of the disagreement may have been that each party had different conceptions of the phenomenon of combustion and, more generally, of the relation between fire and the other three elements.

The cosmogony in Zeno and Chrysippus

The main evidence for Zeno's account of the first stage of the cosmogony comes from three connected texts.

T1: DL *7.142 (*SVF *1.102 and 2.581;* LS *46c;* BS *15.2)*

Γίνεσθαι δὲ τὸν κόσμον ὅταν ἐκ πυρὸς ἡ οὐσία τραπῇ δι᾽ ἀέρος εἰς ὑγρότητα, εἶτα τὸ παχυμερὲς αὐτοῦ συστὰν ἀποτελεσθῇ γῆ, τὸ δὲ λεπτομερὲς ἐξαερωθῇ, καὶ τοῦτ᾽ ἐπὶ πλέον λεπτυνθὲν πῦρ ἀπογεννήσῃ. εἶτα κατὰ μίξιν ἐκ τούτων φυτά τε καὶ ζῷα καὶ τὰ ἄλλα γένη. περὶ δὴ οὖν τῆς γενέσεως καὶ φθορᾶς τοῦ κόσμου φησὶ Ζήνων μὲν ἐν τῷ Περὶ ὅλου, Χρύσιππος δ᾽ ἐν τῷ πρώτῳ τῶν Φυσικῶν καὶ Ποσειδώνιος ἐν πρώτῳ Περὶ κόσμου καὶ Κλεάνθης καὶ Ἀντίπατρος ἐν τῷ δεκάτῳ Περὶ κόσμου. Παναίτιος δ᾽ ἄφθαρτον ἀπεφήνατο τὸν κόσμον.

1 ἐκ πυρὸς om. Φ **2** ὑγρότητα BD : ὑγρὸν PFΦ **3** ἐξαερωθῇ ΦP[4]H : ἐξαραιωθῇ BP[1]QFD **5** γενέσεως τε καὶ φθορᾶς BP[1] **6** μὲν om. F / καὶ Χρύσιππος F **6–7** δ᾽ ἐν τῷ πρώτῳ τῶν Φυσικῶν om. F **8** καὶ Κλεάνθης – κόσμου om. F / δεκάτῳ BPD : del. von Arnim

The cosmos is created when the substance is turned from fire through air into moisture; then the thicker parts of the moisture condense and end up as earth, but the finer parts are thoroughly rarefied, and when they have been thinned still further, they produce fire. Thereafter by mixture plants and animals and the other natural kinds are produced out of these. Zeno speaks of the generation and destruction of the cosmos, in *On the Whole*, Chrysippus in the first book of the *Physics*, Posidonius in the first book of *On the Cosmos*, Cleanthes, and Antipater in the tenth book of *On the Cosmos*. Panaetius, however, declared that the cosmos is indestructible.

T2*: DL 7.135–6 (SVF 1.102 and 2.580; LS 46b; BS 15.3)*

Ἕν τ᾽ εἶναι θεὸν καὶ νοῦν καὶ εἱμαρμένην καὶ Δία· πολλαῖς
τ᾽ ἑτέραις ὀνομασίαις προσονομάζεσθαι. κατ᾽ ἀρχὰς μὲν οὖν καθ᾽
αὑτὸν ὄντα τρέπειν τὴν πᾶσαν οὐσίαν δι᾽ ἀέρος εἰς ὕδωρ· καὶ
ὥσπερ ἐν τῇ γονῇ τὸ σπέρμα <τῷ ὑγρῷ> περιέχεται, οὕτω καὶ τοῦτον
σπερματικὸν λόγον ὄντα τοῦ κόσμου, τοιόνδε ὑπολείπεσθαι ἐν τῷ
ὑγρῷ, εὐεργὸν αὑτῷ ποιοῦντα τὴν ὕλην πρὸς τὴν τῶν ἑξῆς γένεσιν
εἶτ᾽ ἀπογεννᾶν πρῶτον τὰ τέσσαρα στοιχεῖα, πῦρ, ὕδωρ, ἀέρα,
γῆν. λέγει δὲ περὶ αὐτῶν Ζήνων τ᾽ ἐν τῷ Περὶ τοῦ ὅλου καὶ
Χρύσιππος ἐν τῇ πρώτῃ τῶν Φυσικῶν καὶ Ἀρχέδημος ἔν τινι
Περὶ στοιχείων.

1–2 πολλάς τε ἑτέρας ὀνομασίας B **3** πρέπειν B **4** <τῷ ὑγρῷ> add. Marcovich in app. crit. **5** τοιόνδε DP^{4}H : τοιοῦτον B : τοιοῦδε P^{1}F : om. Φ / ὑπολείπεσθαι P^{1}QFΦ : ὑπολίπεσθαι B : ὑπολιπέσθαι von Arnim cum P^{4}HD **6** εὐεργὸν BPFD : ἐνεργὸν Φ/ αὑτῷ von Arnim : αὐτῶ BPD : αὐτὸ F **8–10** λέγει —στοιχείων in mg. super. F^2

[They also say that] god, intelligence, fate, and Zeus are all one, and many other names are applied to him. In the beginning all by itself he turned the entire substance through air into water. Just as the sperm is enveloped in the seminal fluid, so god, who is the seminal principle of the cosmos, stays behind as such in the moisture, making matter serviceable to himself for the successive stages of creation. He then creates first of all the four elements, fire, water, air, earth. They are referred to by Zeno in *On the Whole* and Chrysippus in the first part of the *Physics*, and by Archedemus in a treatise *On the Elements*.

T3*: Stobaeus* Ekl. *(SVF 1.101, 1.497 and 2.471; BS 15.5)*

Ζήνωνα δὲ οὕτως ἀποφαίνεσθαι διαρρήδην· "τοιαύ-
την δὲ δεήσει εἶναι ἐν περιόδῳ τὴν τοῦ ὅλου διακόσμησιν
ἐκ τῆς οὐσίας, ὅταν ἐκ πυρὸς τροπὴ εἰς ὕδωρ δι᾽ ἀέρος
γένηται, τὸ μέν τι ὑφίστασθαι καὶ γῆν συνίστασθαι, [καὶ]
ἐκ τοῦ λοιποῦ δὲ τὸ μὲν διαμένειν ὕδωρ, ἐκ δὲ τοῦ ἀτμι-
ζομένου ἀέρα γίγνεσθαι, λεπτυνομένου δὲ τοῦ ἀέρος πῦρ

ἐξάπτεσθαι, τὴν δὲ μῖξιν <καὶ> κρᾶσιν γίγνεσθαι τῇ εἰς
ἄλληλα τῶν στοιχείων μεταβολῇ σώματος ὅλου δι' ὅλου
τινὸς ἑτέρου διερχομένου."
Κλεάνθης δὲ οὕτω
πώς φησιν· ἐκφλογισθέντος τοῦ παντὸς συνίζειν τὸ μέσον
αὐτοῦ πρῶτον, εἶτα τὰ ἐχόμενα ἀποσβέννυσθαι δι' ὅλου.
Τοῦ δὲ παντὸς ἐξυγρανθέντος τὸ ἔσχατον τοῦ πυρός, ἀντι-
τυπήσαντος αὐτῷ τοῦ μέσου, τρέπεσθαι πάλιν εἰς τοὐ-
ναντίον, εἶθ' οὕτω τρεπόμενον ἄνω φησὶν αὔξεσθαι καὶ
ἄρχεσθαι διακοσμεῖν τὸ ὅλον· καὶ τοιαύτην περίοδον αἰεὶ
καὶ διακόσμησιν ποιουμένου <τὸν ἐν τῇ τῶν ὅλων οὐσίᾳ
τόνον> μὴ παύεσθαι. Ὥσπερ γὰρ ἑνός τινος τὰ μέρη πάντα
φύεται ἐκ σπερμάτων ἐν τοῖς καθήκουσι χρόνοις, οὕτω
καὶ τοῦ ὅλου τὰ μέρη, ὧν καὶ τὰ ζῷα καὶ τὰ φυτὰ ὄντα
τυγχάνει, ἐν τοῖς καθήκουσι χρόνοις φύεται. Καὶ ὥσπερ
τινὲς λόγοι τῶν μερῶν εἰς σπέρμα συνιόντες μίγνυνται
καὶ αὖθις διακρίνονται γενομένων τῶν μερῶν, οὕτως ἐξ
ἑνός τε πάντα γίνεσθαι καὶ ἐκ πάντων [εἰς] ἓν συγκρί-
νεσθαι, ὁδῷ καὶ συμφώνως διεξιούσης τῆς περιόδου.

1 Ζήνωνος lemma add. P **2** δὲ del. Heeren / ante τὴν add. Diels τινὶ **3** τροπῇ FP : τροπὴ Heeren : τραπῇ Meineke (del. γένηται) **4** ὑφίστασται FP corr. Canter / [καὶ] del. Heeren **6** λεπτυνομένου δὲ τοῦ ἀέρος Wachsmuth : ἔκ τινος δὲ τοῦ ἀέρος FP : λεπτυνθέντα δὲ τὸν ἀέρα Usener **7** ἐξάπτειν FP : ἐξάπτεσθαι Diels : ἐξᾴπτειν Meineke / <καὶ> add. Diels **9** τίνας F **10** Κλεάνθου lemma add. P **13** ἐξυγρανθέντος Wachsmuth : ἐξυγραθέντος P **15** τρεπομένου FP corr. Canter **16** τοσαύτην FP corr. Meineke **17–18** τὸν — τόνον (sive τοὺς — τόνους) Meineke : τόνου FP **22** σπέρμα F : σπέρματα P **23** γενομένων F : γινομένων P corr. Meineke **24** τε P^{2} : τι FP^{1} / ἐκ πάντων del. Hirzel / [εἰς] secl. Meineke

And Zeno explicitly argues as follows: it will be necessary that the ordering of the cosmos out of substance that takes place periodically be such that, whenever a change of fire into water occurs through air, one part of it is solidified and becomes earth and, of the remainder, one part stays as water. From the evaporation of the other part, air is generated and, when air is rarefied, fire is ignited. But mixture and blending occurs by means of the reciprocal change of the elements, when a whole body passes through a whole body. Cleanthes, on the other hand, speaks somehow in the following way: once the whole has burnt up, first its centre collapses and, next, the parts contiguous to the centre are completely quenched. And once the whole has been dampened, the last portion of fire, given the resistance of the centre to it, moves away in the opposite direction. And then he says that, when it moves upwards, it grows in size and begins the ordering of the cosmos. The tension that exists in the substance of the whole, being what carries out this kind of periodical process and ordering, never stops.

> As a matter of fact, just as all the parts of a single thing grow at appointed times, so too the parts of the whole, among which there are animals and plants, grow at appointed times. And just as some reasons of the parts, coming together into a seed, are mixed and again separated when the parts are generated, so too all things are generated from one single thing and one single thing is formed out of all things, and the period follows its course methodically and harmoniously.

The theory reported in T1 and T2 is also referred to in lines 1–9 of T3.[2] The three texts identify Zeno as its propounder, which is strong evidence that he was indeed its originator. T1 and T2 also suggest that this theory was subsequently adopted by other major Stoics. T1 mentions Cleanthes, Chrysippus, Antipater and Posidonius, and T2 refers to Chrysippus and Archedemus. However, in contrast with T1, T3 attributes to Cleanthes a theory which happens to be fundamentally different from the theory that the three texts attribute to Zeno. Therefore, if we trust T3 and accept that Cleanthes was indeed the author of this alternative theory, we are forced to conclude that T1 is wrong in saying that Cleanthes accepted Zeno's theory. And if so, Cleanthes' position was unique compared to that of Zeno and the other major Stoics mentioned in T1 and T2. As I shall argue further in this section, there is also good evidence that Chrysippus made an important contribution to the debate. However, unlike Cleanthes, who departed from Zeno on fundamental questions, Chrysippus accepted Zeno's theory by and large.

Let us deal first with Zeno's theory. In order to understand it fully, it is necessary to remember what the conflagration in Stoic cosmology is.[3] The conflagration, or ἐκπύρωσις, is a state of the cosmos in which all bodies are consumed by fire. The cause of this fire is the heavens. Celestial fire is essentially a designing or constructive fire (πῦρ τεχνικόν: *SVF* 1.120), for it is responsible for the heat that sustains the cosmos. This heat, however, gradually desiccates the cosmos until it is completely dry and, therefore, a point is reached where this heat ignites the cosmos and the flames that set upon the cosmos totally consume it.[4]

Thus, some sources, which I quote below, refer to the conflagration as a state where the cosmos is 'fiery through and through' and in which 'nothing remains but fire'.

[2] See also *SVF* 2.555.

[3] Four important works on the Stoic theory of the conflagration are Mansfeld 1979 and 1981, Long 1985 and Gourinat 2002. I have developed my own view on the subject in Salles 2009b.

[4] I have dealt with this issue in Salles 2005.

T4: *Plutarch* Stoic. rep. *1053b–c (*SVF *2.605;* LS *46f)*

λέγει δ'
ἐν τῷ πρώτῳ περὶ Προνοίας· "διόλου μὲν γὰρ ἂν ὁ κόσμος πυρώδης <γένηται>, εὐθὺς καὶ ψυχή ἐστιν ἑαυτοῦ καὶ ἡγεμονικόν· ὅτε δέ, μεταβαλὼν εἴς τε τὸ ὑγρὸν <καὶ τὸ γεῶδες> καὶ τὴν ἐναπολειφθεῖσαν ψυχήν, τρόπον τινὰ εἰς σῶμα καὶ ψυχὴν μετέβαλεν ὥστε συνεστάναι ἐκ τούτων, ἄλλον τινὰ ἔσχε λόγον."

2 γὰρ om. g / ἂν FXg : ὢν Cherniss et Long-Sedley cum AβγNBE **3** <γένηται> add. Pohlenz **4–5** <καὶ τὸ γεῶδες> add. Pohlenz **6** μετέβαλεν Reiske : μεταβάλοι gX^3 : μεταβάλλων codd. alt. / συνιστάναι X^{1}F

In book one of *On Providence* he [Chrysippus] says: 'when the cosmos is fiery through and through it is immediately both its own soul and ruling part. But when, having changed into moisture, into what is earthlike, and into the soul which remains therein, it has in a way changed into body and soul so as to be compounded out of these, it has got a different definition.'

T5: *Cicero* ND *2.118 (*SVF *2.593;* BS *18.2)*

Ex quo eventurum nostri putant id, de quo Panaetium addubitare dicebant, ut ad extremum omnis mundus ignesceret, cum umore consumpto neque terra ali posset nec remearet aer, cuius ortus aqua omni exhausta esse non posset: ita relinqui nihil praeter ignem, a quo rursum animante ac deo renovatio mundi fieret atque idem ornatus oreretur.

As a consequence of what will happen, our philosophers think what they said Panaetius questioned, namely, that the whole cosmos will be ignited, for once the humidity has been consumed, neither will the earth be able to be nourished nor will the air be able to flow, being unable to rise upwards once all water has been consumed; thus nothing will remain but fire. But thanks to it – an animated being and a god – a restoration will take place and the cosmos itself will rise again in an orderly way.

As we know from other sources, one of which will be examined later on, the fire of the conflagration is extinguished when the flames that consume the cosmos run out of fuel and die, and when the mass of heat that is left behind as a result of this extinction gradually cools down.[5] This flameless heat is still part of the conflagration, for in Stoic theory heat is a form of fire and when the flames of the conflagration have died and the cosmos is totally occupied by flameless heat, the cosmos is still 'fiery through and

[5] For the evidence regarding how the conflagration works see notably *SVF* 1.511, 2.593–4, 604–5, 611–12 and 1068. One important passage (missed by von Arnim) is Alexander Lycopolis *Man.* 19.2–12 Brinkmann (quoted below as T11 and discussed in the third section).

through'. According to Zeno the fire of the conflagration, in either of its two forms, flame or heat, is totally extinguished before the cosmogony sets off. This is strongly implied in T2: at the beginning (κατ' ἀρχὰς) the *whole* of matter or 'substance' (τὴν πᾶσαν οὐσίαν) is transformed into water. The matter in question must be the fire of the conflagration, and its complete transformation into water is what determines the beginning of the conflagration, which has as its starting point this mass of water. And this necessarily excludes the possibility that some of the fire of the conflagration persists beyond the start of the cosmogony. As T2 indicates, the whole of matter is transformed into water 'through air' (δι' ἀέρος), a claim that we also find in T1 and T3. The reason for this is not given in any of the three texts. But it may lie in Stoic elemental theory.[6] The four elements change into one another by expansion and contraction of a basic matter. And given that air is denser than fire and thinner than water, any portion of fire that is to be transformed into water must first be transformed into air. Notice that this necessary transition also excludes that some of the fire of the conflagration persists beyond the start of the cosmogony. For, according to Stoicism, air is essentially cold and fire is essentially hot.[7] Hence, no fire can survive the transformation of the whole of matter into air, as is the case in the cosmogony.

Before we look any further into Zeno's theory, one remark is in order. According to T2, when the whole of substance becomes water the four elements are 'generated first' (ἀπογεννᾶν πρῶτον τὰ τέσσαρα στοιχεῖα, πῦρ, ὕδωρ, ἀέρα, γῆν).[8] It is clear that according to T2 they are generated *from* the initial mass of water. I examine below how this generation proceeds in the case of earth, air and especially fire. In the case of water, however, there is a difficulty that we do not find in the other three cases. For how can water, the element, be 'generated' out of the initial mass of water? Either the element water is generated out of a body that is not water, in which case we do not know what this body is and why it is called 'water', or the element water is 'generated' out of something that is literally water, in which case there does not seem to be any generation at all. Some modern scholars have dealt with this dilemma by addressing its first horn.[9] On their view, the initial mass of 'water' is not really made of the element water that currently exists in the cosmos, but of some different substance. However, they fail to provide an adequate explanation of the nature of this

6 See notably *SVF* 2.413, studied in detail in Cooper 2009.
7 See notably *SVF* 2.405, 409, 416, 429, 430–2, 442, 580,664, 787 and 841. 8 Cf. *SVF* 2.555.
9 See notably Cooper 2009: 111–15, to some extent Hahm 1977: 57–90 and Furley 1999: 434–41.

substance. But there is another way to tackle the dilemma, and this is by addressing its second horn. We may argue, in particular, that according to Zeno the original mass of water is literally made of the element water, the water that currently exists in the cosmos, and that the supposed 'generation' of this element out of the original mass is not a qualitative change of the latter – as in the case in the generation of earth, air and fire – but a change of location. Once the extinction of the conflagration is over and the whole of matter has been transformed into air and then into water, the whole cosmos consists of water: no part of it is occupied by something other than this element. But when this large mass of water generates earth, air and fire, and these begin to exist as different kinds of bodies, each possessing its own natural place in the cosmos, then the cosmos ceases to be totally occupied by water: parts of it are taken over by fire, others by air, others by earth and only some of them by water.[10] And this relocation of water in the cosmos – the change from occupying the whole cosmos to occupying only one of its parts – may be what T3 means by the generation of water out of water.

Let us now consider the generation of earth, air and fire. An important aspect of Zeno's theory is that even though god generates these three elements from water, he does not generate them all at once. The great mass of water is first divided into two distinct masses: one that settles down and becomes earth, and the other that remains water. Next, this remaining mass of water is divided into two further masses: one that becomes air, and the other that remains water until the next conflagration. And finally, the mass of air is also divided into two portions: one that remains air until the next conflagration, and the other that becomes fire. This fire is the celestial fire in the uppermost layer of the cosmos. Each of these changes is, ultimately, a qualitative change of a basic matter, the active principle of Stoic ontology, caused by the action of god upon it. Thus, the initial mass of water is this matter in a certain qualitative state and the masses of earth, air and fire that are created out of water are three portions of this same basic matter in a different qualitative state.

Let me complement the account of Zeno's cosmogony by citing three other reports of his theory.

[10] For the distribution of the elements in the cosmos, cf. especially *SVF* 1.99, 2.527–8 and 555. For a thorough discussion of *SVF* 1.99, cf. Mouraviev 2005. Another possibility, as André Laks has suggested to me, is that the difference consists in a change of function: the water currently existing in the cosmos functions as a constitutive element of other bodies; the primeval mass of water, by contrast, did not have this function since at that time no other bodies existed.

T6*: Plutarch* Stoic. rep. *1052f–1053b (*SVF *2.579;* BS *15.8)*

Τὸ βρέφος ἐν τῇ γαστρὶ φύσει τρέφεσθαι νομίζει
καθάπερ φυτόν· ὅταν δὲ τεχθῇ, ψυχόμενον ὑπὸ τοῦ ἀέρος
καὶ στομούμενον τὸ πνεῦμα μεταβάλλειν καὶ γίνεσθαι
ζῷον· ὅθεν οὐκ ἀπὸ τρόπου τὴν ψυχὴν ὠνομάσθαι παρὰ
τὴν ψῦξιν. αὐτὸς δὲ πάλιν τὴν ψυχὴν ἀραιότερον πνεῦμα
τῆς φύσεως καὶ λεπτομερέστερον ἡγεῖται μαχόμενος αὑτῷ.
πῶς γὰρ οἷόν τε λεπτομερὲς ἐκ παχυμεροῦς καὶ ἀραιὸν
γενέσθαι κατὰ περίψυξιν καὶ πύκνωσιν; ὃ δὲ μεῖζόν ἐστι,
πῶς περιψύξει γίνεσθαι τὸ ἔμψυχον ἀποφαινόμενος ἔμψυ-
χον ἡγεῖται τὸν ἥλιον, πύρινον ὄντα καὶ γεγενημένον ἐκ
τῆς ἀναθυμιάσεως εἰς πῦρ μεταβαλούσης; λέγει γὰρ ἐν τῷ
πρώτῳ περὶ Φύσεως· "ἡ δὲ πυρὸς μεταβολή ἐστι τοιαύτη·
δι' ἀέρος εἰς ὕδωρ τρέπεται· κἀκ τούτου γῆς ὑφισταμένης
ἀὴρ ἀναθυμιᾶται· λεπτυνομένου δὲ τοῦ ἀέρος ὁ αἰθὴρ
περιχεῖται κύκλῳ· οἱ δ' ἀστέρες ἐκ θαλάσσης μετὰ τοῦ
ἡλίου ἀνάπτονται." τί οὖν ἀνάψει περιψύξεως ἐναντιώτερον
ἢ διαχύσει πυκνώσεως; <ὡς> τὰ μὲν ὕδωρ καὶ γῆν ἐκ
πυρὸς καὶ ἀέρος ποιεῖ, τὰ δ' εἰς πῦρ καὶ ἀέρα τρέπει τὸ
ὑγρὸν καὶ γεῶδες. ἀλλ' ὅμως ὅπου μὲν τὴν ἄναψιν ὅπου
δὲ τὴν περίψυξιν ἀρχὴν ἐμψυχίας ποιεῖ.

12 πρώτῳ Pohlenz cum g : τρίτῳ codd. cett. **13–17** γῆς et γῆν τῆς et τῆν X¹F **15** περιέχεται codd. corr. Wyttenbach **17** <ὡς> ante τὰ μὲν add. Pohlenz : <ὧν> ante τὰ μὲν add. Reiske **20** παράψυξιν X¹FαA

He [Chrysippus] believes that the foetus in the womb is nourished by nature like a plant but that, at birth, the breath, being chilled and tempered by air, changes and becomes animal and that, hence, soul has not inappropriately been named after this process. On the other hand, he holds soul to be breath in a more rarefied and subtle state than nature; and so he contradicts himself, for how can a subtle and rarefied state have been produced from density in the process of chilling and condensation? What is more, how is it that, while declaring animation to be the result of chilling, he holds the sun to be animate, when it is igneous and the product of an exhalation that has changed into fire? For he says in the first book of *On Nature*: 'The transformation of fire is like this: by way of air it turns into water; and from this, as earth is precipitated, air is exhaled; and, as air is rarefied, ether is diffused round about, and the stars along with the sun are kindled from the sea.' Now, what is more opposed to kindling than chilling or to diffusion than condensation? Thus, the latter produce water and earth from fire and air, and the former turn into fire and air what is liquid and earthy; but nevertheless in one place he makes kindling and in another chilling the origin of animation.

T7: Scholia in Apollonii Rhodii Argonautica *44.4–6* (SVF *1.104;* BS *15.6)*

καὶ Ζήνων δὲ τὸ παρ' Ἡσιόδῳ
χάος ὕδωρ εἶναί φησιν, οὗ συνιζάνοντος ἰλὺν γίνεσθαι, ἧς πηγνυμένης ἡ γῆ στερεμνιοῦται

And Zeno claims that chaos in Hesiod is water, from whose settling down mud is generated, from whose thickening earth is solidified.

T8A: Scholia in Hesiodi Theogoniam *117a* (SVF *1.105*)

Ζήνων δὲ ὁ Στωϊκὸς τοῦ ὑγροῦ
τὴν ὑποστάθμην γῆν γεγενῆσθαί φησι.

Zeno the Stoic claims that the substrate earth was generated from the humid.

T6–8a closely coincide with T1–3, and T7–8a refer to Zeno by name. So here too it is very likely that we are dealing with Zeno's cosmogony. T6 is also extremely valuable because it not only confirms the testimony of T1–2 that Chrysippus followed Zeno, but also identifies a specific work in which Chrysippus developed this theory, *On Nature* Book 1.[11] T7–8a refer to the formation of earth out of water ('the wet'). T6 also mentions this step and, in addition, refers to the generation of water out of the fire of the conflagration (sub-stages 1 and 2 in D1) and to that of air and fire ('ether') out of this water (sub-stages 4 and 5 in D1).

Zeno's theory of cosmogony has significant precedents in Presocratic philosophy. The earliest is probably Thales who, according to Aristotle and Simplicius (*DK* 11B12–13), held that water is the principle of all things in the sense that it is that from which all things first come into being. It is a matter of dispute, however, whether Thales' original idea was cosmological and, in particular, whether it meant that the cosmos as a whole (and not just each individual natural thing) was generated out of water. In this respect a clearer precedent for Zeno's claim is the Presocratic physicist of the fifth-century BCE Hippo of Rhegium. According to the doxographic tradition stemming from Theophrastus (*DK* 38A4 and 10) Hippo was merely a follower of Thales who maintained, like him, that water is the principle and cause of all things. But according to a different doxographic tradition (*DK* 38A3 and 5), independent from Theophrastus, Hippo held the more specific thesis that fire, as the cosmological principle responsible for organising the cosmos, was generated out of water. The main evidence here is Hippolytus:

[11] Cf. *SVF* 2.565 (quoted below as T9) and 2.619.

T8B: *Hippolytus* Ref. *1.16 (*DK *38a3)*

Ἵππων δὲ <ὁ>Ῥηγῖνος ἀρχὰς ἔφη ψυχρὸν τὸ ὕδωρ καὶ θερμὸν τὸ πῦρ. γεννώμενον δὲ τὸ πῦρ ὑπὸ <τοῦ> ὕδατος κατανικῆσαι τὴν τοῦ γεννήσαντος δύναμιν συστῆσαί τε τὸν κόσμον. τὴν δὲ ψυχὴν ποτὲ μὲν ἐγκέφαλον λέγει, ποτὲ δὲ ὕδωρ· καὶ γὰρ τὸ σπέρμα εἶναι τὸ φαινόμενον ἡμῖν ἐξ ὑγροῦ, ἐξ οὗ φησι ψυχὴν γίνεσθαι.

Hippo of Rhegium said that the principles are cold, namely water, and hot, namely fire. When fire was generated by water, it overcame the power of what generated it and it constituted the cosmos. Sometimes he calls the soul the brain, sometimes water. For semen, which manifests itself to us, [scil. comes] from moisture, and it is from this [i.e. semen] that he says that the soul is produced.[12]

In Hippo the cosmogony seems to begin when the amount of fire in the cosmos is large enough to heat it, which is the probable meaning of the idea that fire 'overcomes' the power of water, which is essentially cold. In Zeno, by contrast, the cosmogony begins when water generates the other three elements. But leaving aside this important difference, the parallel between the two theories is strong. In both, water is the origin of fire. And in both fire is, thanks to heat that it emits, an active element playing a central role in the generation and reproduction of natural species. This possible influence of Hippo in Zeno may help to explain an otherwise striking feature: why does fire have to be transformed into air and water and then be created again out of water and air? Surely a simpler account would be that the fire of the conflagration is not totally transformed into air and water, but one part of it is saved from extinction and remains alive from the end of the conflagration until the end of the cosmogony – an idea that, as we shall see, is clearly present in Cleanthes. This alternative and simpler account, however, assumes that fire is a primitive constituent of the cosmos and not, as Hippo suggested, something that was created out of water. Thus, if Zeno followed Hippo on this question, he could not adopt the simpler account of the cosmogony mentioned above and had to put forward the convoluted theory that the sources attribute to him.

Before we move on to Cleanthes, who departed from Zeno, we must bear in mind that Chrysippus, who followed Zeno, made nevertheless a significant contribution to his theory. In fact, there is good reason for thinking that Chrysippus explicitly introduced one crucial idea, namely, that the starting point of the cosmogony is *god* acting on water, rather than water

12 This translation is borrowed (and slightly modified) from Laks and Most forthcoming. My discussion of this text is based on Barney ms.

all by itself. The evidence comes from the basic texts T1–T3. The agency of god only appears explicitly in T1 and T2, which are passages in which the theory is attributed not to Zeno alone, but to Zeno and Chrysippus among others. By contrast, in T3 (lines 1–9) the theory is attributed to Zeno alone and no explicit reference is made to god's agency. This difference strongly implies that the reference to the agency of god was an innovation of Chrysippus. It is not that Zeno denied that god was the agent of the cosmogony. For there is ample evidence that god is a key concept of Zeno's cosmology.[13] The suggestion is just that Zeno did not emphasise this concept in his theory of cosmogony. In any case, it is clear that, from a philosophical point of view, water in Stoicism cannot be, all by itself, the starting point of the cosmogony. In Stoicism water is a passive element and therefore it is not possible for fire, an active element, to proceed from water alone, as Hippo's cosmogony, the antecedent of Zeno's, may suggest.

The cosmogony in Cleanthes

According to T3, Cleanthes' account of the first part of the cosmogony differs substantially from Zeno's. To quote again the relevant lines of the text (Stobaeus *Ekl.* 1.153.7–22):

> Κλεάνθης δὲ οὕτω
> πώς φησιν· ἐκφλογισθέντος τοῦ παντὸς συνίζειν τὸ μέσον
> αὐτοῦ πρῶτον, εἶτα τὰ ἐχόμενα ἀποσβέννυσθαι δι' ὅλου.
> Τοῦ δὲ παντὸς ἐξυγρανθέντος τὸ ἔσχατον τοῦ πυρός, ἀντι-
> τυπήσαντος αὐτῷ τοῦ μέσου, τρέπεσθαι πάλιν εἰς τοὐ-
> ναντίον, εἶθ' οὕτω τρεπόμενον ἄνω φησὶν αὔξεσθαι καὶ
> ἄρχεσθαι διακοσμεῖν τὸ ὅλον· καὶ τοιαύτην περίοδον αἰεὶ
> καὶ διακόσμησιν ποιουμένου <τὸν ἐν τῇ τῶν ὅλων οὐσίᾳ
> τόνον> μὴ παύεσθαι. Ὥσπερ γὰρ ἑνός τινος τὰ μέρη πάντα
> φύεται ἐκ σπερμάτων ἐν τοῖς καθήκουσι χρόνοις, οὕτω
> καὶ τοῦ ὅλου τὰ μέρη, ὧν καὶ τὰ ζῷα καὶ τὰ φυτὰ ὄντα
> τυγχάνει, ἐν τοῖς καθήκουσι χρόνοις φύεται. Καὶ ὥσπερ
> τινὲς λόγοι τῶν μερῶν εἰς σπέρμα συνιόντες μίγνυνται
> καὶ αὖθις διακρίνονται γενομένων τῶν μερῶν, οὕτως ἐξ
> ἑνός τε πάντα γίνεσθαι καὶ ἐκ πάντων [εἰς] ἓν συγκρί-
> νεσθαι, ὁδῷ καὶ συμφώνως διεξιούσης τῆς περιόδου.

Given the obscurity of the text, a full reconstruction of the theory is not possible. But the deviation from Zeno is nonetheless evident. Some

[13] See for example the texts collected by von Arnim under *SVF* 1.152–77.

scholars have tried unsuccessfully to explain it away.[14] In contrast, I should like to emphasise and develop it fully.

As in Zeno, the cosmogony is preceded by a conflagration understood as a state of the cosmos in which it is fully occupied by fire. Once the conflagration is over (ἐκφλογισθέντος τοῦ παντός), the first stage of cosmogony – the one where the four elements are created – begins immediately. It is divided into four distinct sub-stages. The first one is when the centre of the cosmos stops burning and 'sinks' or 'collapses' (συνίζει). It is not immediately clear what this could mean and, in what follows, I devote some space to interpreting this claim.

The key to understanding this claim resides in the meaning of the verb συνίζειν. The only use of συνίζειν attested specifically for Cleanthes is T3. But the term or its cognates appear sometimes in other sources on Stoicism, and there is no reason for thinking that its meaning in T3 is any different from its meaning in these other sources. Now, συνίζειν in these sources is used to refer to the formation of earth. We may consider two examples. One is T7, the scholion on the *Argonautica* collected by von Arnim under *SVF* 1.104 that I quoted earlier. It ascribes to Zeno the belief that mud – a form of earth – proceeds from the sinking of 'chaos' identified by Zeno as water (χάος ὕδωρ εἶναί φησιν, οὗ συνιζάνοντος ἰλὺν γίνεσθαι). The other example is T9, a scholion on verse 115 of Hesiod's *Theogony* collected by von Arnim under *SVF* 2.565:

T9*: Scholia in Hesiodum Theogoniam 115 (SVF 2.565)*

> τρία πρῶτον ἐγένοντο· Χάος, Γῆ, Ἔρως ὁ οὐράνιος, ὃς καὶ θεός· ὁ γὰρ ἐξ Ἀφροδίτης νεώτερός ἐστιν. ἐκ δὲ τοῦ ὕδατος ἐγένοντο τὰ στοιχεῖα· γῆ κατὰ συνίζησιν, ἀὴρ κατὰ ἀνάδοσιν, τὸ λεπτομερὲς τοῦ ἀέρος γέγονε πῦρ, ἡ δὲ θάλασσα κατὰ ἐκμύζησιν, τὰ δὲ ὄρη κατὰ ἐξοστρακισμὸν τῆς γῆς.
>
> Three things were generated at first: chaos, earth and celestial eros, who is also a god. As a matter of fact, the product of Aphrodite is youngest. From water the elements were generated: earth by sedimentation, air by diffusion, and the thin part of air generated fire; the sea, by contrast, [was generated] from earth by extraction and the mountains by expulsion.

The view expressed here belongs to Chrysippus, who is cited by name earlier and after in connection with verses 135b and 459. Here too, the term συνίζησις designates the specific process by which earth is created and, as

[14] Cf. Hahm 1977: 57–90 and 240–8.

we may notice, the term is used in order to set a contrast between this process and the processes by which the other elements were created.[15]

So if this is the meaning of συνίζειν in T3, the first sub-stage in Cleanthes is when the centre of the cosmos stops burning and becomes earth. How should we understand this change? The idea is surely *not* that the mass of fire – either in the form of flame or in the form of heat – is transformed into a mass of earth. This could not be Cleanthes' idea. As we have already seen, Stoic elemental theory allows for fire to be changed into earth, but not immediately as is the case here. It must first change into air, then into water, and *only then* into earth. For elemental change involves a change in density and, according to Stoic elemental theory, the earth is the densest of the four elements, fire is the thinnest and air and water are in between these two extremes. This is well attested for Zeno and Chrysippus, and there is no reason to believe that Cleanthes departed from them on this basic question. So we must think of another possibility, one in which the change in question is not a transformation of fire directly into earth. One option is that the earth generated at the centre of the cosmos is simply a residue left by the combustion of the cosmos during the conflagration: as this combustion progresses it creates a by-product that gradually builds up at the centre of the cosmos. This is not the place to study in detail the theory of the combustion of fire that Cleanthes seems to have followed in his physics. But one basic element of this theory is that fire breaks down complex bodies into, on the one hand, fuel that it immediately consumes and, on the other, an incombustible earthlike residue. On this theory, fire consumes the fuel by causing it to transform into itself by 'assimilation' (ἐξομοίωσις), a concept that Cleanthes himself appears to have used to describe the conflagration (*SVF* 1.510). But the residue is not itself transformed into anything further. This residual earth, like fire itself, is something that remains constant throughout the infinite series of cosmic cycles: when complex bodies are created at the cosmogony, it exists within these bodies as one of its constituents and, at the conflagration, it is released from them when they are broken down by fire. This also applies to the cosmos as whole. The residual earth would be a type of earth that is contained in the material composition of the cosmos and is released when the fire consumes the cosmos at the conflagration. In any case, this residue is not itself combustible, that is, capable of being consumed by fire. And the cosmogony begins precisely when the amount of residual earth

[15] For discussion of this term in this context, cf. Hahm 1977: 245.

accumulated at the centre of the cosmos is large enough to put out the flames that are burning at that place. Note that the residual earth must build up at the *centre* of the cosmos and nowhere else. For, as we saw, this residue is a type of earth, and all earth naturally travels towards the centre of the cosmos in so far as in Stoic cosmology each of the four elements has its own natural place within the cosmos, and the place of earth is at the centre. I referred to this idea at the end of the previous section in connection with Zeno, and it evidently also has a role to play in Cleanthes' account of the cosmogony. Thus, given that this residue builds up at the centre, the centre is precisely the place where the fire of the conflagration is first extinguished.

This takes us to the second sub-stage of the early cosmogony in Cleanthes. According to T3 this step begins when the fire contiguous to the centre is also completely extinguished (εἶτα τὰ ἐχόμενα ἀποσβέν-νυσθαι δι' ὅλου). As a result of this process the whole cosmos is 'dampened' (ἐξυγρανθέντος). Why is this so? One possibility is this: as the mass of residual earth at the centre increases (as it must since large portions of the fire of the conflagration have been burning up until then) one part of this residue becomes water, and one part of this water evaporates and pervades the whole cosmos in the form of vapour. It is worth noticing that this account presupposes the formation of water out of earth, rather than of earth out of water. And this presupposition is well attested for Cleanthes elsewhere, as I shall argue below. The third sub-stage in the early cosmogony according to Cleanthes starts when the centre of the cosmos, which is now occupied by earth and water, 'offers resistance' – the Greek verb employed in T3 is ἀντιτυπεῖν (ἀντιτυπή-σαντος) – to the remaining mass of fire. This last mass of fire is called 'the last portion of fire', τὸ ἔσχατον τοῦ πυρός. The notion that this fire is not generated out of earth, as water is, strongly implies that it is a remnant of the fire of the conflagration. Now, according to T4, this fire, given the resistance it encounters at the centre of the cosmos, travels upwards away from the centre. And when it does so, it forms the heavens: the upper, peripheral, layer of fire that defines the third region of the Stoic cosmos. At this moment, we are told by the text, fire begins the ordering of the whole cosmos (ἄρχεσθαι διακοσμεῖν τὸ ὅλον). This would be the fourth sub-stage in the early cosmogony according to Cleanthes and the beginning of the following stages, which I referred to as '(b)' and '(c)' in my introduction.

In T3 we do not find any explicit reference to the formation of the layer of air. But some information is provided by another source: Hermias, a

Christian apologist of the end of the early third century CE, author of the satirical tract *Derision of Pagan Philosophers*:

T10: *Hermias* Irrisio gentilium philosophorum *14* (DG *26–31;* SVF *1.495)*

ἀλλ' ὁ Κλεάνθης ἀπὸ τοῦ φρέατος ἐπάρας τὴν κεφαλὴν καταγελᾷ σου τοῦ δόγματος καὶ αὐτὸς ἀνιμᾷ τὰς ἀληθεῖς ἀρχὰς θεὸν καὶ ὕλην. καὶ τὴν μὲν γῆν μεταβάλλειν εἰς ὕδωρ, τὸ δὲ ὕδωρ εἰς ἀέρα, τὸν δὲ ἀέρα <ἄνω> φέρεσθαι, τὸ δὲ πῦρ εἰς τὰ περίγεια χωρεῖν, τὴν δὲ ψυχὴν δι' ὅλου τοῦ κόσμου διήκειν, ἧς μέρος μετέχοντας ἡμᾶς ἐμψυχοῦσθαι.

Cleanthes, however, having stuck his head out of the well, ridicules your doctrine and throws up the true principles, god and matter. And earth changes into water, water into air, air travels upwards, and fire moves towards the periphery, whereas the soul permeates the whole cosmos and we are ensouled by participating in part of it.

Thus, according to Cleanthes, air proceeds from water, as in Zeno, but water, by contrast, proceeds from earth which is, therefore, according to Cleanthes, prior to water both logically and chronologically. This explicit reference to air is something that we do not encounter in the report of Cleanthes in lines 7–22 of T3. But there are three important points of coincidence between the two passages: (1) the theories they report are specifically attributed to Cleanthes; (2) whereas water (and air) are created out of the mass of earth at the centre of cosmos, fire is not, which strongly suggests that it is a remnant of the fire of the conflagration; and (3) fire gradually moves away from the centre towards the periphery of the cosmos to form celestial fire.

As in Zeno's cosmogony, at the stage of conflagration there is no differentiation at all in the cosmos; all is fire. At the second stage of the cosmogony the centre of the cosmos is extinguished and is occupied by a large mass of earth. At the third and fourth stages in the early cosmogony the fire contiguous to the centre is also extinguished, and one part of this now larger mass of earth becomes water and Earth is formed. It is at this stage too where, as a result of this, the whole cosmos is dampened and fire is pushed away from the centre towards the periphery. Finally, at the fifth stage, earth, water and air have already been formed and fire, which is now located at the periphery and constitutes the heavens, starts to arrange the whole in order to give the cosmos its current form.

The philosophical grounds of the polemic

We may now compare the views of Cleanthes with those of Zeno and Chrysippus, and bring out the philosophical grounds of the polemic that divided them.

There are three major differences. First, Zeno, followed by Chrysippus, contends that the fire of the conflagration is totally extinguished before the cosmogony begins. This implies that, after its extinction, there is a time in which there is no fire at all in the cosmos, either in the form of flame or in the form of heat. It is only after a number of processes have been completed that fire is generated again out of water and air. Cleanthes, in contrast, maintains that the fire of the conflagration is *never* totally extinguished. The fire that currently exists in the cosmos – and especially celestial fire – is a remnant of the fire of the conflagration. It does not proceed from water or from any of the other elements, as is the case in Zeno. This difference is far-reaching. It reveals that Cleanthes departed from Zeno–Chrysippus on the question of when the conflagration ends. According to the latter the conflagration ends once its fire is totally extinguished, that is, once the large mass of heat that is left by the flames of the conflagration cools down and is transformed into cool air. But this is not so according to Cleanthes. On his view, the conflagration is over not when the fire of the conflagration is totally extinguished, for he claims that this never really happens, but rather when the fire at the *centre* of the cosmos dies out. The extinction of the central fire, as we have seen, is caused by an increasingly large amount of residual non-combustible earth that builds up at the centre as a result of the combustion of the fire of the conflagration. Cleanthes' idea that fire is something primitive that endures throughout the cosmic cycle – which is a conception that we do not find in Zeno or in Hippo – also has a clear Presocratic precedent, namely Heraclitus and his description of cosmic fire as 'everliving' (*DK* 22B30: ἀείζωον). However, Heraclitus himself – apparently contradicting this description – also refers to the 'death' of cosmic fire (*DK* 22B76: θάνατος) and to the periodic extinction of the sun (*DK* 22B6), which are notions that are alien to Cleanthes. The second major difference between Zeno and Cleanthes turns on the origin of earth. Zeno and Chrysippus hold that earth is created out of water, whereas Cleanthes maintains that water is created out of earth, and earth out of this residual earth. And this leads us to the third major difference: in Zeno, but not in Cleanthes, all the matter in the cosmos is combustible and, in consequence, the fire of the conflagration leaves no residue. As I argue below, this third difference is the most basic because it entails the other two.

To appreciate this third difference, we must enquire into why Zeno held that the fire of each conflagration must come to an end and be totally extinguished. A set of reasons emerges from a report of the third-century Neoplatonist Alexander Lycopolis.

T11: *Alexander Lycopolis* Man. *19.2–12* (LS *46i;* BS *18.1*)

καλῶς γὰρ δὴ πρὸς τὸν Ζήνω-
νος τοῦ Κιτιέως εἴρηται λόγον, ὃς "τὸ πᾶν ἐκπυρωθή-
σεται" λέγων "πᾶν τὸ καῖον ἔχον <ὅτι> καύσῃ ὅλον καύσει,
καὶ ὁ ἥλιος πῦρ ἐστιν καὶ ὃ ἔχει οὐ καύσει;" ἐξ οὗ
συνήγετο, ὡς ᾤετο, τὸ πᾶν ἐκπυρωθήσεσθαι·

2–3 ἐκπυρωθήσεσθαι C **3** πᾶν τὸ καῖον ἔχον καύσῃ ὅλον καύσει codd : πᾶν τὸ καῖον ἔχον <ὅτι> καύσῃ ὅλον καύσει Mansfeld and Van der Horst : πᾶν τὸ καῖον <ὅ> ἔχει καύσαι, ὅλον καύσει *vel* πᾶν τὸ καῖον ἔχον [καύσῃ], ὅλον καύσει *vel* πᾶν τὸ καῖον ἔχον <ὅ> καύσει, ὅλον καύσει Brinkmann in app. crit.

For it has been rightly argued against the argument of Zeno that 'the whole will undergo a conflagration' saying [that if] 'everything which burns and has something to burn will burn it completely and the sun is fire, will it not burn what it has?' and from which he inferred, as he thought, that the whole will undergo a conflagration.

The aim of the argument is to prove that the cosmos will undergo a conflagration, and that this conflagration will consume *everything* because the sun is made of fire and the cosmos is combustible.[16] But the argument also explains why the fire of a conflagration must be totally extinguished. The explanation may be reconstructed as follows:

(1) Any fire burns as long as it does not run out of fuel and as long as it is not put out by some body distinct from the fire itself.
(2) There is no body, either inside or outside the cosmos, distinct from the fire itself, that could put out the fire of the conflagration (no internal body since at the conflagration the whole cosmos is occupied by fire, and no external body since there are no bodies in the extra-cosmic void).
(3) Given that the cosmos is finite, there is necessarily a finite amount of matter in it and, therefore, a finite amount of possible fuel for the fire of the conflagration.

[16] An account of the place of this passage in Alexander's treatise is provided by Mansfeld and Van der Horst in 1974: 74 nn. 293–6 and by Mansfeld in 1979: 147–55. The term 'the whole', τὸ πᾶν, sometimes used by the Stoics to mean the sum of the cosmos and the infinite extra-cosmic void (see notably *SVF* 2.522–5), must refer here simply to the cosmos.

Therefore, from (1) and (2):

(4) The fire of the conflagration will not be put out by some body.

And from (1), (3) and (4):

(5) The fire of the conflagration will run out of fuel and, in consequence, it will be totally extinguished at some point.

There is no evidence that Cleanthes denied any of the three premises of the argument. So how did he manage to avoid its final conclusion while accepting the premises? One hypothesis is that he subscribed not just to the three premises of Zeno's argument but also to three further theses that are fully consistent with these premises. The three further theses are: (i) celestial fire is never totally extinguished, though not because it has an unlimited amount of fuel (which would contradict the third premise of Zeno's argument), but because the extinction of the fire at the centre and its subsequent transformation into earth and then water provides *new* fuel to celestial fire; (ii) no body is fully combustible and, therefore, the action of fire upon any body necessarily releases some non-combustible residue; and (iii) at the conflagration, the residue left by the combustion of the cosmos gradually builds up at its centre.

These three theses allow Cleanthes to avoid the conclusion of Zeno's argument while retaining its premises. For if no body is fully combustible, as is stated in (ii), then all combustion is necessarily incomplete and the conflagration – which consumes all bodies – is bound to leave a residue. But this residue, which is a type of earth, will have to build up at the centre of the cosmos because earth naturally travels towards its centre. And finally, as (i) indicates, a portion of this residue, when it is transformed into water, provides new fuel to the peripheral fire, which explains its endurance from the conflagration to the new cosmogony. On this reconstruction, the root of the difference between Zeno–Chrysippus and Cleanthes – that which ultimately explains why their views on the cosmogony are so radically different – is that they had different conceptions of how bodies react to the action of fire and, especially, different ideas of how much of them is combustible.

Conclusion

I claim to have shown that the first stage of cosmogony – the stage where the four elements are created and the cosmos acquires its basic structure – was a controversial subject in early Stoic cosmology. There were two competing accounts that radically departed from each other on central questions and, notably, on the question of the origin of celestial fire: according to Zeno it issued from a large mass of water that occupies the totality of the cosmos once the fire of the previous conflagration is completely extinguished, whereas according to Cleanthes celestial fire is a remnant of the fire of the conflagration, which is therefore never completely extinguished. As I have argued, the ultimate basis of this polemic may have been that Zeno, followed by Chrysippus, and Cleanthes had different views on the relation between fire and its fuel at the conflagration: according to Zeno–Chrysippus the combustion of the latter by the former is complete and leaves no residue; according to Cleanthes, by contrast, this combustion is not complete, and a residue is necessarily left.

One important question that must be addressed is how Cleanthes' account of the cosmogony is consistent with his theory on the formation of the sun. The sun, being the largest celestial body, is given special attention in Stoic accounts of the cosmogony. In Zeno, the origin of the sun is the large initial mass of water and air that was formed as the heat left by conflagration gradually cooled down. The sun, according to Zeno, was created out of an exhalation – ἀναθυμίασις – from this water,[17] by which he means that the sun is the result of the physical expansion of part of this water into air and of this air into fire. There is also evidence, however, that Cleanthes too thought that the sun was created out of an exhalation from the sea.[18] And this poses a problem. For how, according to Cleanthes, can the sun proceed from water if, on his view, celestial fire proceeds, not from water, but from the fire of the conflagration? Therefore, either Cleanthes' position is inconsistent or some explanation must be given of how the two claims may be consistent with each other. This is an exegetical problem that I cannot tackle here but which, I believe, can be satisfactorily solved.

[17] This view is attested for Zeno individually in *SVF* 1.121 and for Chrysippus, also individually, in *SVF* 2.579 (T6 above) and 652.
[18] See notably *SVF* 1.501.

CHAPTER 2

Plotinus' account of demiurgic causation and its philosophical background

Riccardo Chiaradonna

Demiurgy and causation

Plotinus' account of the sensible world is based on two assumptions:

1. The sensible cosmos is rationally ordered, and its order depends on the activity of a prior cause.
2. This order does not reflect any rational design on the part of the cause, since the cause has no reasoning or calculation in it.

Plotinus therefore rejects intelligent design theology, while at the same time maintaining that our world has an ordered structure, which is the effect of a superior cause.[1] Here I aim to set this theory against its background.[2] I will argue that the debate between Platonic and Aristotelian philosophers during the second century CE played a prominent role in the genesis of Plotinus' account.

A crucial passage to assess Plotinus' view of demiurgic causation is the opening chapter of *Enneads* VI.7 (treatise 38).[3] It contains an exegetical section on the *Timaeus*, where Plotinus considers Plato's account of the making of the cosmos and the fashioning of the human body. Plato's text

I wish to thank Pierre-Marie Morel and Francesco Verde, who were kind enough to read a first draft of this chapter. Thanks are also due to David Sedley and Christopher Noble, for their extremely valuable remarks, and to Brian Prince, for checking my English. All mistakes are my own.

[1] Here I will not go into Plotinus' complex attitude to teleology. Suffice it to say that he rejects a horizontal account of natural teleology according to which the sensible world is arranged in a particular way for the sake of certain good ends (see VI.7.1 ff. and Plotinus' criticism of Aristotle's account of motion in VI.1.16 (treatise 42)). Instead, Plotinus accepts a vertical account of teleology, as it were, which is connected to his views about emanation and conversion, according to which each thing is in need of, and is directed towards, what is higher and better (see III.8.7.17–18: τέλος ἅπασιν ἡ ἀρχή). See Thaler 2011; Chiaradonna 2014a.

[2] For an in-depth discussion of Plotinus' arguments against divine planning, see Noble and Powers' contribution in this volume (Chapter 3).

[3] All translations of Plotinus are taken from Armstrong's Loeb edition of the *Enneads*, with some slight changes (see Armstrong 1966–88). References to the Greek text follow Henry and Schwyzer's Oxford edition (*editio minor*: see Plotinus 1964–82).

raises a puzzle for Plotinus, for it describes the activity of the demiurge in terms of discursive reasoning and calculation (λογισμός: *Tim.* 30b4, 33a6, 34b1). Plotinus, however, denies that god – whom in this section Plotinus equates with the Intellect – produces our cosmos like a human craftsman. This view often occurs in Plotinus, who generally claims that true and intelligible causes do not deliberate. Therefore order in the sensible world derives from a superior nature, but this fact does not require any process of reasoning (λογισμός) or foresight (προόρασις) on its part (VI.7.1.28–32). Plotinus maintains this view in his treatises on providence, where he distinguishes providence (based on the causation of universal *logos*) from a kind of foresight based on a process of reasoning (see III.2.2.8–9, III.3–4, 14.1–2). The same holds for the world soul, whose thought activity Plotinus opposes to that of discursive and inferential reasoning (IV.4.11.11–17), and for the demiurgic contemplation of nature: again, Plotinus separates the causal activity of nature from that which depends on reasoning and research (III.8.3.13–17).

Interestingly, the agent is not the same in all these passages: in VI.7 Plotinus focuses on Intellect, in III.2 on universal *logos* (the status of *logos* in this treatise and its relation to Plotinus' usual metaphysical hierarchy is famously debated),[4] in IV.4 on the world soul and in III.8 on the lowest productive part of the world soul, i.e. nature. It may actually be difficult to define the position of the demiurgic cause in Plotinus' metaphysics, and this fact reflects a certain distinctive fluidity in Plotinus' gradualist metaphysical hierarchy.[5] Be that as it may, the distinction between the causation of intelligible substances and a kind of craftsmanlike causation based on calculation or discursive reasoning is a recurring aspect of passages where Plotinus focuses on how true intelligible causes act on the physical world. In fact, this thesis is deeply rooted in Plotinus' philosophy and is connected to a key aspect of his theory of knowledge: that intelligible beings should be conceived of adequately and according to the principles proper to them (see VI.5.2), whereas discursive and inferential thinking is typical of our embodied souls (see IV.3.18.1–7; IV.4.6.10–13; IV.4.12.5–48).[6]

[4] See Armstrong 1940: 102–5 (the account of *logos* in III.2 and III.3 conflicts with Plotinus' usual theory of metaphysical principles, since *logos* comes to be something like a fourth hypostasis). Criticism in Rist 1967: 90–7.

[5] See Opsomer 2005a. I dwell on this issue in Chiaradonna 2014a.

[6] On the position of *logismos* in Plotinus' account of the soul see Karfik 2011–12. Plotinus' views have been taken to show a certain inconsistency: for in demarcating the soul's activity from that of the Intellect, Plotinus sometimes does not refrain from ascribing a kind of transitional and incomplete thought activity to the universal soul (see III.7.11.15–17), and this conflicts with what he says elsewhere about its non-inferential thought activity (see the discussion in Karfik 2012).

In the background of this view lies Plotinus' account of emanative causation, based on the so-called double *energeia* theory.[7] The central idea of this theory is that real causes act without undergoing any affection and in virtue of their own essence (the first *energeia*, i.e. the internal act that constitutes their own nature). According to the first *energeia*, real causes are what they are and 'abide in themselves' (see Plato *Tim.* 42e5). However, an external act (the second *energeia*) flows from them in virtue of their very nature, as a sort of by-product, without entailing any transformation or diminution on their part. The secondary act can never be separated from its origin and is like an image of it, whereas the first activity stands as a paradigm. Plotinus' favourite images of fire emanating heat through its environment and of light propagation are intended to convey these features of causation. It is this model of gradualist or emanative causation that replaces that of artisanal causation in Plotinus' thought.

Rather than exploring Plotinus' theory of causation, I wish to focus more narrowly on his attitude to Plato in the opening part of VI.7 and try to spell out the background of his position.[8] There Plotinus aims, so to speak, to neutralise Plato's account in so far as it suggests that god's causality is an activity based on provident calculation. Accepting such an account without qualification would entail that god is conceived of anthropomorphically, something Plotinus does his best to avoid. His solution is as simple as it is radical: he reads Plato's words as a metaphor suggesting that our sensible world is ordered *as if* it were produced by the rational plan of a provident craftsman (λογισμός; πρoόρασις: VI.7.1.29–32); but this is not what happens in reality, since our world is nothing but a lower and spatially extended image, which unfolds what exists 'all together' at the intelligible level. This process of derivation implies no planning or foresight on the part of god: what depends on god derives somewhat automatically from his very nature, so that the same essential content that exists without succession or deficiency in god is split and comes into existence at the level of the corporeal and extensional world (VI.7.1.54–7). Accordingly, Plotinus reads the artisanal model of causation set out in Plato's *Timaeus* as a metaphor expressing the derivation of the sensible world from its higher principles.

[7] Plotinus sets out this theory in a number of passages, esp. V.1.6.28–53, V.2.1.12–18, V.3.7.13–34, V.4.2.21–7, V.9.8.11–19. Furthermore, the theory is alluded to in many other texts. There is a vast debate on Plotinus' 'double activity' and its sources. Here I only refer to Emilsson 2007: 52–68.

[8] The literature on VI.7 is abundant. I have especially profited from both the commentary by Hadot 1988 and the annotated translations by Tornau 2001 and Fronterotta 2007.

Plotinus' approach to demiurgic causation has puzzled interpreters. It is worth quoting some remarks by Jean-Marc Narbonne, who has recently explained Plotinus' metaphorical reading of Plato as a reaction against Gnostic cosmology:

> The opposition appears to be categorical, even literal, between the Platonic statement according to which the Demiurge proceeds through reasoning ... and Plotinus' solemn declaration stating that the universe was not produced 'as the result of any process of reasoning [οὐκ ἐκ λογισμοῦ γενομένου] ...' (47 [III.2], 3, 4). How might this declaration be interpreted? Undoubtedly, for Plotinus, these are distinctions between the different types of reasoning, such as that which is simply a way of expressing or manifesting the intelligence at work in the eternal and stable generation of things, and that which serves as a pretext for the introduction of contingency, change, and even conflict in the world. It is only with this second type of reasoning that Plotinus in fact disagrees and not with the first – as long as it is correctly interpreted. The problem with the second type is real, however, precisely because an exegesis of the *Timaeus* did exist at Plotinus' time, which depicted Plato's reasoning demiurgy as a form of contingency, by emphasizing its arbitrary character. These exegetes were, of course, none other than the Gnostics who became so problematic for Plotinus that he was driven to open controversy with them in Treatise 33. (Narbonne 2011: 118–19)

This reading is unpersuasive. Certainly Plotinus' account of demiurgic causation conflicts with a literal reading of the *Timaeus*, and this is a crucial fact to be taken into account when assessing his interpretation of Plato. *Pace* Narbonne, however, this situation is not unique. Plato's *Timaeus* indeed plays a prominent role in Plotinus' philosophy and references to this dialogue are ubiquitous in the *Enneads*.[9] Yet, Plotinus' interpretation is opinionated to say the least. For example, he neglects the mathematical background of the dialogue, to the extent that he virtually ignores the atomic triangles. Plotinus' account of bodies is actually based on a creative reinterpretation of Aristotle's hylomorphism, whereas Plato's geometrical atomism finds no place in this account.[10] The same holds with the mathematical structure of the soul: while Plotinus often refers to Plato's account of the composition of the world soul (*Tim.* 35a ff.), he basically ignores its harmonic structure. Therefore, Plato's emphasis on the

[9] References to the *Timaeus* cover no less than seven columns in Henry and Schwyzer's *Index fontium* (see Plotinus 1964–82, III: 361–4). This list is certainly not complete: for additions see e.g. Riedweg and Gritti 2010. Also, see D'Ancona 2012: 948–9.

[10] *Pace* Magrin 2010, Plotinus' reference to Plato's σχήματα in III.6.12.12 is no exception: see Chiaradonna 2012: 196–8.

mathematical *composition* of the soul is simply left out.[11] The overall picture to be drawn from this scenario is that Plotinus' interpretation of the *Timaeus* is extensive but very selective, and that his reading is profoundly shaped by his own agenda.

The example of mathematics is particularly clear. Plotinus departs significantly from what we find in the *Republic*, where the dianoetic thought of mathematics is set out as 'mediator (μεταξύ) between opinion and understanding' (*Rep.* 6.511d). This passage probably lies behind Plotinus' view that the discursive self has a middle position between sense-perception and the Intellect (V.3.3.36–40). Plato, however, famously connects 'geometry and the sister arts' to dialectic (the upper sections in the line analogy: *Rep.* 6.509d), in that they differ from opinion and are relative to the intelligible; hence the crucial position of abstract mathematical disciplines in the *curriculum* of philosophers in the ideal city (*Rep.* 7.525d–531e). None of this is to be found in the writings of Plotinus, who is instead inclined to bring together mathematics and the visible world of bodies (the *lower* part of Plato's line), since both involve a kind of quantitative, extensional multiplicity which differs from that of intelligible substances (see IV.3.2.24–9).[12] The conflict with Plato's words is, in this case, no less striking than the conflict concerning demiurgic causation. And indeed, Plato's overall thesis about the philosophical and ethical significance of mathematics finds no echo in Plotinus.[13] To sum up: Plotinus' metaphorical reading of demiurgic causation patently conflicts with the literal meaning of Plato's *Timaeus*, but this fact fits with Plotinus' overall approach to this dialogue. As we shall see below, in developing his peculiar reading of Plato, Plotinus was actually deeply influenced by the philosophical debates of the second century CE, and especially by some Peripatetic criticisms of Plato and his followers.

Plotinus and the second-century-school debate

Narbonne's reading is also questionable in one other respect. At the end of the passage quoted above he asserts that the reason why Plotinus departs

[11] See IV.1 (treatise 21), IV.2.1–2, IV.3.19, IV.9.2. I follow the interpretation given by Schwyzer 1935. The criticism addressed against Schwyzer by Phillips 2002: 245–6 seems unconvincing to me. Also, see Mesch 2005.

[12] See Chiaradonna 2014b for further details.

[13] Plotinus' reference to 'mathematical studies' (μαθήματα) as preparatory to philosophical thought in I.3.3.5 is too cursory and conventional to provide a genuine counterexample. On Plato's view see Burnyeat 2000.

from a literal reading of Plato's demiurgic causation is to be found in his reaction against Gnostic cosmogony. This statement needs substantial qualification. Certainly, Plotinus rejects the Gnostic account of cosmogony as involving a gross misinterpretation of Plato's demiurge, which emphasises the arbitrary and anthropomorphic character of his activity (see II.9.6).[14] Indeed, Plotinus' theory of intelligible causation (as opposed to that which involves reasoning and calculation) plays an important role in his critical discussion (see II.9.2 and II.9.8). Yet there is simply no reason to infer from this fact that Plotinus' account was *determined* by his polemic against the Gnostics.[15] Rather, it seems that the anti-Gnostic passages in II.9 refer cursorily to philosophical views which Plotinus develops elsewhere in more detail. As a matter of fact, allusions to the Gnostics are absent or marginal in passages where Plotinus delves into intelligible causation and argues in favour of his theory. Instead, these passages often display very interesting allusions to Peripatetic theories, which have completely escaped Narbonne. Some recent contributions have convincingly downplayed the Gnostic aspects in Plotinus' account of nature in treatise III.8 while at the same time emphasising the importance of the Aristotelian background.[16] As we shall see below, the same applies to Plotinus' discussion in VI.7. It is arguably this Peripatetic background that makes it possible to understand something more about Plotinus' theory and its genesis (and in particular the idea that demiurgic causation does not entail *logismos*).

A recently (2008) discovered fragment from Diogenes of Oinoanda's inscription (NF 155 = YF 200) sheds new light on the ancient debate about Plato's demiurge (Hammerstaedt and Smith 2008: 24–6). This text is one of the 'Maxims' from the Epicurean inscription, and it provides a most

[14] The interpretation according to which Plotinus II.9 is part of an anti-Gnostic *Großschrift* (see Harder 1936), including treatises 30–3, has mostly been discarded. A *status quaestionis* can be found in D'Ancona 2012: 905–6. See also Narbonne 2011: 1–4.

[15] Narbonne appears to take Plotinus not to be denying that there is reasoning on the part of the demiurge, but only that there is the sort of arbitrary demiurgic reasoning endorsed by the Gnostics. But Plotinus is pretty clear that he is denying reasoning (as usually understood), not just some peculiarly erratic form of reasoning (I owe this remark to Christopher Noble).

[16] See D'Ancona 2009 and Morel 2009. As D'Ancona (2009: 365) rightly notes, 'Nul ne doute que cette thèse [i.e. Plotinus' view of productive contemplation], une fois établie, puisse servir aussi pour contrer les doctrines de ces platoniciens d'emprunt que sont les Gnostiques aux yeux de Plotin: l'action non délibérée et toujours parfaite de la nature, qui se révèle être une forme de θεωρία, s'oppose en effet diamétralement aux doctrines de ceux qui soutiennent que le démiurge du cosmos est méchant et que le cosmos aussi est mauvais . . . Pourtant le traité III, 8 [30] ne semble pas viser ces derniers, si ce n'est dans la mesure où, eux aussi, sont les tenants d'une fausse conception de la causalité des principes.'

interesting criticism of Plato's cosmology. This is the translation of the fragment given by Hammerstaedt and Smith: 'Although Plato was right to acknowledge that the world had an origin, even if he was not right to introduce a divine craftsman of it, instead of employing nature as its craftsman, he was wrong to say that it is imperishable.'[17] The existence of an Epicurean polemic against Plato's *Timaeus* is well attested. Until now the evidence included Epicurus' criticism of the theory of elements in Περὶ φύσεως book 14 (*PHerc* 1418) and Velleius' objections against Plato's demiurge in Cicero's *De natura deorum*.[18] Velleius criticises both Plato's account of craftsmanlike causation and the unacceptably asymmetrical view that the world had a beginning but will have no end (*ND* 1.18–20). The newly discovered fragment from Diogenes supplements the evidence from Cicero and gives further details on the Epicurean rejection of Plato's demiurge. Furthermore, the objections against Plato are consistent with the extant evidence concerning Diogenes' criticism of Stoic cosmology as entailing that god created the cosmos for his own sake and that of human beings (see NF 126/127 VI 14, fr. 20 I.10, II.12). Diogenes (or rather his source) indeed takes Plato to be correct in assuming that our world is generated. *Pace* Plato, however, this does not entail that it is the work of a divine craftsman: the only true craftsman is nature, as Diogenes emphasises. And, of course, Plato is wrong to say that the world is imperishable while being generated.

Diogenes' inscription dates from (possibly the second half of) the second century CE. The parallel with Cicero shows that Diogenes draws from the earlier Epicurean tradition, and there is little if any evidence for implicating him in contemporary debates. Yet it is interesting to set Diogenes' new fragment in parallel with some fragments of Atticus,

[17] Hammerstaedt and Smith 2008: 25. The Greek text is as follows:
καλῶς Πλάτων
ὁμολογήϲαϲ γε-
νητὸν εἶναι τὸν
κόϲμον, ᵛ εἰ καὶ μὴ
καλῶϲ ἐδημιούρ-
γηϲεν αὐτόν,
τῇ φύϲει δημι-
ουργῷ μὴ χρη-
ϲάμενο{ι}ϲ, ᵛ κα-
κῶς ἄφθαρτον
ᵛᵛᵛ εἶπεν.
vacat

[18] On Epicurus' Περὶ φύσεως book 14 see the recent discussion in Verde 2013: 333–45. The text is edited by Leone 1984.

a Platonist philosopher of the same period.[19] We know that Atticus vehemently rejected Aristotle's philosophy and the conciliatory efforts made by those who attempted to use Aristotle as a sort of guide for interpreting Plato. Atticus devoted a whole work to the refutation of those Aristotelising opponents (whatever their identity may be), and a number of fragments of this work can be found in Eusebius' *Preparation for the Gospel*.[20] One of the most striking features of Atticus' criticism is the parallel he draws between Epicurus and Aristotle (*apud* Eusebius *Prep. Evang.* 15.5.1–14 = fr. 3 des Places), as both of them deny god's concern for our world and ultimately reject providence. Atticus' defence of Plato is based on a close and literal reading of the *Timaeus*. He takes the world as having been 'really' generated in time, and regards the demiurge's activity as that of a craftsman who is capable of producing things and cares for his products by intervening on them.[21] It is on account of his view of demiurgic causation that Plato 'makes all things connected with god, and dependent on god' (εἰς θεὸν καὶ ἐκ θεοῦ πάντα ἀνάπτει: fr. 3.16 des Places, trans. Gifford), a claim immediately substantiated by Atticus with references to Plato's *Laws* (4.715e–716a and 10.885b7), *Timaeus* (29e1–3, 30a4–5) and *Phaedo* (62b1, 97c4).

Famously, Aristotle conceives of nature as teleologically ordered and regards god as the cause of motion. Furthermore, Aristotle claims that ordered astral motions ensure regularity in the sublunary region (see *De gen. et corr.* 2.10). Atticus takes these aspects of Aristotle's philosophy as efforts to develop a notion of providence while at the same time removing all intervention from a divine and artisanal cause external to nature. On his view, however, this is just a mischievous attempt to dissimulate a complete denial of providence. Atticus' parallel between Aristotle and Epicurus is meant to substantiate this claim: 'For if according to Epicurus providence

[19] On Atticus see Baltes 1983; Moraux 1984: 564–82; Zambon 2002: 129–69; Karamanolis 2006: 150–90. Fragments in des Places 1977.

[20] Atticus' work is referred to as Πρὸς τοὺς διὰ τῶν Ἀριστοτέλους τὰ Πλάτωνος ὑπισχνουμένους *apud* Eusebius *Prep. Evang.* 11.1.2 = fr. 1 des Places, but it is controversial whether this was the title of the treatise: see Karamanolis 2006: 151. On the identity of Atticus' adversary see Zambon 2002: 138 (who argues in favour of a Platonist opponent) and Karamanolis 2006: 153–7 (who argues in favour of an Aristotelian opponent). Part of Atticus' fr. 3 des Places is translated by Sharples 2010: 202 (= 22 N).

[21] The fragments about the temporal generation of the world are preserved by Proclus' *In Timaeum* and probably derive from Atticus' lost commentary to this dialogue (see frr. 19–25 des Places). On Proclus *In Tim.* 1.276.31–277.7 Diehl (fr. 19 des Places) and 1.381.26–381.12 Diehl (fr. 23 des Places) see Dörrie (†) and Baltes 1998: 112 and 414–15 (*Bst.* 137.5 and 137.6). On the preservation of the world through the demiurge's artisanal intervention see Eusebius *Prep. Evang.* 15.6.11–14 = fr. 4.64–109 des Places.

disappears, although the gods according to him employ the utmost solicitude for the preservation of their own goods, so must providence disappear according to Aristotle also, even if the heavenly motions are arranged in a certain order and array' (fr. 3.66–71 des Places, trans. Gifford).[22] Very interestingly, Atticus contrasts the administration of human affairs by Aristotle's nature (φύσει τινί) with that conducted by god's reasoning (θεοῦ λογισμῷ: fr. 3.85 des Places). Only a provident god, who is capable of both producing this world and intervening in it, can ensure order and providence. This cannot be achieved by Aristotle's nature. A further crucial aspect of Atticus' criticism is his emphasis on the existence of an animated power (δύναμις ἔμψυχος) that pervades the whole, binding and holding all things together (Eusebius *Prep. Evang.* 15.12.3 = fr. 8.18 des Places). This is precisely the role of the world soul: Atticus once again emphasises the contrast between this idea and Aristotle's nature (fr. 8.8–10: πρὸς οὐδὲν τούτων ἡμῖν Ἀριστοτέλης ὁμολογεῖ. οὐ γὰρ εἶναι τὴν φύσιν ψυχήν). The criticism of Aristotle's cosmology is distinctive of Atticus' approach, but his reading of the *Timaeus* is not unparalleled.[23] Although the issue of the generation of the world was debated in Middle Platonism, and interpreters of Plato were famously split among those who read Plato's *Timaeus* literally and those who favoured an allegorical interpretation, there was seemingly no disagreement about the fact that God acts on the world as a craftsman.[24] Pre-Plotinian Platonist philosophers argued that natural order reflects the reasoning of the demiurge. Thus according to Alcinous the demiurge 'proceeds through a most admirable providence and administrative care (δίαιταν) to create the world, because "he was good" (*Ti.* 29e)' (*Didask.* 12.167.13–15, trans. Dillon).[25] Galen's Platonising teleology too is based on his account of providential nature as a benign craftsman (see esp. *De us. part.* 3.507 Kühn).[26] Atticus' emphasis on the *logismos* of the demiurge, then, simply reflects what was then the current (and indeed

[22] εἴπερ γὰρ καὶ κατ' Ἐπίκουρον τὸ τῆς προνοίας οἴχεται, καίτοι τῶν θεῶν κατ' αὐτὸν πᾶσαν κηδεμονίαν ὑπὲρ τῆς σωτηρίας τῶν οἰκείων ἀγαθῶν εἰσφερομένων, οὕτως ἂν οἴχοιτο καὶ κατ' Ἀριστοτέλην τὸ τῆς προνοίας, εἰ καὶ τὰ κατ' οὐρανὸν ἐν τάξει τινὶ καὶ κόσμῳ διοικεῖται. The polemical remark about Epicurus' denial of providence is current in ancient authors and is also to be found in Plotinus (II.9.15.8–10).

[23] Atticus' criticism was actually followed by Harpocration, probably a pupil of his (see Proclus *In remp.* 2.377.15–378.6 Kroll = Atticus fr. 25 des Places, Harpocration fr. 13 Dillon / 21 T. Gioè, *Bst.* 137.7 Dörrie (†) and Baltes), and Galen (see Philoponus *Aet. mun.* 600.1–601.16 Rabe, from Galen's lost treatise *On Demonstration*, book 4).

[24] On the debate about the generation of the world, see the excellent survey in Dörrie (†) and Baltes 1998: 84–146 (texts) and 373–465 (commentary). Further discussion can be found in Noble and Powers' contribution in this volume (Chapter 3).

[25] See the parallels in Whittaker 1990: 110 n. 224.

[26] See Sedley 2007: 239–43.

most natural) way of reading Plato's *Timaeus*, although the criticism of Aristotle and the parallel between Aristotle and Epicurus are distinctive features of his approach.

The parallel between Epicurus and Aristotle can be taken to confirm Atticus' superficial knowledge of Aristotle's philosophy.[27] Comparing Aristotle and Epicurus (to Aristotle's detriment) could also be a mere rhetorical tactic for denigrating Aristotle with no genuine philosophical interest. Yet at the time of Atticus there still was a living Epicurean tradition, and Atticus' polemical tone does not rule out the possibility that his objections have some real ground. Prudence is necessary, but the least one can say is that the recently discovered fragment of Diogenes' inscription fits very well with Atticus' polemical argument. As a matter of fact, the similarity between Diogenes' criticism of Plato and Atticus' *Aristoteles interpretatus* is striking, as both entail a replacement of Plato's demiurge with a non-craftsmanlike *physis*. Atticus has, therefore, some good reason for regarding Epicurean and Aristotelian philosophers as forming a common front against the supporters of divine artisanal causation in nature – that is, Plato and the Stoics – and this despite the obvious differences between Epicurus and Aristotle's accounts of nature.[28]

This background helps explain some features of Alexander of Aphrodisias' cosmology. In a number of well-known texts Alexander challenges Plato's views on demiurgic causation. Yet at the same time Alexander carefully defends the existence of order in nature, while arguing that order in the sublunary region does not depend on craftsmanlike reasoning. Furthermore, in *Quaestiones* 2.3 (a work possibly directed against Atticus, or against someone who had levelled criticisms against Aristotle closely reminiscent of those of Atticus), Alexander develops a problematic account of the 'divine power' (θεῖα δύναμις) in nature, derived from celestial bodies: this power acts on the composition of bodies, and on account of it bodies acquire a certain psychic principle (*Quaest.* 2.3 49.3 Bruns).[29] Probably a

[27] It is very difficult to detect any serious textual work on Aristotle's treatises behind Atticus' objections: the hypothesis that he was not very familiar with Aristotle's works is plausible. See Moraux 1984: 570–1, 580.

[28] The disputed philosophical allegiance of the anti-Stoic philosopher Diogenianus (known through Eusebius) confirms further the analogy between the Epicurean and the Peripatetic views on fate: *status quaestionis* in Sharples 2010: 234.

[29] Unfortunately, the Greek text is far from being clear: see Fazzo and Zonta 1999: 209 n. 36. On this difficult and aporetic *Quaestio* see the discussions in Moraux 1967, Donini 1996 (= Donini 2011: 125–38) and Rashed 2007: 288–91. English translation in Sharples 1992: 93–8; Italian translation in Fazzo and Zonta 1999: 195–217. On Alexander's likely polemical allusions to Atticus see Sharples 1990: 90–1.

Platonist opponent would not have been persuaded by Alexander's defence of Aristotle.[30] However, Alexander's combined attempt to reject demiurgic causation while maintaining the existence of providence marks a crucial step in post-Hellenistic accounts of causation, and can plausibly be seen as a source of Plotinus' account. Alexander's rejection of craftsmanlike causation is indeed so radical that he does not refrain from defining nature as an 'irrational power' (ἄλογος . . . δύναμις, *apud* Simplicius *In phys.* 311.1 Diels; see Alexander of Aphrodisias *In metaph.* 104.3 Hayduck).[31] Furthermore, in order to illustrate how natural motion is transmitted from the first mover, Alexander adopts the mechanical analogy of a marionette whose parts are connected with strings (see Alexander of Aphrodisias *apud* Simplicius *In phys.* 311.8 ff.).[32] No choice or plan is involved in this process, and Alexander develops the distinction between nature and art both in his treatise *On Providence*, preserved in two Arabic versions (see *De prov.* 79 Ruland), and in his commentary on the *Metaphysics* (*In metaph.* 104.3–10 Hayduck).

Alexander certainly maintains that rational structures and regularities exist in the sublunary region. This holds especially for natural species, which exist in virtue of their hylomorphic forms (human beings beget human beings: see *In metaph.* 103.33 Hayduck); the eternal and regular character of these forms is connected to the cyclical motions of celestial bodies. What Alexander rejects is the Platonic view that such rational structures should be seen as depending on a demiurgic and reasoning cause that produces cosmic order by contemplating an external paradigm. This paradigm instead simply coincides with the hylomorphic immanent form, in so far as it is taken as the goal of natural motion (*In metaph.* 349.6–16 Hayduck). Predictably enough, Alexander (*De prov.* 33.1 ff., 87.5 ff. Ruland; *Quaest.* 1.25 41.8 ff. Bruns, 2.19 63.15 ff. Bruns) regards natural regularities (such as the processes of coming to be and perishing, and the continual existence of sublunary natural species) as connected to (and depending on) cyclical celestial motions. Therefore, the hylomorphic structure of the sublunary region, and the cosmological relation between this region and regular celestial motions, make it possible to account for natural order without having to conceive of nature as a demiurgic and

30 See the remarks in Donini 2011: 133.

31 See on this Sharples 1982; Genequand 1984 (who suggests that Alexander reacts against Galen's account of nature); Accattino 2003 (with a criticism of Genequand); Adamson 2007. On Alexander's *On Providence* see the translations in Fazzo and Zonta 1999 and Thillet 2003. The discussion in Rashed 2007: 278–85 and 294–304 is fundamental. Sharples 2010: 196–210 provides a survey of the Peripatetic debate about providence.

32 On this analogy see Rashed 2007: 278–85 and Rashed 2011: 151–2, whose interpretation I follow.

reasoning power. It is extremely interesting that Epicurus' name crops up again in the discussion: Alexander argues that the Epicureans denied finality in nature since they mistakenly connected it to choice and reasoning (κατὰ προαίρεσιν ... καὶ λογισμόν, *apud* Simplicius *In phys.* 372.9–15 Diels), (correctly!) noting that things in nature do not happen this way. Yet, as Alexander immediately adds, the situation is different, since nature produces for the sake of some end, although this happens without reasoning (ἡ φύσις ἕνεκά του μὲν ποιεῖ, οὐ κατὰ λόγους δέ).[33]

As other scholars have already remarked, Alexander's criticism of demiurgy in his treatise *On Providence* is similar to Plotinus' position.[34] Certainly, Plotinus repeatedly argues that the material world results from 'action' or 'production' on the part of intelligible principles (see Plotinus' usage of the verbs ἐργάζεσθαι and ποιεῖν in II.7.3.9, IV.4.12.29–41, VI.3.15.28 etc.), and this conflicts with the Peripatetic theory. As noted above, however, 'production' here should not be taken to mean demiurgic or artisanal production, for Plotinus' primary concern is to strip intelligible causality of any anthropomorphic connotation. True principles do not actually engage in any reasoning or calculation, and their causal action merely depends on their essential nature, without involving any deliberation or choice between different alternatives. Both in Alexander and in Plotinus λογισμός and cognate expressions refer to that which does *not* explain order in nature. Both authors reject the idea that teleology should be explained anthropomorphically, as a plan devised by nature. And both Alexander and Plotinus regard nature as an 'irrational' power, although Plotinus adds the crucial qualification that the irrational power of nature is nonetheless a kind of contemplation.[35] Indeed, the differences here are also very significant. Alexander rejects Plato's demiurge in favour of a

[33] Note that Alexander's treatise *On Providence* opens with a criticism of the Atomist position: see Alexander of Aphrodisias *De prov.* 1.5 ff. Ruland. On Alexander's reading of Epicurean philosophy see now Rashed 2011: 110–13, 356–7 and *passim.*

[34] See Thillet 2003: 46–54, esp. 49. On Alexander's account of providence and its posterity among late antique and Arab philosophers see Adamson 2007. More details in Chiaradonna 2014b. Needless to say, Plotinus does not mention Alexander in his treatises (no philosopher later than Epicurus is mentioned in the *Enneads*), but Plotinus was certainly familiar with Alexander's works (see Porphyry *Vit. Plot.* 14.10–13; further details in Chiaradonna 2008 and D'Ancona 2012: 973–5). The least one can say is that Plotinus and Alexander display a common attitude to demiurgic causation, and this fact points to a common school background. Yet I am inclined to think that Plotinus was familiar with Alexander's discussion.

[35] See Alexander of Aphrodisias *apud* Simplicius *In phys.* 311.1 Diels (ἄλογος ... δύναμις) vs. Plotinus III.8.3.12–14: Πῶς δὲ αὕτη ἔχει θεωρίαν; Τὴν μὲν δὴ ἐκ λόγου οὐκ ἔχει· λέγω δ' ἐκ λόγου τὸ σκοπεῖσθαι περὶ τῶν ἐν αὐτῇ. On Plotinus' attitude to the Peripatetic account of θεωρία in III.8 see D'Ancona 2009. I should emphasise again that Narbonne 2011: 122–7 neglects these parallels. His account of Plotinus' contemplation is, then, one-sided and unconvincing.

cosmological account of teleology which does not involve separate Ideas of the Platonic sort and is based on the theory of the immanent specific form. This is certainly not the case with Plotinus, who rather develops a distinctive account both of Plato's metaphysical essential causes and of the theory of participation. Plotinus' Platonist views on natural kinds and on their derivation from intelligible causes are certainly very different from Alexander's hylomorphic essentialism.[36] Yet Plotinus' correction of Plato's account of demiurgic causation in the *Timaeus* can be seen as incorporating some features of the Peripatetic position; and the background of second-century CE school debates between Platonic and Peripatetic philosophers helps explain why Plotinus discarded the most obvious reading of Plato's text.

From demiurgy to gradualism: Plotinus' account of the nature of man

The suggested reading is further confirmed by Plotinus' detailed discussion of Peripatetic views in the same texts where he focuses on demiurgic causation. It is as if Plotinus felt it necessary to demarcate his position from two equally misleading views: the Middle Platonic reading of the *Timaeus*, involving a kind of anthropomorphic activity on the part of true causes, and the Peripatetic view, which rejects craftsmanlike causation but also discards Plato's paradigmatism altogether. Plotinus' own middle path between these alternatives leads precisely to his theory of ontic gradualism and emanative causation. As noted above, some recent studies have elucidated the Peripatetic background of Plotinus' views about nature, contemplation and demiurgy in III.8 in interesting ways.[37] I do not wish to dwell on this issue here, but rather to focus again on VI.7. In VI.7.2 Plotinus moves on to Aristotle, after his initial discussion of the demiurge, and in particular after having just qualified the nature of the Intellect by using the term 'cause' (αἰτία) (VI.7.1.57).[38] In order to explain the Intellect's causal nature, Plotinus adopts the characteristic Peripatetic distinction between ὅτι and διότι or διὰ τί,[39] while arguing that this distinction can only

[36] For further details on Plotinus' theory and his attitude to Aristotle's hylomorphism see Schiaparelli 2010; Wilberding 2011b; Chiaradonna 2014a and 2014b.

[37] See above, note 16.

[38] The bibliography on VI.7.2 is abundant. See, in particular, D'Ancona 1992; Schroeder 1992; and the annotated translations by Hadot 1988 and Fronterotta 2007. I am especially indebted to Schiaparelli 2010. What follows is just a sketchy account of this chapter. Further details can be found in Thaler 2011 and Chiaradonna 2014a.

[39] See Aristotle *An. post.* 2.2.90a15 and *Metaph.* 8.4.1044b14. At VI.7.2.12 Plotinus refers to the Aristotelian example of the eclipse.

acquire an adequate foundation within a Platonist account of reality. In fact, Plotinus grants the Peripatetic view that at least some things 'here below' can ultimately be seen as identical with their essence and that form is the 'cause of being' for each sensible thing (VI.7.2.11–17). However, he claims that this is not sufficient to understand how something may be genuinely identical with its cause.[40] For in order to adequately grasp this identity we cannot start from things here below, whose structure is necessarily extensional. Extensional things cannot but form a plurality whose parts are external to each other, although they are certainly connected to the other parts of the same ordered whole (see VI.7.2.30). But a thing and its form can never be completely identical within such an extensional context. In order to account for the unity between things and their causes, then, we must ultimately abandon the extensional world of bodies and focus on Form itself (αὐτὸ τὸ εἶδος: VI.7.2.18, i.e. not as the form of a certain body). But in so doing we must go beyond hylomorphism and examine the very nature of the Intellect. 'There' things are perfectly unified, and it is only in that context that each object can genuinely be seen as being identical with its cause.

At the end of this argument Plotinus claims that 'when you state the cause, you state everything' (VI.7.3.13–14: αἰτιολογῶν πάντα λέγεις). Although the details are open to debate, Plotinus' general account conveys the idea that everything in our world (with the exception of spatial extension) is ultimately the effect (or rather the extensional unfolding) of intelligible causes.[41] Accordingly, all features in our material world derive from intelligible principles, so that we actually know these beings as a whole by considering their cause. This conclusion, however, raises the obvious puzzle that the intelligible realm should pre-contain features that appear to be peculiar to the corporeal world, such as perception in human beings. This puzzle is raised in VI.7.3.22–33 and it sets the context for Plotinus' discussion about the nature of the human being in the following four chapters (VI.7.4–7). Here I cannot provide a full interpretation of this long and difficult section, which includes Plotinus' celebrated discussion about the 'three men' (i.e. the three ontic and cognitive levels that

[40] καὶ πειρωμένοις οὕτως τὸ τί ἦν εἶναι λαμβάνειν ὀρθῶς συμβαίνει. Ὃ γάρ ἐστιν ἕκαστον, διὰ τοῦτό ἐστι. Λέγω δὲ οὐχ, ὅτι τὸ εἶδος ἑκάστῳ αἴτιον τοῦ εἶναι—τοῦτο μὲν γὰρ ἀληθές—ἀλλ᾽ ὅτι, εἰ καὶ αὐτὸ τὸ εἶδος ἕκαστον πρὸς αὐτὸ ἀναπτύττοις, εὑρήσεις ἐν αὐτῷ τὸ «διὰ τί» (VI.7.2.14–19). P. Hadot 1988: 89 and 201 and Schiaparelli 2010: 481 argue (in my view convincingly) that the generic πειρωμένοις refers in fact to Aristotle and the Peripatetics. This would not be an isolated case: see VI.7.4.26–8 and VI.1.1.29–30.

[41] For further details and discussion I would refer again to Chiaradonna 2014a.

characterise the nature of human beings: see VI.7.6). I would only like to emphasise again how Plotinus' reading of Aristotle shapes his account of the Platonic hierarchy of being and, in particular, his view about the relation between the soul and enmattered form. It seems to me that this account provides Plotinus' answer to the exegetical puzzle about Plato's *Timaeus* raised in the first chapter of the treatise.[42]

Plotinus starts his discussion by focusing on man 'here below' (VI.7.4.3) and mentions three hypotheses about the nature of 'this man': '(1) Is this man a *logos* other than the soul which makes this man and provides him with life and reason? (2) Or is the soul of this kind the man? (3) Or the soul which uses the body of such a kind?' (VI.7.4.7–10).[43] Plotinus' use of τοιαύτη (as referring to the soul) and τοιόνδε (as referred to the body) is interesting. In his hypotheses (2) and (3) Plotinus suggests that man is identical with his soul: man can in fact either be a soul 'of this kind/of a certain sort' or 'the soul which uses a body of this kind/of a certain sort'. Both views are clearly indebted to Plato's *Alcibiades* (129e–130a, 130c) and Plato's anthropology in the *Alcibiades* shapes Plotinus' discussion in VI.7.4–7 (see, in particular, VI.7.5.24).[44] Yet the use of τοιαύτη/τοιόνδε as referring to both soul and body cannot be traced back to Plato and points to Aristotle's account of the soul in *De anima* 2.1.412a16–28, where the soul is said to be the actuality of a body 'of such and such a kind' (τοιόνδε: 412a16; τοιοῦτο: 412a21), that is, a body having life potentially within it. According to the same pattern outlined above in relation to nature and providence, Plotinus integrates into his psychology Aristotle's view about the hylomorphic relation between soul and body (organic bodies are alive in virtue of their formal component), while at the same time profoundly transforming it: for the aspect in virtue of which the body 'of such a kind' is alive, is, according to Plotinus, a mere trace or shadow produced by the soul, whereas soul proper (i.e. the ultimate source of the body's life) is not part of the hylomorphic compound (see IV.4.18.4–10, 18.30).[45]

[42] What follows is a very cursory account. Further details can be found in the notes *ad loc.* by Hadot 1988 and Fronterotta 2007. Horn 2012: 226–7 provides a reading of VI.7.4–5, whose conclusions are different from my own.

[43] ἆρα ὁ ἄνθρωπος οὗτος λόγος ἐστὶ ψυχῆς ἕτερος τῆς τὸν ἄνθρωπον τοῦτον ποιούσης καὶ ζῆν αὐτὸν καὶ λογίζεσθαι παρεχομένης; Ἢ ἡ ψυχὴ ἡ τοιαύτη ὁ ἄνθρωπός ἐστιν; Ἢ ἡ τῷ σώματι τῷ τοιῷδε ψυχὴ προσχρωμένη. Here I will mostly leave the term *logos* transliterated. The exact meaning of it is a *crux* of Plotinus scholarship and I cannot dwell on it. For further details I would only refer to Hadot 1988: 217–20; Kalligas 2011; Wilberding 2011b; Gerson 2012.

[44] On Plotinus' reading of Plato's *Alcibiades* see Aubry 2007.

[45] See on this Noble 2013.

Plotinus, therefore, does not merely consider the hypothesis that man is a soul, or a soul that uses a body, but adds the further remark that soul and body must be of such-and-such a kind, in other words that soul and body must be structured in such a way as to constitute a human being and not some other living being. How can this peculiar structure be accounted for? Plotinus' answer to this problem leads him to transform the Peripatetic notion of 'composite' in chapter VI.7.5: for unlike Aristotle he claims that man 'here below' is in fact the composite not of soul plus a body of such a kind, but the composite of soul (i.e. the intelligible principle which provides man with life and knowledge) plus a *logos* of such a kind (i.e. the intelligible forming principle that accounts for the features of bodies).[46] Here, again, Plotinus' critical allusion to *De anima* 2.1 is evident. In fact, he incorporates the Peripatetic idea that human beings are composite and that the soul alone does not suffice to account for their nature. Yet what must be added to the soul is not a certain body, but a certain intelligible forming principle which accounts for the features of that body, i.e. *logos*, which is, in its turn, an activity of the soul (VI.7.5.3–4).[47] In so doing, Plotinus transfers, so to speak, Aristotle's hylomorphic account into a different framework of thought, according to which even the nature of human beings 'here below' is completely constituted at the intelligible level (where 'intelligible' should be taken to mean not merely the world of Forms, but the different degrees or levels that constitute Plotinus' intelligible realm). It is extremely interesting to examine in some detail how Plotinus comes to this conclusion, for a careful reading of his argumentation sheds further light on his approach to Aristotle's hylomorphism.

Plotinus starts (VI.7.4.12–31) by considering the first hypothesis on his list, i.e. that man is a *logos*. First he equates *logos* with some kind of definition or 'formal determination' that should express the nature of its object (i.e. man, a living being composed of soul and body). The standard example for this definition is 'rational animal', and this is indeed the definition mentioned in VI.7.4.12.[48] A *logos* conceived of in this way, however, does not show at all what man here below really is (i.e. his essence or nature), but simply describes the factual structure of concrete beings composed of body and soul. This kind of *logos* or definition, then,

[46] See VI.7.5.2–3: Τί κωλύει συναμφότερόν τι τὸν ἄνθρωπον εἶναι, ψυχὴν ἐν τοιῷδε λόγῳ.

[47] According to Plotinus, features in the intelligible *logos* include 'corporeality': see II.7.3, where there is a further critical allusion to the Peripatetic account of essence and definition (II.7.3.8–10).

[48] This definition is famously later than Aristotle, and it plays a crucial role in Alexander of Aphrodisias' account of essence and definition (e.g. Alexander of Aphrodisias *In Top.* 46.6–14 Wallies): see Rashed 2007: 153–5.

has no explanatory power. Plotinus' polemical reference to Aristotle's view of definition is evident here (see the parallel in II.7.3.8–10): for Plotinus argues that even if we grant that we should focus on forms in matter (i.e. even if we provisionally accept Aristotle's position), this kind of definition is nonetheless insufficient, since it accounts for beings composed of matter and form (τόδ' ἐν τῷδε, VI.7.4.22–3; see Aristotle *Metaph.* 7.5.1030b18, 7.11.1036b23), whereas it is incapable of grasping form alone. But *according to the Peripatetics themselves* definition should be able to grasp the essence of things (VI.7.4.26–8). Their account of *logos* as definition is therefore insufficient even according to Peripatetic standards, since it is incapable of adequately grasping the essence of man, even if one regards enmattered forms as the proper object of definitional *logos* (VI.7.4.24–5). With his usual philosophical acumen, Plotinus thus points to a major difficulty in Aristotle's theory of form and definition in *Metaphysics* 7, namely the issue of whether a definition of form should include material features or not.[49] Plotinus exploits this puzzle to his own advantage, since he suggests that the only way of solving the problematic status of Peripatetic definitions points to a different meaning of *logos*: not as the definition of concretely existing human beings, but as a principle that produces human beings and accounts for their nature; this is what Plotinus calls τὸν λόγον τὸν πεποιηκότα at VI.7.4.25. The *logos* itself is 'one which *makes* the rational living being' (VI.7.4.31).

Next Plotinus dwells on the status of the productive *logos* and, in doing so, introduces the soul as the essential constituent of the nature of man (VI.7.4.31–VI.7.5.8). Plotinus suggests that in order to define the real nature of man we should replace the derivative term ζῷον with the original form ζωή: Ζωὴ τοίνυν λογικὴ ὁ ἄνθρωπος (VI.7.4.33). According to Plotinus 'life' is necessarily connected with 'soul', so that the two may even be conceived of as identical to one another (see VI.7.11.9–12). Therefore, replacing 'living being' with 'life' ultimately leads to the identification of the nature of 'man' either with an activity of the soul or with the soul itself (VI.7.4.35–6). But this raises a further problem (VI.7.4.37), since the same soul can pass from man into different living beings (Plotinus admits reincarnation and claims that human souls can pass into irrational animals and even plants: see III.4.2.11–30). If this is the case, the soul alone cannot be identical with man: for example, if the soul were the same thing as man, we could not regard human beings as being essentially bipeds, since human souls can reincarnate as animals with four legs. At the beginning of VI.7.5

[49] See the contrasting views of Frede 1990 and Peramatzis 2011.

Plotinus solves this predicament by claiming that man is not soul alone, but something composed of soul plus a *logos* 'of this kind' (VI.7.5.3). As noted above, in this way Plotinus incorporates and transforms Aristotle's hylomorphic account of man as an ensouled living being: for the nature of human beings includes not the body, but the formative principles that are responsible for the bodily structure of human beings.[50] These formative principles are nothing but activities (ἐνέργειαι) of the soul, and could not exist without the soul which acts within them (VI.7.5.4–5).[51]

Plotinus distinguishes between different levels in the soul: the nutritive soul (φυτική, VI.7.5.10); a higher soul that 'makes a living being' (VI.5.10) by entering into proximate matter and 'making shapes in body according to itself' (VI.5.14); and finally a higher and 'more divine' (θειότερα, VI.5.21) soul. This tripartition is reminiscent of the Aristotelian distinction between nutritive, sensitive and rational soul. Yet unlike Aristotle, Plotinus does not distinguish between different levels in the soul based on their function or kind of activity. The whole section actually aims to show that in some sense the highest and reasoning soul *can* be said to perceive. Furthermore, in VI.7.6.19–20 the formative and sensitive soul (what Plotinus here calls the second man) also appears to include the activity proper to the discursive and embodied soul (as opposed to the higher soul that does not leave the Intellect). What distinguishes each level of the soul is not so much its kind of activity as the degree of 'clarity' or luminosity of its activity. This view is characteristically Plotinian in so far as Plotinus sometimes sets out the 'vertical' hierarchy of intelligible principles (Intellect, world soul and nature) as an intensive hierarchy of degrees of life with different levels of 'clarity' (see III.8.8.16–24). In other words, while different levels of the soul may well perform the same life activities, they will do so in different ways and with different degrees of intensity. Hence the soul that makes the living being is 'clearer and more alive' than the nutritive one (VI.7.5.10–11), whereas the divine soul has 'clearer sense-perceptions' than the lower soul (VI.7.5.22–3); in turn, the perceptions of the lower soul are 'dimmer' than those of the higher soul. It is indeed somewhat difficult to make sense of this view, but Plotinus' analysis of the various levels of perception (from the lowest and unconscious *petites perceptions* to the highest and clearest perceptions, which are actually intellections) provides a good example of it.

Plotinus dwells on the different ontic degrees of man and on their mutual relation in VI.7.6, where he sets out his famous distinction between

50 This is parallel to what Plotinus argues about the relation between soul and *logoi* in V.7.1.8–10.

51 The connection between *logos* and *energeia* is characteristically Plotinian: see Gerson 2012: 20.

three 'men' and their cognitive activities. Interpreting this exceedingly difficult section is a task that lies beyond the scope of the present chapter. Suffice it to say that Plotinus' gradualism emerges there as the positive side of both his metaphorical account of Plato's demiurge and his criticism of Peripatetic hylomorphism. At the end of his discussion Plotinus argues again that those features that characterise human beings 'here below' (including perceptions) are nothing but lower manifestations of a higher and intelligible nature: 'these sense-perceptions here are dim intellections, but the intellections there are clear sense-perceptions' (VI.7.7.30–1).

The sketchy account just provided shows that Plotinus' criticism of Aristotle in VI.7.2–5 is connected with his account of demiurgic causation in VI.7.1. Plotinus agrees with the Peripatetic criticism of craftsmanlike causation, but does his best to demarcate his position from hylomorphic essentialism. His sophisticated discussion of essence and definition, therefore, completes his account of demiurgic causation. In a way, Plotinus' attitude towards Peripatetic philosophy is similar to Alexander's attitude towards the Epicureans. Alexander does his best to differentiate his position from that of Epicurean philosophers precisely because the Peripatetic and Epicurean views could be seen as similar (since both entail the replacement of Plato's demiurge with nature – though Peripatetic and Epicurean philosophers obviously conceive of nature in very different ways from each other). Likewise, Plotinus' metaphorical interpretation of Plato's demiurge and his account of causation are very similar to Alexander's view. Plotinus, however, certainly does not wish to replace demiurgy with hylomorphism and his critical discussion of Aristotle's essentialism in VI.7.2–5 shows that his metaphorical account of demiurgic causation holds within a fully Platonic framework according to which things 'here below' derive from intelligible causes and are just lower images of extra-physical essences. Moreover, Plotinus argues that hylomorphism presents certain internal difficulties and inconsistencies, which can only be solved via the assumption of 'Platonic' intelligible causes.

I do not wish in any way to deny the existence of a Gnostic background to Plotinus' discussion. Plotinus' sections on demiurgic causation and the 'three men' can indeed be seen as a response to Gnostic speculations. It may well be the case that Plotinus conceived of the first section of VI.7 as his alternative (or implicit response) to the Gnostic views on the cosmos and man.[52] But it cannot in any way be proven that Plotinus developed his theories against a Platonic–Gnostic background: as I hope to have shown,

[52] For this reading see Corrigan 2000: 160, 176–7.

much goes against this hypothesis. Rather, Plotinus makes creative and sophisticated use of the philosophical school debates that raged in the second and early third centuries and develops his distinctive 'version' of Platonism against this philosophical background. In his anti-Gnostic polemic (see esp. II.9) he simply employs his views about demiurgy and causation as a resource against Gnostic cosmology (since Gnosticism, in his view, is nothing but a misleading form of Platonism: see esp. II.9.6). It might well be the case that Plotinus' 'spiritual experience' was close to that of the Gnostics.[53] However, Plotinus' philosophical arguments point in a different direction.

[53] See Narbonne 2011: 151.

CHAPTER 3

Creation and divine providence in Plotinus

Christopher Isaac Noble and Nathan M. Powers

Introduction

According to Plato's *Timaeus* our world was constituted and arranged in the best possible way by a divine craftsman (the demiurge). Plato represents the demiurge as being guided by his own planning or calculation (*logismos*: *Tim.* 30b, 34a–b) about how to bring about maximally good order in the cosmos he produces. Indeed, it seems that the demiurge's causal power with respect to our cosmos is labelled 'foresight' or 'providence' (*pronoia*: *Tim.* 30b–c) precisely to reflect the claim that benevolent divine planning is explanatory of various features of the physical world.[1] Among Platonists of the early empire, planning was widely understood (on Plato's authority) to be an integral aspect of the demiurge's creative activity. Plotinus, however, dissents sharply from the earlier Platonist tradition on this point. He argues that divine reason (*nous*) cannot make plans of any sort concerning the physical cosmos, or indeed entertain any thoughts whatsoever about physical realities. This leaves Plotinus with the challenge of developing an alternative account of how divine intellectual activity determines and guarantees – as he affirms it does – a 'providentially' optimal world order.

The bulk of this chapter will discuss Plotinus' case against the view that the cosmos is the product of demiurgic planning and the alternative account of divine providence that he proposes in that view's stead. But it will be useful to begin by briefly focusing attention on the picture of divine planning presented in the *Timaeus*, because Plotinus' dissenting story about providence can plausibly be taken to be

We wish to thank audiences at the LMU-München and at a conference at the CNRS Centre Jean Pépin in Paris (organized by Meryem Sebti and Daniel de Smet) for valuable discussion of this chapter, and Peter Adamson for helpful written comments on an early draft.

[1] Cf. *Tim.* 44c, 45a–b.

motivated by a desire to resolve tensions internal to Platonism that arise in connection with this picture.

In the *Timaeus* the demiurge is initially confronted with pre-cosmic stuff in a disorderly state; and since he is good, he desires to make everything that exists (including this stuff) as good as possible. To this end, he decides to construct a cosmos endowed with soul and intelligence (30b–c); and since the cosmos is to be a living being he selects the 'complete living being' as the appropriate intelligible model after which to fashion the cosmos (27d–29a, 30c–31b).[2] In making these and other arrangements for the structure and constitution of the world and its parts, both the demiurge and the 'lesser gods' – to whom he delegates the creation of mortal beings – are guided by their thoughts about how to bring about the best cosmic order and well-designed organisms. In Plato's own account there is some obscurity about the respective contributions of the intelligible model and of god's planning activity to particular features of the sensible cosmos.[3] Still, it is clear that basic to Plato's conception of the divine planning are (i) that it is thought about the sensible cosmos; (ii) that this thought involves the discovery of means to ends; and (iii) that it is directed towards prospective states of affairs (as its description as 'forethought' suggests). Plotinus, however, does not accept that the divine thought responsible for cosmic order could have any of these features, and accordingly denies that the cosmos is the product of planning.

Plotinus' case against divine planning rests upon a set of assumptions that hail in part from the *Timaeus* itself and its tradition of interpretation. Among these, of particular importance are the Platonist identifications of the demiurge with a divine intellect and of the intelligible model that the demiurge looks to with the Platonic Forms.[4] These two interpretative claims appear to have been widely regarded as compatible with the thesis

[2] The 'complete living being' that the demiurge uses as a model is apparently the Form of Animal, which is 'complete' by virtue of being the whole of the kind animal, rather than any of the specific animal kinds that are its parts (*Tim.* 30c). The *Timaeus* leaves unresolved how the design of the physical cosmos might reflect the population of the intelligible world more generally.

[3] In the *Timaeus* some features of the cosmos evidently reflect the demiurge's desire to instantiate what is already given in his model, but planning is also required to work out how features of the model are to be best approximated in a physical medium; other features here may well not represent attempts to copy any feature of the model. This left wide room for variation in ancient interpretations of the demiurge; at one extreme, the early Stoics were inspired by the *Timaeus* to identify the benevolence of their divine craftsman entirely with his providential reasoning about the best disposition for the cosmos, even though they dismissed the notion that the cosmos has an intelligible model (see Powers 2013).

[4] The identification of the Forms (collectively) with the demiurge's model finds support in the causal role other Platonic texts accord to a wide array of Forms.

that god plans certain features of the sensible cosmos.[5] By contrast, Plotinus reaches the conclusion that, once other Platonic commitments are taken properly into account, the identification of the demiurge with a divine intellect and of the model for the sensible cosmos with the Forms leaves no place for divine planning in cosmology (or, indeed, for the demiurge to have any intentions whatsoever regarding the sensible world). Faced with this result, Plotinus proposes that all references to planning in the *Timaeus* are to be taken metaphorically, as Plato's way of indicating that the cosmos, since it is as good as it can be, accords with the arrangement at which a divine designer would have arrived, had the cosmos in fact been the product of planning.[6]

The framework for Plotinus' views about providence

Before turning to Plotinus' arguments against the view that the demiurge engages in planning the cosmos, it will be helpful to briefly orient ourselves to the place of the demiurge and his creation in Plotinus' metaphysics, and to articulate the further commitments that serve as the basis for those arguments.

Plotinus maintains that the whole of reality derives from a perfectly simple first principle identified as 'the One' or 'the Good'. This first principle, the One, generates a divine Intellect (*nous*) that thinks the Platonic Forms, which also constitute its substance, so that Plotinus' divine Intellect is, like Aristotle's, a self-thinker. This Intellect in turn generates Soul. Finally, one particular soul, the so-called world soul (and

[5] Middle Platonist authors widely adopt the *Timaeus*'s characterisations of the demiurge as planning the cosmos and its soul, often repeating Plato's own examples without any evident qualms: see e.g. Philo of Alexandria *De op. mund.* 16 ff.; Apuleius *De Plat.* 1.8, 1.10; Atticus fr. 6.6–9 des Places; Plutarch *De an.* procr. 1025a–f, 1027a; Plutarch *De sera* 550d–e; and Alcinous *Didask.* 12.2–3, 14.4; cf. Galen *Hipp. et Plat.* 7.5.21. The state of our evidence makes it difficult to determine whether the absence of explicit references to planning in other authors reflects dissent on this point. For providence in Middle Platonism see Dragona-Monachou 1994; Mansfeld 1999; and Sharples 2003. For Platonists on the demiurge see Opsomer 2005b.

[6] III.2.14.1–6, V.8.7.36–44, VI.2.21.32–7, VI.7.1.28–32, VI.8.17.1–4; cf. III.2.3.6–9. Plotinus also appears to interpret the claim that the demiurge desires to create because he is without envy (*phthonos*, *Tim.* 29e) metaphorically. True to the letter of the *Timaeus*, he describes the productive activity of intelligible entities as a manifestation of their lack of 'envy' (IV.8.4.1–10, IV.8.6.1–18, V.4.1.34–6). But since, in the cases of his first principle (the One) and of *nous*, this production involves no desire to produce their product (III.2.2.10–12), Plotinus construes this language in non-intentional terms; cf. his qualifications of the claim that they lack envy (*hoion*, IV.8.6.12; *hôsper*, V.4.1.35), his application of this description to a radiant light source (IV.8.4.3–5), and more generally his use of inanimate entities (e.g. fire's warming and snow's cooling) to illustrate the productive activity of intelligible entities (cf. e.g. IV.5.7, V.1.3, V.1.6, V.3.7, V.4.1).

specifically, its nutritive faculty, Nature), generates the sensible cosmos by producing matter and imparting form to it. At each stage of this causal sequence the producer timelessly generates its product (whole-cloth, as it were) as a necessary outcome of its own essential activity – a causal pattern typically labelled 'emanation' (from Plotinus' images of the product 'flowing from' its cause) or 'double activity' (with reference to Plotinus' distinction between the *energeia* that is intrinsic to the cause and the *energeia* that constitutes the product). In this scheme the demiurge is (for Plotinus as for Platonists before him) identified with the divine Intellect; and his role in creating the physical cosmos is mediated by the activity of the world soul he creates.

This Plotinian interpretation of the broad structure of Platonist metaphysics represents in certain ways a departure from earlier mainstream Platonist views. Most conspicuously, Plotinus' view that the highest god is a principle prior to divine *nous* (the One) and his theory of 'emanation' are exceptional. Yet the basic commitments that underlie Plotinus' arguments against divine planning each enjoyed some measure of acceptance among his Platonist predecessors, and the battery of arguments that proceed from them, taken together, present a forceful challenge to the compatibility of divine planning with other Platonist principles. These commitments are as follows:

C1: *Two-worlds epistemology*

> Only Forms can be objects of knowledge; physical realities, with which we are acquainted through sense-perception, are objects of an inferior mode of cognition: mere belief or opinion (*doxa*).

The *locus classicus* for this two-worlds epistemology is Plato's account of knowledge and belief in *Republic* 5–6. According to the argument in *Republic* 5, knowledge (*epistêmê*) has as its object Forms, whereas opinion (*doxa*) has as its object sensibles. This epistemic schema is further refined in the simile of the divided line in *Republic* 6, where Plato designates the cognitive state consisting of the apprehension of intelligible objects as *noêsis*.[7] As we shall see below, this epistemological picture raises a difficulty about how divine *nous* can reflect upon how sensible reality should be disposed; for the epistemic states most plausibly attributed to divine *nous*, knowledge (*epistêmê*) and understanding

[7] See *Rep.* 508d, where the soul's apprehension of intelligible objects is designated *nous* in contrast to *doxa* of sensible items; and *Tim.* 27d ff. Cf. v.9.7.

(*noêsis*) are directed solely towards intelligible reality and do not concern the sensible realm at all.[8]

C2: *The perfect actuality of divine* nous

Divine *nous* is a fully actual self-thinker.

This view of *nous* owes an obvious debt to the account of divine *nous* in *Metaphysics* 12, where Aristotle argues that god is a fully actual self-thinking intellect. But Plotinus cannot reach his conception of divine *nous* via the strategy Aristotle pursues there, because Aristotle deduces that god has these attributes in a series of arguments built around the common premise that god is the very best thing (a status that does not apply in Plotinus' view to *nous*, but rather to his highest god, 'the One' or 'the Good').[9] Instead, like Aristotle in *De anima* 3.5, Plotinus proceeds from the assumption that there must be some *nous* that knows in full actuality in virtue of its essence; the existence of a *nous* of this sort is necessary as a cause of the actualisation of knowing in merely potential knowers.[10] While there is some dispute about what the *nous* of *De anima* 3.5 knows and how it relates to the divine *nous* of *Metaphysics* 12, for Plotinus the actual *nous* that is the cause of knowing in us is the divine *nous* that thinks the Forms, and it is able to serve as a cause for the actualisation of knowledge in potential knowers (i.e. souls) precisely because it contains all knowledge in actuality (v.9.4). Moreover, Plotinus contends that *nous*'s possession of knowledge in virtue of its essence implies that the beings it knows are internal to it and, indeed, that *nous*'s knowledge and the beings it knows must be one and the same. For if *nous* were other than its objects of thought, it would not know in virtue of its essence, but instead be a potential knower brought to actuality by something else (v.9.5.1–10). In this way Plotinus reaches the view that *nous*'s thought consists in the necessarily complete and comprehensive theoretical contemplation of its own substance (i.e. the Forms).[11] As

[8] Although Aristotle likewise accepts that *nous* has as its object *noêta* rather than *aisthêta* (*De an.* 3.4, 429a13–18), this thesis does not imply a two-worlds epistemology, as the *noêta* acquired by scientific inquiry are attained via perception and concern aspects of sensible reality (*An. post.* 2.19).

[9] Aristotle's arguments in *Metaphysics* 12 are, however, an effective critique against Platonists who identify as the highest god a demiurge who engages in cosmic planning. Aristotle argues that, since god's activity must be the very best activity, he will engage exclusively in an activity of theoretical contemplation (*theôria*), and that this activity is wholly devoid of change. Such considerations provide reasons to doubt that the planning of mundane affairs befits the highest divinity; cf. Sedley 2007: 169–71.

[10] For Aristotelian influences on Plotinus' theory of *nous* see Menn 2001: 235–9.

[11] Plotinus also has epistemological grounds for identifying *nous* with its objects: if they were not the same, *nous* would not grasp the intelligible things themselves, but rather have mere 'impressions' of them (v.3.5.21–9; cf. v.5.1.50–68); see Emilsson 2007: ch. 3.

we shall see below, this conception of *nous* will be an important basis for Plotinus' denial of planning to *nous*. For Plotinus takes planning to involve consideration of non-actual states of affairs; but since *nous* necessarily thinks only actual beings, it is impossible for it to consider such objects.

C3: *The eternity of the cosmos*

The physical cosmos had no beginning in time, but is rather the eternal product of *nous*'s activity.

According to the *Timaeus* the cosmos had a beginning in time, and there was controversy among ancient Platonists as to whether this claim is to be interpreted literally or not. 'Literalists' about creation maintain that there is a temporal beginning for the cosmos, whereas 'non-literalists' about creation maintain that the cosmos is eternal and that Plato speaks as if it had a temporal beginning solely for expository purposes.[12] This debate between literalists and non-literalists has implications for divine planning. If planning involves the discovery of means to ends, and the resulting plan can be implemented only *after* such discovery has occurred, then only literalists about creation in time are well positioned to claim that the permanent features of the cosmos are planned – whether this planning is conducted by divine *nous* or by some other divine entity (e.g. the world soul).

The case against divine planning

Plotinus occasionally appeals to C3 as sufficient to establish the claim that the cosmos is not a product of divine planning: since the cosmos has always existed, there was no time before it existed when it could have been planned out in advance (III.2.1.15–26, VI.8.17.1–9). But in the opening chapters of *Ennead* VI.7, 'How the multitude of Forms came into being and On the good', Plotinus offers a far more ambitious and thoroughgoing series of arguments against divine planning.[13]

Plotinus' discussion takes its point of departure from the case of sight and eyes, which figure prominently in the *Timaeus* as examples of the kinds of abilities and organs that divine reason confers on animals to ensure that they can survive and flourish.[14] The clear implication of the

[12] For these 'literalist' and 'non-literalist' interpretations see Baltes 1976–8 and Sorabji 1983: 268–83.

[13] For a complementary treatment of *Enn.* VI.7 that focuses on its engagement with the earlier Platonist and Peripatetic traditions see the chapter by Riccardo Chiaradonna in this volume (Chapter 2).

[14] 'Did the God [i.e. the demiurge] or one of the gods [i.e. the 'lesser gods'], when sending souls to birth, put "light-bearing eyes" in the face and give them the other organs for each of the senses because he foresaw that safety would be ensured in this way, if one saw and heard beforehand and by

Timaeus's account is that divine reason, in determining that animals are to have features such as eyes, took into consideration which features would be useful to animals in negotiating their physical surroundings.[15] Plotinus, however, sharply rejects the implication that the demiurge literally engages in planning out eyes or any other features of the sensible cosmos. After raising a preliminary query about whether the demiurge should be thought to confer just sense-organs or also the sense-perceptual capacities that are active through them (VI.7.1.10–21), Plotinus proceeds to offer a set of complementary arguments for the claim that the demiurge's creative activity cannot involve anything like means–end reasoning about the sensible cosmos.

The first argument (VI.7.1.21 ff.) attacks the idea that divine *nous* is in a position to think about sensible reality at all. Plotinus begins by granting, for the sake of the argument, that divine *nous* could engage in discursive reasoning (*logismos*). This is not (as we shall see below) a supposition that Plotinus himself accepts; but even if it were true, Plotinus says, divine *nous* could not arrive by such reasoning at conclusions about perceptible, physical items. This is because a reasoner must start from premises about perceptible items in order to reach conclusions about perceptible items; but *nous*, since it lacks perception, has no access to such premises.[16] Instead, *nous* only has intelligible items (*noêta*) as premises and can, accordingly, only draw conclusions about *noêta*. In this argument Plotinus assumes, first and explicitly, that *nous* does not have the capacity for sense-perception (which is posterior to *nous*; cf. 'not yet', VI.7.1.24); and, second (albeit implicitly), that *noêta* do not have as their contents facts about sensible reality. Neither of these assumptions is here defended by Plotinus, but both derive support from (C1), since that view denies to *nous* epistemic states associated with perception, and holds that *nous*'s proper objects (*noêta*) are Forms *as contrasted with* sensible reality.

touching could avoid one thing and pursue another?' (VI.7.1.1–5) Given that Plotinus goes on to deny that the god or gods did in fact confer eyes for the safety of organisms here, it seems preferable not to translate this sentence (with Armstrong) as an assertion.

[15] Although the explicit rationale in the *Timaeus* for giving human beings eyes is above all to facilitate the soul's attainment of wisdom (*Tim.* 47a–c), the proposal that eyes are given for survival is perfectly consonant with the *Timaeus*'s general outlook that bodily structures are designed with a view to the preservation of embodied life (cf. *Tim.* 33a, 45d–e, 72e–74e).

[16] Thaler 2011: 163–4 takes the argument to be that *nous* lacks '*aisthêsis*' in the sense that it fails to have a general concept of sense-perception, and so cannot reason about a particular sense modality (e.g. hearing or sight). But Hadot 1988: 197–8 is surely right to take Plotinus' point to be that *nous* lacks a capacity for sense-perception and, accordingly, lacks appropriate starting points for drawing conclusions about sensible reality (cf. VI.7.1.25–8).

Having contended that divine *nous* cannot make plans concerning the physical world because it cannot even entertain thoughts about sensible reality, Plotinus goes on to offer a series of three arguments for the further claim that *nous* cannot indeed engage in planning (*logismos*) and foresight (*proorasis*) about anything whatsoever, since its nature is incompatible with these modes of thought. The series begins with a brief and elliptical argument (VI.7.1.32–5) for the claim that god could have no need of planning and foresight. Planning, he suggests, is useful only for those who are in the epistemically deficient position of not knowing what is to be done.[17] Similarly, foresight is a cognitive power that would be superseded by something better.[18] The thrust of these suggestions is that if divine *nous* had to plan out the cosmos and (in doing so) to foresee future events, this would imply that it is in the position of a practical agent who must have recourse to deliberation; which in turn implies (unacceptably) not only that divine *nous* is in an inferior epistemic position, but also that it exercises powers whose utility is altogether obviated by its comprehensive theoretical knowledge (see C2 above).

This second argument prepares the way for two further arguments directly contending that divine cognition is incompatible with the sort of reasoning involved in planning. According to the third argument (VI.7.1.36 ff.) planning and forethought necessarily involve consideration of alternative states of affairs; as Plotinus puts it, planning necessarily involves deliberative thoughts of the form 'X rather than Y' (or 'X in order that not Y'). To arrive at (or possess) such a reasoned preference entails the availability of the two alternatives, X and Y, for consideration. Plotinus contends, however, that in any case where planning might seem appropriate, there will be only one object (say, X) present within the ambit of divine apprehension. Therefore *nous* cannot plan. The argument as it stands is severely compressed, and there is a question as to how best to construe the basis for the contention that for any two genuine alternatives X and Y, there is in *nous* only X or Y (and not both). On one prominent reading, *nous* faces no such alternatives because everything in *nous* is necessary rather than contingent.[19] We agree that there are no

[17] This is a familiar Plotinian point: III.2.1.10–13, IV.3.18.1–7, IV.4.6.10–13, IV.4.12.1–29, IV.8.8.13–16, V.8.4.36–8.

[18] Presumably, what is meant is contemplation (*theôria*); cf. the claim that nature governs the cosmos by contemplation: III.8.1 ff.

[19] For this proposal see Bréhier 1924–38 vol. VI.2: 44; Thaler 2011: 164–5; and Hadot 1988: 197–9. Hadot takes Plotinus to exclude the existence of (future) contingents in *nous* by way of the argument below (VI.7.1.45 ff.) for the thesis that all things *nous* knows are simultaneously present to it

contingencies or open possibilities in *nous* (indeed, this much is implied by Plotinus' denial of alternatives in *nous*). However, Plotinus does not deny the existence of alternatives in *nous* by adverting specifically to modal considerations, and we favour a reading that does not require an additional, implicit appeal to the necessity of *nous*'s thought-contents. Recall that on Plotinus' theory of *nous*, *nous*'s objects of thought are the actual beings it knows (see C2). So *nous*'s thoughts will consist just in thought of what is actual. On this view, *nous* cannot think 'X, so that not Y' or 'there would have been Y, if not X' for the simple reason that unrealised alternatives (Y's) – precisely in so far as they are unrealised – would be what potentially was (or is or will be); and *nous* thinks only actual beings (X's). The actuality of divine thought thus precludes its entertaining alternatives and so precludes planning.

In his fourth argument (VI.7.1.45 ff.) Plotinus continues to develop the contrast between *nous*'s thought and deliberative forethought by observing that all the beings of which *nous* can be aware are wholly and simultaneously present to it. So there are, for *nous*, no prospective (i.e. not yet realised) states of affairs to be *fore*-seen or pre-planned; this much is implied by the complete actuality of *nous* (see C2). Nonetheless, Plotinus vindicates a special sense in which *nous* might be said to exercise foresight (as he affirms it does, lines 42–3), claiming that 'the future' is already present in *nous* in the sense that it comes to be later in time not in *nous*, but 'in another' (i.e. the sensible world) 'as if it had been planned in advance' (lines 49–52). Thus, *nous* 'foresees' the future just in so far as it knows once and for all the intelligible counterparts of those things that come into existence earlier and later in the sensible world.[20]

Finally, Plotinus presents a fifth argument (VI.7.3.1 ff.) against the proposal that certain features of the sensible world might be the products

('[L]e Dieu ne peut raisonner parce qu'il n'y a pas pour lui de choix possible, c'est-à-dire parce que tout est en lui présent de manière simultanée'). Hadot's construal of the development of the passage relies on reading the sentence at VI.7.1.42–5 as the question of whether *nous* exhibits foresight rather than as the assertion that it does (cf. Gurtler 2002: 112). But this is linguistically difficult, given the adversative force of the phrase (οὐ μὴν ἀλλὰ) that introduces the next argument (see Denniston: 28–30; Smyth §2767 ['nevertheless']; Kühner-Gerth II, §534.7 ['jedoch']) and serves to announce Plotinus' intention to qualify the sense in which *nous* can be said to exhibit foresight.

20 One might be tempted to read Plotinus' claim that the 'the future' is already present in *nous* as the stronger claim that *nous* actually has knowledge of future states of affairs here. Plotinus' epistemological commitments however tell against this stronger reading: given that *nous*'s knowledge is not of sensible reality (see C1), it cannot strictly speaking know future states of affairs in the sensible world. Contrast e.g. Boethius' God, whose knowledge is completely and eternally present to him, and embraces all states of affairs in the sensible world – past, present, and future (*De cons.* 5.6.1–17).

of planning that 'supplements' *(prostheinai,* VI.7 2.54) what is already given in the Forms, on the grounds that such supplemental planning is incompatible with the eternity of the world (C3).[21] As noted above, Plotinus elsewhere points out the difficulty in thinking that an eternally existing cosmos was pre-planned, since there was no time before the cosmos existed when it could have been planned out. Here he builds on this difficulty by confronting the proponent of a planned cosmos with two further problems as to the circumstances in which the alleged planning occurred. On the one hand, if the cosmos has always been optimally arranged in accordance with an already existing divine plan, then a divine craftsman could only engage in planning – that is, in the formulation or discovery of a plan for the cosmos – if he had somehow forgotten that previously implemented plan.[22] But if on the other hand the divine plan is a plan to remedy pre-existing deficiencies in the cosmos, then we will be unable to explain why this improvable cosmos had not previously been improved, since there was (*ex hypothesi*) a beneficent cause of good already present to it. Thus, god could relate to an eternal cosmos via planning only at the cost of being cognitively imperfect or imperfectly good; neither of these alternatives is acceptable. It is worth noting that this final argument also excludes the possibility that some divine entity other than *nous* (e.g. the world soul or the 'lesser gods' of the *Timaeus*) could be responsible for planning permanent features of the cosmos; for the same problems can be posed for any such entity.[23]

Plotinus' five arguments together constitute a robust case against divine planning of the sensible cosmos. Divine *nous*'s nature implies that it (i) lacks appropriate starting points for thinking about the sensible world; (ii) is not in a defective cognitive condition where planning would be

[21] Thaler 2011: 165–6 rightly notes that the argument targets the idea that creation of sensible kinds involves supplementing features given by the Forms with 'additions or alterations'. The opening lines of the chapter (ll. 1–3) may well adumbrate a further criticism of supplemental planning in particular, namely that such planning of structures here would be counterproductive, as it could only serve to add or subtract features given in the Form, and so, reduce the product's conformity to its intelligible model.

[22] That is, if he forgot the plan specifying how Forms should be instantiated; contra Hadot 1988: 206, the problem envisaged is not that of the fallen demiurge (à la the Gnostics) who must plan because he has forgotten the Forms. See preceding note.

[23] Indeed, Plotinus elsewhere stresses that the world soul (and nature) do not deliberate or plan, but rather reflect on thought-contents that derive ultimately from divine *nous* (IV.4.10.6–18, IV.4.11.1–13, IV.4.12.1–29, IV.8.8.13–16), a fact that underlies Plotinus' decision to characterise their mode of thought as *theôria* (III.8.1 ff.). Armstrong obscures this point in his translation of III.9.1 by construing the mental activity (*dianoeisthai*, cf. *Tim.* 39e9) whereby a soul undertakes to reproduce what it sees in *nous* in the sensible cosmos as 'planning'.

necessary or appropriate; and cannot engage in the consideration of (iii) alternatives or (iv) prospective states of affairs as such, both of which are essential to planning. Furthermore, (v) the eternality of the world's structures is incompatible with these features being planned by gods.[24] Plotinus accordingly suggests (VI.7.1.28–32) that the terms *logismos* and *proorasis* are properly used of divine *nous* to indicate that things here are arranged as someone who reasoned about them *would have* arranged them, not to indicate that god actually engages in *logismos* or literally 'fore-sees' what happens here. This comment seems to be intended to legitimise Platonic ways of talking.

Providence without planning

What room is left for providence by Plotinus' claim that divine planning is impossible (and so *a fortiori* that it has no causal role in the ordering of the physical world)? Plotinus gives his clearest explanation of what he takes providence to be in III.2–3, *On Providence*, which comprises a continuous essay in theodicy. The main purpose of this text is to rebut the claim (put in the mouths of Epicureans and Gnostics) that our cosmos is not arranged in the best possible way for a thing of its kind. We shall have something to say below about his strategy for this rebuttal; what is of immediate interest is the 'position statement' that Plotinus stakes out near the outset of his discussion. He asserts that providence (*pronoia*) exists, but warns against understanding this assertion to entail that the cosmos has been ordered in accordance with divine deliberation or *logismos*. For the term '*pronoia*' when applied to *nous* as the cause of the cosmos does not imply that *nous* is prior to (*pro*) the cosmos in time (and so that it gives *fore*-thought to the cosmos); indeed, the term does not imply divine thought about the cosmos at all. Rather, the term designates the metaphysical priority of *nous* to the cosmos (III.2.1.21 ff.; cf. VI.8.17.9–12); or, as he puts it in what appears to be his official definition of *pronoia*, 'providence [for the physical cosmos] is its being in accordance with *nous*' (III.2.1.21 f.).[25]

It is worth pausing to appreciate just how radically austere an account of providence this is. In asserting that divine providence has brought about the best physical cosmos possible, Plotinus explicitly refuses to ascribe any

[24] These considerations may not exhaust Plotinus' reasons for denying planning to *nous*. For example, since Plotinus' *nous* is, like Aristotle's, unchanging, it cannot engage in a step-by-step process of reasoning (Opsomer 2005b: 84).

[25] In line with this definition, the world soul exercises providence in a derivative way, by communicating the contents of *nous* to the sensible world (III.3.4, IV.8.2).

planning or supervisory capacity to divine *nous* with regard to that cosmos. For him, to speak of *nous* as *pronoia* is simply to point to its causal role vis-à-vis the cosmos; one might just as well say that the term properly refers to the katanoetic nature (as it were) of the physical world, that is, to the fact that the contents and features of this world are causally determined by the Forms that *nous* contemplates.

It is in the very nature of the Forms to emanate forth from their own being, and in so doing to generate the physical world as a reflection and image of themselves. On the austere Plotinian picture, whatever exists here in our world is simply a necessary result of that emanation. Accordingly, the only form of explanation Plotinus will admit for the world's containing some given formal feature is '[in order] that there shall be everything' (VI.7.3.14 f.), that is, because intelligible reality necessarily gets expressed as fully as possible here in its image (the sensible realm). So all things here are present here because they are already present there, among the Forms; but, again, they are not present either here or there through divine planning. The benefit and welfare of the inhabitants of the physical universe were not (and could not possibly have been) taken into account in its creation.[26]

Problems for Plotinian providence

This view of providence, according to which all formal features of the sensible cosmos have counterparts in *nous*, faces certain difficulties. To begin with, Plotinus himself identifies two *aporiai* to the effect that *nous*'s possession of such counterparts is incompatible with the perfection of its thought. It is moreover unclear how to explain the fact that *nous* contains counterparts of structures here in the sensible world that seem optimised for the survival of living creatures, given that (on Plotinus' theory) *nous* is precluded from giving any thought to the conditions of embodied life.

[26] It has been proposed that Plotinus' austere theory of providence may owe a debt to Alexander's. For Alexander providence is exercised over the sublunary region by the heavens, whose orderly motions ensure regular processes of generation and destruction as well as the permanence of species kinds (*Quaest.* 1.25 41.4–19 and 2.19 63.10–28 Bruns; *De prov.* 33.1 f., 87.5 ff. Ruland). Alexander's view that providence involves no planning anticipates a key aspect of Plotinus' position, but his view as stated accords no prominent role to divine thought. In particular, it is not clear whether, for Alexander, divine *nous* itself (i.e. the Prime Mover) exercises providence (for a negative answer see Sharples 1982: 200–4); or whether Alexander takes the contents of divine thought to determine the structure of the cosmos or the natural kinds within it, as Plotinus does. For Alexander see Sharples 1982, and, for his possible influence on Plotinus, Thillet 2003: 46–54, and the chapter by Riccardo Chiaradonna in this volume (Chapter 2).

Let us begin with the two *aporiai* Plotinus himself discusses. First, since *nous* has counterparts of all things here, there must be a counterpart of sense-perception in the Form of Man. But if sense-perception, a non-rational cognitive capacity, itself has a counterpart in *nous*, then this threatens to involve *nous* with a cognitive activity inferior to the *theôria* which alone is proper to it (VI.7.3.22–33, VI.7.4–7). Plotinus addresses this difficulty by insisting that the noetic counterpart of sense-perception is not a capacity to apprehend sense-objects. As designations such as 'fire' or 'horse' are not applied synonymously to the Forms and their participants, Plotinus can reasonably deny that the existence of 'perception' in *nous* straightforwardly implies its possession of a capacity to apprehend sense-objects. Moreover, since the capacity to perceive sense-objects would be incompatible with *nous*'s nature as a pure intellect, Plotinus has a positive reason to insist that this feature of 'perception' here does not apply to the noetic counterpart of the same name. Plotinus thus accepts that there must be a counterpart of sense-perception in *nous* that serves as an analogue for the cognitive faculty here that is essentially related to sense-objects, *without* itself standing in a cognitive relation to sense-objects. To explain how this might be possible, Plotinus proposes that the relation of our non-rational faculty of sense-perception to perceptible bodies has its analogue in our rational soul's cognitive relation to the noetic counterparts of 'bodies' or 'sensibles' – presumably, what is meant here are those Forms that correspond to sensible bodies here, e.g. the Form of Fire[27] – whereas the rational soul's cognitive relation to Forms in *nous* has as its analogue the cognitive relation of the Form of Man to itself (VI.7.6.1–12, VI.7.7.19–31).[28] With this proposal, Plotinus suggests that counterparts for non-rational faculties can be found in *nous* without *nous* itself being saddled with non-theoretical capacities.

Second, there is the *aporia* that the perfection of *nous* will apparently be compromised if it contains noetic counterparts of inferior sensibles (e.g. 'worthless' [*eutelês*] non-rational animals: VI.7.8.15–18, VI.7.9.1–5). Plotinus' initial line of response to this difficulty is to point out that the counterparts of non-rational animals must (in accordance with his theory of *nous*) be intellects and so, far from being worthless, very fine indeed.

[27] For the suggestion that 'bodies' and 'sensibles' in *nous* are to be so understood see Emilsson 1988: 360–3.

[28] There is a question here about how the cognitive relation of the Form of Man to itself serves to prefigure the rational soul's cognitive relation to a plurality of Forms. The answer may lie in Plotinus' view that each Form, in thinking itself, also thinks other Forms, apparently because its comprehensive self-knowledge requires an understanding of how it relates to other Forms.

But he concedes that these intellects are presumably inferior to those intellects within divine *nous* that correspond to rational animals. This response shifts Plotinus' burden to showing that the existence of such inferior intellects in *nous* is consistent with its optimal condition overall. And Plotinus thinks this burden can be met by contending that the inclusion of these inferior intellects in *nous* contributes to its perfection by making it complete ('*teleios*', '*pantelês*': VI.7.10.4–7, VI.7.12.1–4).[29]

This appeal to completeness might look like an ad hoc solution to the problem at hand, but in fact the view that the completeness of *nous* involves an extensive plurality of inferior Forms has a firm basis in the principles governing *nous*'s constitution.[30] On Plotinus' theory there are relations of priority and posteriority within the contents of *nous* (e.g. each genus is prior to its subordinate species), and inferior Forms are a necessary outcome of successive stages in the articulation of *nous*'s contents (VI.7.9.15–20). However, even though these articulations bring with them some Forms that are (taken individually) inferior to others, Plotinus has a principled explanation of why such articulations are intrinsic to *nous*'s nature. He contends that an object of thought must have some complexity, and so, that some degree of composition in the object of thought is a minimal condition on *nous*'s thinking. But what is crucial for present purposes is that Plotinus *also* takes the minimal differentiation required for thought to presuppose a principle responsible for maximal differentiation within *nous*.

> If it [i.e. *nous*] has no change in it, and no 'difference' wakes it to life, it would not even be an activity. . . . Now certainly what has come to be has come to be from the same and the different; and not just any different, but universal difference; for its same is also universal. But since it is all that is same and all that is different, there is no one of the others that it leaves out. Its nature therefore is to become different in every way. (VI.7.13.11–12, 21–5)

In this passage Plotinus alludes to his interpretation of the five 'greatest kinds' at Plato's *Sophist* 254b–55e – being, sameness, difference, motion, rest – according to which these generic 'kinds' serve as primary principles explaining the nature and constitution of *nous* (II.4.5, V.1.4, VI.2.7–9). On this interpretation, 'difference' is the cause not just of the minimal differentiation in the object of thought required for thinking, but also – apparently, because 'difference' is *universal* in scope – of comprehensive differentiation in *nous*'s object. So, on Plotinus' theory, it follows from the

[29] Cf. the characterisation of the model as a '*complete* living being' at *Tim.* 31b.
[30] For these principles see Rappe 2002; Emilsson 2007: 160–4; and Thaler 2011.

basic principles of divine thought that there will be comprehensive differentiation among *nous*'s objects of thought. Thus, Plotinus can reasonably insist that maximal internal diversification, together with the existence of inferior Forms it implies, is necessary to the completeness of *nous*'s nature.[31]

Thus far, we have focused on a pair of difficulties for the perfection of divine thought that Plotinus sees as arising from the thesis that all formal features of the cosmos are prefigured in *nous*. But this thesis might be thought to involve a further problem, one that presents a significant obstacle to regarding Plotinus' theory as a respectable alternative to the natural teleology of Aristotle or the Stoics. According to those models of natural teleology, the perfection of individual species and the good order of the sensible cosmos as a whole are to be explained by reference to the choiceworthy ends their features subtend. Plotinus by contrast maintains that although the sensible cosmos and the species within it are optimally constituted, their features are fully determined by the contents of a divine mind that cannot take consideration of the natural world or of ends internal to that world. How then can Plotinus explain the fact that *nous* possesses those very thoughts that ensure a well-ordered physical cosmos and well-adapted organisms, given that it cannot (*ex hypothesi*) give any thought to the sensible cosmos and the conditions of embodied life? If we concede to the Platonist that divine *nous* cannot think about, for example, the utility of eyes and horns for organisms here, can he still tell a plausible story about how *nous* entertains thoughts that guarantee the existence of such useful structures here? Plotinus is clearly optimistic that *nous*'s thoughts ensure an optimal cosmos, albeit without fully explaining the grounds for his confidence. In what follows we will argue that his presentation of his theory does suggest one important reason for this optimism.

It will be helpful to begin by drawing a distinction between two different possible ways of conceiving of the noetic counterparts of things here. On the one hand, Plotinus might think that *nous* contains more or less detailed blueprints for each of the structures here; for example, the noetic counterpart of 'horn' in the Form of Bull might be a specification for a pointy keratinous protuberance of the skull of a certain shape and relative size (Option 1). But this suggestion is problematic. It is hard to see

[31] Compare Thaler 2011: 177–9, who proposes that the diversity of living Forms serves the teleological purpose of supplying all necessary parts of 'the general notion of life' (178). He rightly calls attention to evidence that, for Plotinus, each Form contributes to the 'completeness' of *nous*'s life; less clear is whether Plotinus means by this (as Thaler proposes) that life is a genus whose existence is dependent on that of all its sub-species.

how *nous* could be in a position to conclude that just this sort of structure is necessary for the completeness of the Form of Bull, given that *nous* is (*ex hypothesi*) precluded from thinking about the conditions here that make the desirability of a structure with just these features intelligible (e.g. defence against predators or competition with other members of the herd). On the other hand, Plotinus might think that the noetic counterpart of 'horn' does not contain a detailed articulation of the features of horn we find here; rather, 'horn' in *nous* is something that has a function in *nous* analogous to that of horn here, while the structural features of horn here only emerge in the process of emanation (Option 2). We will now adduce some textual evidence that lends some support to Option 2, and then consider how Option 2 helps Plotinus avoid the problem raised for Option 1.

As mentioned above, Plotinus appears to hold that the noetic counterpart of the sense-perceptual apparatus in humans is constituted simply by the cognitive relation of the self-thinking Form of Man to itself. Similarly, when Plotinus discusses the noetic counterparts of horns and claws, he asserts that these are not for defence (a Form, after all, doesn't need to defend itself), but rather serve to make the Form in question self-sufficient (*autarkês*) and complete (*teleios*) as the kind of 'living being' (*zôon*) it is, namely as the theoretical intellect in question (VI.7.9.39–10.7). The notion that 'horns' and 'claws' make the Forms in question complete is neutral as to what the Form thinks in having these elements. But there is no suggestion that 'horns' and 'claws' are otherwise like the structures of the same name in the sensible world, and Plotinus' treatment of the noetic counterparts of sense-perception (discussed above) tends to count against the suggestion that they are. So what we seem to find in *nous* corresponding to claws and horns are features that play the functional roles of providing intelligent life, self-sufficiency and completeness to the Forms, rather than blueprints for whatever bodily structures or psychic faculties might best serve to realise the analogues of these excellent divine attributes in the sensible world. If this is right, then Plotinus is not committed to the implausible position (Option 1) that *nous* works out in detail structures that will turn out to have utility here without taking into consideration the conditions and ends here with reference to which these structures have utility.

But if the physical structures of such features as 'eye' and 'horn' are not already worked out in *nous*, as Option 2 would have it, then how do features of *nous* ensure the existence of such useful structures here? Presumably, the answer must be that these features in the sensible world come into existence automatically (as it were) as images of the Forms generated though the process of emanation. This is not as uninformative a solution as it might seem. Once we accept the assumption that the Forms cause (but do not contain detailed specifications of) optimal physical structures, the emergence of such structures in the sensible cosmos is secured by the Platonist assumption that Forms cause their sensible participants to approximate to the Forms' own perfections. Plotinus can be confident that every species here will turn out to be a successful design, since it will necessarily possess whatever physical structures replicate (as far as is possible in the physical medium) the completeness, self-sufficiency and vitality of the Form of which it is a likeness.

A theory of *providence*?

Earlier we observed that Plato characterises the demiurge's thought in relation to the physical cosmos as 'providence' (*pronoia*) because the organisation and features of the cosmos are planned out in advance by the demiurge. Given Plotinus' categorical denial that the Platonic demiurge in fact has any plans or concerns whatsoever regarding the cosmos (let alone any with temporal priority to it), one might reasonably wonder how he manages to take his theory about the causal relationship between divine *nous* and the physical world to constitute an account of *providence* at all.[32] Plotinus no doubt feels some obligation to maintain that the world is under the control of providence simply in order to remain exegetically faithful to Plato's assertions about *pronoia*, and this represents one motivation for his efforts to show that the application of the *term* '*pronoia*' to divine *nous* is justified. Thus far, we have seen that Plotinus legitimises Plato's ascription of *pronoia* to the demiurge in two ways: (1) *pronoia* is *nous*'s being causally (but not temporally) prior to (*pro*) the cosmos; and (2) *nous* can be said to have foresight in the sense that it eternally knows the noetic counterparts of what exists both earlier and later in time here in the sensible world. That is not the end of the matter, however; Plotinus has a further (and more substantial) rationale for framing some of his claims about *nous* as claims specifically about providence.

Divine *pronoia* is typically invoked by Plotinus' Platonist predecessors (as well as by the Stoics, who were also in part inspired by the *Timaeus*) precisely as an explanatory basis for the claims (a) that the cosmos is

[32] Cf. Atticus' complaint that Aristotle, like the Epicureans, effectively does away with providence by denying that the gods are interested in human affairs (fr. 3 des Places).

optimally organised and (b) that there is divine justice (i.e. the wicked do not go unpunished, nor the virtuous unrewarded).[33] So if Plotinus can show that divine intellection (as he characterises it) guarantees the truth of both (a) and (b), then he can fairly claim to have delivered a story about 'providence' to rival those traditional theories that appeal to divine planning in order to support the same claims.

We have already seen above how Plotinus takes himself to have established (a): all of the formal features of natural entities (including those that enable animals to survive and flourish, each in its own way) are given in the intelligible content of *nous*, which is also ultimately responsible for their generation (via emanation). It is worth emphasising that Plotinus will say the same thing about the entire physical cosmos itself, taken as a whole: it is the grand product of the emanation of the Forms taken together as a whole, and this is precisely what ensures that it is optimally organised. In III.2.1 Plotinus characterises the Forms collectively as constituting the 'first and truest' cosmos on account of the fact that the relations that they bear to one another are completely harmonious, with no trace of conflict or disunity. This harmonious unity is instantiated in the relations among the various parts of the physical world – in so far as such unity can be instantiated in the latter's imperfect material medium – earning it too the title of 'cosmos' (III.2.2; cf. VI.8.14.25–8).

This justification of (a) amounts to a reinterpretation of the Platonic demiurge: given that formal causes in *nous* are alone what determine the best possible state of affairs in the cosmos, the theoretic activity of divine intellect yields results that are just the same as – or perhaps even better than – those that would have been brought about by optimal planning.[34] In effect, the self-propagating nature of the good precludes the need for divine benevolence understood as goodwill directed towards the world as such; the benevolence of *nous* on Plotinus' view just *is* its self-propagating nature, *qua* good.

As for (b), Plotinus pursues in III.2–3 a theodical strategy that derives in its broad contours from book 10 of Plato's *Laws*. This strategy is too complex to discuss in detail here,[35] but of special note for present purposes

[33] The association of these claims with the assertion of divine providence was not confined to philosophical circles, but enjoyed somewhat wider currency among educated Greeks. See, e.g., Theon *Progymnasmata* (126.3–4), where (a) is invoked as a commonplace that an orator could draw upon in support of the thesis that 'the gods have forethought (*pronoousi*) for the cosmos' 126.3–128.2 (Patillon) and Plutarch *De sera* 549b–e, where doubt of (b) is presented as perennially leading some people to believe that there is no providence.

[34] See note 6 above.

[35] Plotinus borrows liberally from Stoicism in his theodicy as well; see Bréhier 1924–38, III: 17–23.

is Plotinus' claim (following *Laws* 904b–e) that the soul of every mortal after death is reincarnated or transposed into a new situation that reflects its 'worth' (*axia*), and that this new situation itself represents an appropriate reward or punishment for its character and conduct in life (III.2.13). Plotinus emphasises that this process of justice via reincarnation is built into the orderly arrangement of the world; it does not require special divine attention or intervention.[36] So souls, which are themselves each co-eternal with the cosmos, are fully implicated in the cosmic order; every soul after death comes to be disposed in just the way necessitated by that soul's past lives. The career of every soul is in effect the expression of a law that is both unique to itself and part of a natural order that is as just as possible.

Conclusion

Plotinus rejects the traditional Platonist conception of the demiurge of Plato's *Timaeus*, on which divine *nous* is understood to cause the physical world to be well ordered (to be a fine *kosmos*) through deliberation or planning; as we have seen, he argues this conception of the demiurge to be incompatible with certain important commitments (C1–3) that he thinks all Platonists should share. He offers an alternative account of divine providence that is interesting because of its very austerity. The physical world exists because the Forms necessarily produce it by emanation; and the physical world is well ordered because it is necessarily the physical expression of the perfections of the Forms, in all their variety and mutual harmony. Plotinus urges us to interpret all Platonic talk of divine benevolence in terms of the former claim, and all Platonic talk of divine craftsmanship in terms of the latter.

Plotinus' account of the demiurge's providential activity was to enjoy an influential but mixed legacy in later antiquity.[37] While his position that the cosmos is not the product of divine planning would become the predominant view, his denial that the providential deity thinks about the sensible cosmos would prove to be too radical a thesis for many of his successors. On these points Proclus is illustrative of broader trends. Proclus

[36] Claims about justice through reincarnation appear frequently in the Platonic myths (e.g. those in *Gorgias*, *Phaedrus*, *Phaedo* and *Republic* 10), but Plotinus is apparently interested in a feature unique to the myth of *Laws* 10: the process of soul relocation is there claimed to occur automatically and in accordance with general rules, so that the divine administration of the universe is accomplished, as Plato puts it, with 'marvellous ease' (904a).

[37] For providence in later Neoplatonism see Schibli 2002: ch. 3; Adamson 2015; and the introductions to Steel 2007 and Opsomer and Steel 2012.

wholeheartedly accepts Plotinus' verdict that divine providence involves no deliberation or planning, and to this extent he endorses Plotinus' decisive turn away from the traditional Platonist view.[38] On the other hand, Proclus is not prepared to follow Plotinus in his insistence that there is no knowledge of the physical world on the part of the providential deity. Rather, Proclus contends that providence entails knowledge of what falls within the scope of its oversight, and so, knowledge of physical reality, at the level not just of kinds but even of particular items and contingent events.[39] Here Proclus' attempt to develop an epistemology capable of reconciling the transcendence and immutability of divine cognition with the idea that providence does involve thought about our cosmos reflects – at least implicitly – a partial acceptance and a partial rejection of Plotinus' remarkably austere and revisionary account of divine *pronoia*.

[38] Proclus contends that gods and lesser divinities exercise providence without calculation (*logismos*: *El. Th.* § 122; *De dec. dub.* 65).

[39] For Proclus, although the demiurge proper continues to be situated at the level of divine *nous* (specifically, he is identified with Proclus' third triad therein, the 'intellective intellect'; for details see Opsomer 2000: 116–19), providence is identified in the first instance with the beneficent activity of the henads, which are gods posterior to the One, but prior to *nous* (*El. Th.* §§ 120, 134; *De dec. dub.* 66); see Proclus' etymology of '*pronoia*' as 'prior to *nous*' (*pro nou*) (*El. Th.* § 120; cf. Prov. 13). Lesser divinities also exercise providence, thanks to their participation in the henads (*El. Th.* § 120; *De dec. dub.* 62 ff.). For the thesis that the henadic gods know particulars and contingents, albeit timelessly and non-contingently, see *El. Th.* § 124; *De dec. dub.* 2–8 and *In Parm.* 4.952 ff., with Opsomer and Steel 2012: 5–16. For divine foreknowledge of particulars in Neoplatonist authors see Sorabji 1983: 255–6.

CHAPTER 4

Waiting for Philoponus

Richard Sorabji

Part I Three Gazan Christians waiting for Philoponus

The fifth-century CE pagan Neoplatonist Proclus wrote eighteen arguments for the eternity of the cosmos which were to be recorded and attacked one by one in the next century by the Christian Philoponus in his *Against Proclus on the Eternity of the World.*[1] By 'the world' (*kosmos* in Greek) was meant the entire universe envisaged as an orderly system, and they agreed with each other that the universe was an orderly system, even when they used an expression, 'the universe' (*to pan* in Greek), that did not actually include that in its meaning. Their disagreement was on whether it had a beginning and an end, as the Christians said. Proclus died in 485 CE, and Philoponus wrote his reply, the first of a series against the pagan philosophers, in 529. In the period of almost fifty years between 485 and 529 three Christians from Gaza had already attempted to attack the dominant pagan philosophy of Platonism. They had some limited successes, but I shall argue that it was only Philoponus who had the ability to carry the attack home into the pagan camp.[2]

Three Christian attacks on Neoplatonism

Aeneas of Gaza was the first of the three Gazans. His dialogue *Theophrastus*, written between 485 and 490,[3] was named after his fictitious Platonist

1 For modern editions and translations see Philoponus 1899, Philoponus 2004; Philoponus 2005; Philoponus 2010; and Philoponus 2006.

2 This chapter is a revised version of a paper in Hansberger et al. 2012. Both versions form a sequel to the extensively updated second edition of Sorabji 2010. I should rectify a couple of oversights in that updating. I was aware at the time of only one of the writings, discussed below, of Edward Watts on the Alexandrian school. I further regret that in describing the research of Clemens Scholten, I ascribed one part of it to his reviewer, Christian Wildberg, who had intended only to endorse what Scholten had himself said (1996: 129–43) in his response to Verrycken's hypothesis on the chronology of Philoponus.

3 See Aeneas of Gaza 1958. For the date see Aujoulat 1986 and Segonds 1989. A date earlier than 485 when Proclus was still working would render incomprehensible the claim at Aeneas of Gaza 1958: 3.4–8 that there was no longer any proper study of philosophy being conducted in Athens.

interlocutor who represented the case for Platonism and who supposedly lost to the fictitious Christian Euxitheus. The main subject of the *Theophrastus* was the human soul and its fate before birth and after death, including resurrection. But it overlapped in one part with the subject of the other two Christian texts, the Christians' creation of the world from a beginning, as opposed to the Neoplatonists' eternal creation of the world. Moreover, the overlap turns out to be greater, because the second Christian also finishes his text by discussing resurrection.

The second Christian, Zacharias, born in the port of Gaza, Maiuma, had gone to study rhetoric and philosophy in Alexandria in the 480s. After moving on to study law in Berytus around 487, he wrote in the 490s a dialogue caricaturing the great Neoplatonist philosopher of Alexandria, Ammonius, and named it after him as *Ammonius* (or *On the Creation of the World*). This dialogue represented Ammonius as being refuted and silenced by the arguments about creation of a Christian student identified in the preface as Zacharias himself, even though he warns that when he is in dialogue with the two pagan professors he will switch his designation from 'A' to 'Christ.' (short for 'Christian'), as he does several times.[4] It cited Aeneas a couple of times, and reused some of his arguments, although it added many more, often naïve, but occasionally better than those of Aeneas. Zacharias was to feature as the hero in another of his own treatises, *The Life of Severus*, this time as the most important figure in a riot of about 486 that ransacked the pagan temple of Isis outside Alexandria, and led to the flight from the city of three pagan teachers, the torture of another, and the death in hiding of a fifth. Both dialogues added a later insertion. The *Ammonius* inserted the supposed refutation by Zacharias of Gessius, a younger medical colleague of Ammonius, and it has been suggested that this 'refutation' was added after 525 CE, when Ammonius was no longer alive himself to provide the straw man.[5] The dialogue is amusingly unreal in its boasts, but also sociologically interesting, as it reveals much about the practices in the Alexandrian school, which has recently been excavated,[6] and about the feelings of a zealous and

[4] Preface, lines 8–13. The truth is that Ammonius at some time made a concession, but not the one that Zacharias wanted. He extended Aristotle's argument that God's infinite power is needed to produce eternal *motion* by saying that Aristotle actually *recognised* the point (already made by Proclus) that it was also needed to produce eternal *existence* (see Sorabji 1988: 253–4). But the effect remained, contrary to Zacharias, *eternal*.

[5] Edward Watts has illuminated the situation in a series of four writings to date: 2005; 2006: 204–61; 2009: 113–32; 2010.

[6] By the Polish Centre of Archaeology, Cairo, under Grzegorz Majcherek. I have drawn conclusions from the excavation in the second edition of Sorabji 2010: Introduction, part 1, as above.

dangerously disruptive sub-group of Christian students. Zacharias finished up as the bishop of Mytilene.

As regards a third Christian, Procopius of Gaza (d. *c.* 538), it has been drawn to my attention[7] that there are two pages in Procopius' *Commentary on Genesis*[8] that argue against the eternity of the world, and this battery of summarised arguments includes a few that are more skilful than those that we find in Zacharias. Procopius himself had a brother called Zacharias, but it is not known whether that Zacharias was ours.

The anti-Platonist arguments of the last two Gazans very occasionally anticipate those found in Philoponus. But we shall see that they cannot be compared. One important factor is that Philoponus does repeatedly what the Gazans do only occasionally. He takes on his Platonist opponents on their own terms, using their premises to arrive at his conclusion. Not even Proclus had done that in writing against the Christians, and Philoponus often complains in his *Against Proclus* that Proclus is relying on Platonist dogma which he, Philoponus, finds unpersuasive and sometimes absurd.

This is only one of the reasons why I have described the Christian opponents of Neoplatonism as 'waiting for Philoponus'. But the situation has been illuminatingly nuanced by Edward Watts. It is not that the three Gazans were trying to do the same thing as Philoponus and failing. On the contrary, Watts has argued, in discussing the first two of the three Gazans, that they had more parochial aims (Watts 2005: 218, 222, 225). On this account, it turns out to be only natural that they did not feel it necessary to refute the Platonists on their own terms. Aeneas was a rhetorician, and Watts argues that the readers he wanted to impress were in the literary salons especially of Alexandria and regarded themselves as being on equal, and not unfriendly, terms with the Neoplatonists. Zacharias, by contrast, was writing to impress certain Christian students in Alexandria and to undercut the personal authority of their pagan teachers in religious matters. Neither questioned the value of Neoplatonist cultural knowledge, and Aeneas was interested in displaying his rhetorical acquaintance with it. Both might presumably have achieved their objectives, without showing the Neoplatonists themselves that their philosophical position was untenable, which is what Philoponus set out to do.

[7] I am indebted to Michael Champion for drawing this to my attention in his dissertation (2010). I had the benefit of reading the dissertation along with his draft translations of Aeneas and Zacharias, which were not included in it.

[8] *PG* 87.1, that is, Migne 1857–66: cols. 29B–33A.

Watts offers a hypothesis about Philoponus' motivation, which must inevitably remain conjectural (Watts 2006: ch. 9). But it would, if true, explain the fact that Philoponus opposed the Neoplatonists on their own terms. Up to about 525 CE Philoponus had edited by far the largest number of commentaries on Aristotle based on Ammonius' lectures, with innovative contributions of his own.[9] From 529 onwards, while writing at least one more commentary on Aristotle, he embarked on a series of attacks on the pagan world-view of Aristotle and Proclus. Watts' suggestion is that, on finding that he had been passed over for the chair of philosophy upon the death of Ammonius' immediate successor around 525, in favour of the very much less experienced Olympiodorus, he decided that he could attract Christian students away from Olympiodorus to his own better exposition of pagan thought, accompanied by a critique of it.

I shall now look more closely at the successes and limitations of the three earlier efforts to combat pagan Neoplatonism.

Aeneas on the food-chain objection to resurrection

I shall start with an example from Aeneas' defence of Christian views on the soul and the resurrection of the body. This makes the overlap with Zacharias complete, because Zacharias tacks the same problem about the resurrection onto his treatise on the creation of the world. Like Philoponus, Aeneas is aware of an objection to the resurrection of the body, that our bodies will be eaten after death and will have passed through a food chain, but he is unaware of the sharpest formulation of the problem given by Origen (d. 253/4 CE), which motivates the solution offered by Origen and was apparently followed by Philoponus.[10] The crucial point in the sharpest formulation is that the food chain includes other individual humans.

The food-chain problem was known at least as early as Athenagoras in the second century CE. Christians wanted us to be given back the very same bodies in the resurrection, and the usual view was that this would require us to be given back some of the same particles we had had before. But if the food chain through which our dead bodies had passed included other *human* bodies, the question arose how those particles would belong

[9] Watts does not commit himself on Koenraad Verrycken's hypothesis that the anti-Aristotelian arguments were added later, but that hypothesis has not on the whole found favour.

[10] I am drawing on my discussion in Sorabji 2006: 70–7. The evidence on Philoponus is in fr. 32 in Syriac with French translation in van Roey 1984: 135–8.

to one human individual rather than to another. Origen solved the problem, but at a price that most Christians other than Philoponus did not like to pay. He takes an unusual passage of Aristotle which uses the notion of form (*eidos*) in the sense of the structure of an individual organism, and compares it with an elastic tube, preserving the organism's structure and identity however much its *matter* may change and grow.[11] Origen accordingly suggested that it would count as the very same individual body if the unique structure (*eidos*) of the former body was restored in the resurrection. There would be a positive advantage if the matter was *not* the same, because the Christian promise is of an indestructible new heaven and a new earth – this point Philoponus certainly accepted. Flesh is inevitably destructible. What we need is a body made of a spirit (*pneuma*), whether of air or fire or a mixture, which cannot be sliced, cut or otherwise damaged, but one having the very same unique *structure* as we had at some point in our lives. We might think of the structure displayed in a photograph taken in our most characteristic prime.

The way Aeneas puts the food-chain problem is different. He considers only a *single* link, and not a chain. He puts it not as a question about recovering the *individual* body, but as an apparently easier question about recovering a body of the right – human – *type* after a *single* devouring.[12] It is not obvious why he thinks this a problem. If the idea is merely that our particles get mixed up with particles from an animal of another *type*, why should God not retrieve them for the resurrection? But perhaps he thinks they are *converted* into particles of the other animals' type, so that the particles of *our* type would be lost. That would justify his speaking of *types*.

However, Aeneas occasionally half-recognises, but only briefly, that even a *single* devouring calls also for a discussion of *individual* bodies, *not* types. He does so when he says, 'God despatches each soul to its own and separates and distributes them, just as, in our world, good shepherds easily marshal even one of their myriad sheep in the right place, and there is no risk of any confusion.'[13] He returns a little later to the example of the bronze statue of an *individual*, Achilles, being melted down and turned into gold. Aeneas is following the discussion in Gregory of Nyssa's *On the Soul and Resurrection*, which talks merely of bodily elements being scattered, not of their being *devoured*. Nonetheless Gregory's talk of elements (*stoikheia*) seems to be referring, even if a little ambiguously, to *individual*

[11] Aristotle *De gen. et corr.* 1.5, 322a28–33. I have analysed this reference to form as applied to the Christian resurrection by Origen in Sorabji 2006: 59–60 and 76–7.

[12] Aeneas of Gaza 1958: 54.20. [13] Aeneas of Gaza 1958: 59.3–6.

elementary particles, rather than to *types* of element, when he says that each human soul has to remember for the sake of bodily resurrection the original elements and the order in which they were originally placed. When he adds that the soul remains diffused amongst these elements and remains with them, this is what suggests *individual* elementary particles. The soul recognises the original particles by means of the signs (*sêmeia*), peculiar characteristics (*idiotêtes*), and distinguishing structure (*idiazousa sustasis*) proper to them (*oikeion*, *idion*). Some of the analogies offered by Gregory concern individuals, but Gregory may be referring to types, when he gives the analogy of colour hues, and imagines a painter who can recover the original separate hues out of a mixture.[14] Gregory has occasion to focus again on *types*, when he goes on to discuss certain distinct questions. Criticising the view of some (not all) Platonists that individual humans can be reincarnated as animals of different *types*, he says that this obliterates the peculiar characteristics (*idiotêtes*) of their nature.[15] Again, he takes up St Paul's resurrection analogy that the death of a seed of a given *type* is needed for the life of a plant of a given type.[16] Aeneas repeats this sort of example, without any warning that it concerns types, in his answer to the food-chain problem.[17]

The food-chain problem is raised again in Zacharias' *Ammonius* as a final supplementary question on resurrection, but again with only a *single* devouring considered. He shows no appreciation of the dangers involved in further stages of the food chain.[18]

Platonists for and Christians against an eternal world

I shall now come to the debate between Platonists and Christians on whether the world is eternal.

(i) Platonists Porphyry, Sallustius, Proclus: the Creator's eternal good will implies an eternal world created

The pagan arguments for the world's eternity, like the difficulties raised against resurrection, go back well before Proclus. Proclus in the fifth century is said to have been drawing on Porphyry in the late third century,

[14] Gregory of Nyssa 1857–66c: cols. 73A–85A. Similarly Gregory of Nyssa 1857–66a: ch. 27, cols. 225A–229A.
[15] Gregory of Nyssa 1857–66c: col. 109, lines 5–26.
[16] Gregory of Nyssa 1857–66c: cols. 153, lines 33–40, 156, 1–6, citing 1 Corinthians 15:35–44.
[17] Aeneas of Gaza 1958: 56.9–17. [18] Zacharias 1973, lines 1421–86.

and was also anticipated by the later fourth-century Platonist Sallustius, when he argued that the Creator's eternal good will implied an eternal world created.[19]

(ii) Proclus: otherwise he would sometimes be a potential, so imperfect, creator

Proclus offers a further argument: that if the world that God created was not eternal, he would sometimes be a merely potential, and so imperfect, creator.[20] God then must be an *actual* creator, and in that case the world he creates must be *actual*.[21]

(iii) Christians: Zacharias, Procopius, Philoponus: being a creator depends on internal powers, not external products

I shall start with the Gazans' best replies. Aeneas did not contribute significantly to the first topic. But Zacharias offered some good arguments directed not against Proclus, but against Proclus' pupil Ammonius, under whom Zacharias studied in Alexandria. The pagan view makes God dependent for his being on what he creates.[22] But in fact God is a creator not because of what he creates, but because of what he has within him: creative principles (*dêmiourgikoi logoi*),[23] so that he creates by will and freely (*thelein, eleutheros*), not by any necessity.[24] In appealing to creative principles, Zacharias may deliberately have been using a favourite concept of his opponent Ammonius. At any rate, one of Ammonius' more philosophical pupils, Asclepius, wrote a commentary based on Ammonius' seminars on Aristotle's *Metaphysics*, and the commentary uses the term 'creative principles' of the Platonic Forms in God's mind no less than ten times.[25] The idea of creative principles goes back earlier to Proclus, who quotes it from Porphyry, who based it in turn on a discussion in different wording by his own teacher, Plotinus.[26] But Zacharias might

[19] Proclus, recorded by Philoponus, *Against Proclus*. Proclus' first argument is missing in Greek, but preserved in the Arabic of Ishāq ibn Hunayn, and translated into English by Peter Adamson in Philoponus 2004: 34.3–6. See further Sallustius, *On the Gods and the World*, ch. 7 and Proclus' sixth argument, translated in Philponus 2005; this is Philoponus 1899: 224.18–225.6.

[20] Proclus recorded by Philoponus *C. Proc.* 56.10–12 and 18–20, 82.17–25, 93.8–10.

[21] Proclus recorded by Philoponus *C. Proc.* 42.1–43.23.

[22] Zacharias 1973, lines 740–3.

[23] Zacharias 1973, lines 762–7.

[24] Zacharias 1973, lines 389–94.

[25] In Sorabji 1988: 279, n. 123 I recorded Asclepius, from the voice of Ammonius, 44.32–7, 68.17–27, 75.27–8, 167.14–34, 183.14–16, 233.38–40, 363.1–5, 393.34–394.2, 441.27–31, 442.1–2.

[26] Proclus 1903: 396.22, in a passage said at 391.4–5, to be drawn from Porphyry; cf. Plotinus *Enn.* III.2.2.18 ff. (1951–73).

also – probably not instead – have found the idea of creative principles in the *Christian* tradition. At any rate, without the *word* 'creative', Origen, perhaps inspired by the still earlier Jewish thinker Philo of Alexandria, sees principles (*logoi*; Latin: *rationes*) in the mind of God as being used as a model for creation. More exactly, he thinks these principles are within God the Son in his role as God's Wisdom.[27]

Zacharias continues by saying that human professionals are analogous to God as creator. A doctor does not depend for being a doctor on patients being treated continuously, but on the medical principles (*iatrikoi logoi*) within him, and the same goes for a craftsman or rhetorician.[28] He adds a telling objection, that the pagan argument about God being eternally good would prove too much. For God is surely the creator of individuals such as Socrates and Plato, but they do not exist always. Does that detract, however, from the goodness of God?[29]

The other Christians agree. Procopius says that if God's creatures were necessary for his being *by nature* a creator, they would also be needed for completing the divine essence (*ousia*).[30] But a building or ship is not co-terminous (*sunhuparkhei*) with a builder or shipwright. For the creating is *within* the creator, whereas the thing created comes *after* him.[31]

Philoponus repeats some of the same arguments as Zacharias before going beyond him. God creates by means of creative principles, analogously to a shipwright or builder, so what he creates need not exist at the moment he thinks of it.[32] He repeats also that an individual human is not forever being created by God.[33] But he takes the argument about potentiality further, and treats it in a more systematic way. It was Aristotle who had first introduced the distinction between potentiality and actuality, but he had recognised an intermediate level between two extremes. A human embryo may be potentially a builder. A fully trained wakeful human may *actually* be building. But the trained human, when sleeping or otherwise occupied, is in an intermediate state, which can equally be called a level of second potentiality, or of *first actuality*. It is to be contrasted with the level of *second* actuality occupied by someone who is building at the moment.[34] Philoponus' point is that, although the second actuality of building may

[27] Origen (partly in the Latin translation of Rufinus) *De princ.* 1.2.2; *Commentary on John* 1.19.(22), 111–15; *C. Cels.* 5.39, discussed in Sorabji 1983 (2nd edn 2006): 250–2.
[28] Zacharias 1973, lines 372–4. [29] Zacharias 1973, lines 180–207.
[30] Procopius *Commentary on Genesis*, *PG* 87.1, col. 29D.
[31] Procopius *Commentary on Genesis*, *PG* 87.1, cols. 32D–33A.
[32] Philoponus *C. Proc.* 36.18 ff., 79.5–6, 84.14–15. [33] Philoponus *C. Proc.* 92.1–93.14.
[34] Aristotle *De an.* 2.5; Philoponus *C. Proc.* 47.18–48.7.

call for (some of) a building to exist, the first actuality does not require there to be any building in existence. The same goes for God, who is an *actual* creator, so long as he possesses the *first* actuality of creating. The point is a very general one that takes in Zacharias' examples of doctor, craftsman and rhetorician, and brings these and many other examples under a very general principle supplied by that very systematic philosopher, Aristotle.

(iv) Christians Zacharias, Philoponus: God's will for a beginning is not a change of his beginningless will

Zacharias has another good argument that uses his opponents' premises. He responds to the objection, ascribed to his opponent Ammonius, that if God gave the world a beginning, he will at that time have changed his will (*boulêsis*). Zacharias makes Ammonius' Christian student reply that God had a *beginningless* will for this creation.[35] Zacharias could once again be taking or adapting an idea from his opponent Ammonius, in order to turn it against him. At any rate, Ammonius' pupil Philoponus later repeatedly uses the argument that God could have a beginningless and changeless will for a creative change to occur,[36] and he goes on to compare God's changeless will for change with his unitary knowledge of the diverse and timeless knowledge of the temporal.[37] Possibly the point about will, and certainly the point about knowledge, had been used by Ammonius. The point about knowledge had first been made been made by Iamblichus and was then repeated by Proclus and by Ammonius himself in application to timeless knowledge of the temporal and definite knowledge of an indefinite future.[38]

(v) Christians Maximus, Procopius, Philoponus: even Aristotle allows instances of white to come into being out of nothing

The other good argument that deserves to be mentioned is made by Procopius, but this one had already been anticipated by an earlier Christian. It is that the creation of the world out of nothing in which Christians believe was also accepted by Aristotle for the special case of individual forms. This argument, applied to qualities (*poiotêtes*) rather than forms,

[35] Zacharias 1973, line 1087. [36] Philoponus *C. Proc.* 563.12, 56.6–21.
[37] Philoponus *C. Proc.* 658.21 ff. [38] Ammonius *In de int.* 136.15–25.

had been ascribed by Eusebius in the fourth century to a still earlier Christian called Maximus. When Philoponus applies the argument to individual forms, he gives Aristotle's example of a *particular* instance of whiteness as a form that comes into being out of nothing.[39] This explains why Maximus could have called it a quality. It is, of course, a *particular* form or quality, and most *particular* forms would be of finite duration. Aristotle says that white and certain other forms start existing without undergoing a process of coming into existence.[40]

(vi) Some Platonists: must not an endless world be beginningless? Reply: only if *by nature* endless

From the Platonist side, Augustine reports the case made by many Platonists for treating the world as beginningless.[41] Augustine, like Aristotle before him, and like the Platonists Plutarch and Atticus, understood Plato differently, as assigning a beginning to the orderly cosmos, while having the Creator spare it from coming to an end. This is just what Augustine believed to be true of the human soul, that it had a beginning, but was spared destruction. But many Platonists insisted that what has no end can have no beginning. This reason for making the world eternal had neither Plato's support, in Augustine's view, nor his own.

Proclus used the same Platonist view that Augustine had criticised. He claimed Plato's authority for the view that the world was imperishable and that therefore it must be beginningless, and hence eternal.[42] But Lindsay Judson has shown that Philoponus outmanoeuvred him and took the subject further. Not only was Philoponus more careful in specifying the relations between imperishable and ungenerable, and between unperishing and ungenerated, but he also stressed that Plato's *Timaeus* made the world imperishable only through God's overriding its perishable nature. It is only if it is *by nature* imperishable, which Plato denies, that we can infer that it is ungenerable.[43]

[39] Maximus is cited by Champion 2010. See, for Maximus, Eusebius *Prep. Evang.* 7.22.12 ff.; for Procopius, *Commentary on Genesis*, *PG* 87.1, cols. 29B–C; for Philoponus, *C. Proc.* 340, 347, 365.3.

[40] Aristotle *Metaph.* 1039b26, 1043b14, 1044b21. For other things besides forms of which this is true, see Sorabji 1983 (2nd edn 2005): 11, n. 5.

[41] Augustine *Civ. Dei* 10.31.

[42] Proclus *apud* Philoponus *C. Proc.* 120.5–14.

[43] Philoponus *C. Proc.* 225.21–226.19; Judson (2010) points out that while Philoponus rightly records Plato as making God override the world's perishable nature, he has no sound argument for his further assumption that even God could not override its natural ungenerability.

(vii) Christians Basil, Zacharias, Procopius: would not an eternal world be co-eternal with God, and so have the same honorific status as its Creator? Replies: why same in all respects? And is God eternal in the same sense?

Zacharias and Procopius, like Basil of Caesarea before them, raise the further complaint that a world co-eternal with its Creator would share God's honorific status.[44] Zacharias has a better version in which he argues, more modestly, that if the world were co-eternal with God, there would be *one* respect (eternity) in which God was not superior.[45] But why should the world have the same status in *all* respects?

There is an even more glaring oversight. The Neoplatonists drew on Plato's denial in the *Timaeus* that 'was' or 'will be' apply to Platonic Forms or that the Forms grow older, and in the *Parmenides* that the One postulated by Parmenides is in time or has a share of time. Although Plato's own description of Forms wobbled and his description of the One ended in paradox, Plotinus and subsequent Neoplatonists were emphatic in describing not only Forms, but also Plato's Creator God – normally equated with the divine Intellect – as outside time, and the supreme God, the One, as indescribable. Neither God, then, is in time. Time already in Plato's *Timaeus* was only an image of eternity, and the Neoplatonists deny eternity any duration. This means that the physical world, even if eternal in the sense of having everlasting temporal duration, does not have the kind of eternity that sets the Neoplatonic deities outside time altogether.[46]

Nonetheless, the Christians' mistake did not consist in putting their point in terms of co-eternity, because the Platonists themselves often speak that way. In arguments (i) and (ii) above they speak of the Creator's good will, of his status as Creator and of the created world all in the same terms as being eternal (*aïdion*), existing always (*aei*), or in the Arabic of Proclus' first argument as being everlasting (*abadī*).[47] Only at the very end of the Arabic discussion is it explained that God is everlasting in a different sense

[44] Basil of Caesarea, *Homilies on Hexaemeron* 2.2.10, cited by Champion 2010; Zacharias 1973, lines 516–17, 856; Procopius, *Commentary on Genesis*, *PG* 87.1, col. 29B. Philoponus' Platonist enemy Simplicius had heard analogous complaints raised against the idea of matter as a principle on a par with God: Simplicius *In phys.* 256.14–15.

[45] Zacharias 1973, lines 833–5.

[46] I have assembled the textual evidence in Sorabji 1983 (2nd edn 2006): ch. 8. I thank Sebastian Gertz for stressing its relevance here.

[47] *Aei*: Sallustius *On the Gods and the World* ch. 7; Proclus in Philoponus *C. Proc.* 87.17–25, 225.3–10, *Aïdion*: 56.18–20, *Abadī*: first argument of Proclus preserved in Arabic 34.3–35.8, translated by Peter Adamson in Philoponus 2004.

of being eternal and timeless. The Platonists do not mind the talk of co-eternity, because what interests them is what the timeless and the everlasting have in common: they are beginningless and endless. But the Christian conclusion that they have the same status requires more: that they are eternal in the same sense, and this the Platonists do not grant.

(viii) Christians Procopius, Philoponus: is not effect chronologically later than cause? Reply: or only causally posterior?

On the Christian side, we have noticed Procopius insisting that a ship or building comes *after* (*meta*) the shipwright or builder. But Philoponus makes the claim quite general. Whenever anything is generated, the cause precedes and the effect comes later.[48]

The Platonists, however, had made use of a distinction of Aristotle's between types of posteriority, in order to say that some things are posterior only in the order of causation, and not chronologically. Augustine reports that some Platonists had used this distinction in the present context to say that Plato intended to give the world a causal origin, but not a temporal origin.[49] But Aeneas and Zacharias seem not to distinguish the two types of posteriority, temporal and causal. Thus Aeneas mistakenly cites Plotinus as on his own side in insisting that the Creator is older (*presbuteron*) than the matter he creates, and cannot be at the same level, or on the same footing (*hama*). He does not consider whether the priority and sameness of level and footing in Plotinus is chronological or causal.[50] Zacharias repeats the mistake over whether *hama* means chronologically or causally at the same level. In a passage to which I shall return, he thinks Ammonius will be forced to agree that, since co-eternal things are *chronologically* at the same level – *hama* – one cannot be the cause of the other. But this will follow only if the meaning of *hama* is switched to being *causally* at the same level, so that one is not *causally* prior to the other.[51] It is interesting that Zacharias is really hoping here to argue from the opponent's premises, because he introduces his point by having the Christian student say that he will furnish *also* proofs that are derived from premises familiar and dear to Ammonius.[52] But the word 'also' acknowledges the point that he has not thought it necessary to confine himself to premises accepted by Ammonius, but sometimes plays only to the Christian gallery.

[48] Philoponus *C. Proc.* 14.18–20; cf. 17.22–5, 41.14–16. [49] Augustine *Civ. Dei* 10.31.

[50] Aeneas of Gaza and Colonna 1958: 46.11–16, referring to Plotinus *Enn.* II.4.7.2–10 (1951–73).

[51] Zacharias 1973, lines 1027–40. [52] Zacharias 1973, lines 1025.

(ix) Platonist examples of effect with the same duration, but not status, as its cause: a shadow, light, a footprint

To the Christian objections that an effect must be chronologically later than cause the Platonists had a set of analogies to offer by way of further reply: a shadow; light from a light-source; and a footprint. Thus Taurus in the second century CE explained the sense in which the world is 'generated' in terms of the analogy that there never was a time when the light of the moon was not being generated by the sun.[53] Plotinus in the third century spoke of Aristotle's prime matter, the ultimate subject of properties that has to be endowed with form if it is to constitute bodies, as 'a shadow upon a shadow'. It was not even a being at all.[54] We shall later have to consider whether that would prevent it being an effect. It is the last item in the chain of things that proceed co-eternally from the supreme deity, the One. If it is a shadow upon a shadow, this may imply that the penultimate items in the chain, bodies, are themselves regarded as a shadow. The first thing that proceeds from the One, the thing that is above all a being, is the divine Intellect. For its proceeding forth Plotinus uses the analogy of light. It proceeds from the One not by the One's willing, but as a shining round about the One (*perilampsis*) like the brightness (*to lampron*) that is always being generated (*gennômenon*) round the sun.[55] The fourth-century Platonist Sallustius argues that the world is as ever-existent as its Creator, in the way that light coexists with fire or the sun, and a shadow with a body.[56] Proclus says that the world is always being brought into being (*ginetai*), and brought forth (*paragetai*), and always simultaneously is being generated (*gignomenos*) and has been generated (*gegenêmenos*), just as solar light emanates from its source.[57]

Christians began to notice these Platonist analogies, which they disliked, at least as early as Basil of Caesarea (329–79), who ascribes to his Platonist opponents a shadow and radiance that are spontaneous (*automaton*) and not generated by choice (*prohairesis*),[58] while Augustine's mentor, Ambrose (339–97), ascribes to them a co-eternal world existing like a shadow spontaneously (*sua sponte*), and not through God's will (*ex voluntate*), although both Christians recognise that the Platonists allow God to be the *cause*.[59] Augustine himself in the early fifth century ascribes to

[53] Taurus *apud* Philoponus *C. Proc.* 147.7–9. [54] Plotinus *Enn.* VI.3.8.34–7 (1951–73).
[55] Plotinus *Enn.* V.1.6.25–39. [56] Sallustius, *On the Gods and the World* 7.
[57] Proclus *In Tim.* 1.290.29–291.2 Diehl.
[58] Basil of Caesarea, *Homilies on Hexaemeron* 1.7, *PG* 29, cols. 19–28.
[59] Ambrose *Hex.*, CSEL vol. XXXII, Day 1, ch. 5, paragraph 18.

certain Platonists the analogy of an eternal footprint.[60] Aeneas objects to the shadow analogy,[61] Zacharias allows his Platonist straw man, Gessius, to defend Ammonius with the analogy of a shadow that is simultaneous with the body that casts it, but does not have the same status.[62]

Augustine's example of a Platonist co-eternal cause and effect was of a foot implanted without beginning or end in the dust, causing an eternal footprint. In discussing this and other examples elsewhere,[63] I suggested substituting a springy cushion for the dust, in order to justify treating the foot as a cause. To call it a cause would be to imply that, for example, if the foot *were* to be removed – which it need not be – the indentation *would* spring up.

Shadow and light would not be accepted by modern physics as examples of effects exactly simultaneous with their causes, since light takes time to travel. But although time for travel has been used as a criterion of causation in certain special scientific experiments,[64] the case for excluding a time lapse from the *ordinary* concept of causation is that for ordinary purposes it is useful to be able to take shadow or light as genuine effects, without having first to discover whether there was such a time lapse. Augustine's Platonist example of the footprint in any case involves a different physical mechanism.

(x) Christian objections to the analogy of shadow and light: Basil, Ambrose, Aeneas, Zacharias

It might seem that Christians would be in a poor position to object to the Platonist appeal to light from a source as an example of an effect with the same duration as its cause. For the Christians themselves, ever since the Letter to the Hebrews 1:3, ascribed to St Paul, had repeatedly used the example of a light-source and its radiance (*apaugasma*) to illustrate the begetting (*gennêsis*) for all eternity of God the Son by God the Father.[65]

60 Augustine *Civ. Dei* 10.31. 61 Aeneas of Gaza and Colonna 1958: 46.1–9.
62 Zacharias 1973, lines 522–6. 63 Sorabji 1983 (2nd edn 2006): 310 and n. 16.
64 The simultaneous production of two opposite electric charges at a distance too great for light to traverse has been confirmed in the so-called non-locality phenomenon, and has been taken to rule out one charge affecting the other, but at the same time it has created a serious challenge to current physical theories.
65 In Sorabji 1983: 310, n. 16 I cited, besides the Letter to the Hebrews 1:3, Origen *De princ.* 1.2.4; a fragment cited *ad loc.* in Origen, Koetschau et al. 1973 (namely, fr. 1 of *Homily on the Letter to the Hebrews*, in Origen 1831–48); Dionysius of Alexandria *apud* Athanasius *De sententia Dionysii*, *PG* 25, cols. 501C–504B, 508A–B, 512C, 516C; Theognostus *apud* Athanasius *De decretis Nicaenae Synodi*, *PG* 25, col. 460C; Alexander of Alexandria *Epistolae de Ariana haeresi*, *PG* 18, cols. 557C, 565C, 576A–B; Athanasius, *Contra Arianos*, *Oratio* 1.8, 1.12, 2.34–6, 4.2, *PG* 26, cols. 25C–28C, 37A, 220A–225A,

But the Christian Nicene Creed of 325 CE insisted that the Son was *begotten, not created*, being of *one substance* with the Father. This may have made the relationship too close to count as one of cause and effect. We shall see Zacharias and Philoponus drawing the same conclusion for the analogy of sunlight from the sun as applied to the different case of the creation of the world.

The Christian objections to the analogies of a shadow and of light do raise some questions of interest, but they virtually all make the mistake of supposing that the casting of a shadow or light would need to be analogous to the creation of the world in *all* respects, instead of just the *one* respect of producing an effect exactly simultaneous with its cause. Even Philoponus did not entirely escape this defect. Sometimes too Christians were simply addressing their co-religionists and deploring the analogies as not fitting their own beliefs – which, of course, they were not meant to do. This would be true of Basil and Ambrose, who were complaining in the texts already cited that shadow and light are generated spontaneously, not by will, as Christians require. Ambrose in addition casts around for disanalogies and points out that God has no body and his splendour is not corporeal, something that had not been disputed.

Aeneas repeated Basil's displeasure at the idea that a shadow is spontaneously generated (*automaton*), for we must recognise what a shadow is. It occurs when a body in front of the sun does not allow light to fall behind it, and this is why the shadow is drawn with the same shape as the body. Zacharias was to ascribe this account of shadow by Aeneas to 'one of the sages (*sophoi*) of our time'.[66] Aeneas immediately added a series of disanalogies. There is nothing bodily about God.[67] In any case, who would wish to create or destroy his own shadow?[68]

Zacharias similarly repeated much of what was displeasing to Christians. A shadow is unchosen, unwilled and spontaneous (*aprohaireton, aboulêton, ek t'automatou*). He, like the others, looked for disanalogies: to cast a shadow, a light, unlike God, needs a third thing, a body in between. A shadow would follow God, even if he did not wish it (*thelein*).[69]

Procopius did not add to the case, but only repeated that a co-eternal world is due not to will (*boulêsis*), but to the necessity of nature (*anangkê*

469C, 480C–D; Basil of Caesarea *Homilies on Hexaemeron* 1.7, col. 17B–C; cf. Basil of Caesarea *De Spiritu Sancto*, *PG* 32, col. 186B–C; Gregory of Nyssa *De fide*, *PG* 45, col. 140A–B; Gregory of Nyssa *Oratio de deitate Filii et Spiritus Sancti*, *PG* 46, col. 560C–D; Ambrose *Hex.* Day 1, ch. 5,.

66 Zacharias 1973, l. 543, as pointed out by Sebastian Gertz.

67 Aeneas of Gaza and Colonna 1958: 46.1–6.

68 Aeneas of Gaza and Colonna 1958: 46.8.

69 Zacharias 1973, lines 526–40, 758–61.

phuseôs).[70] Even Philoponus makes this point about an unwilled necessity of nature (*aboulêtos anangkê phuseôs*) in a related discussion.[71] But he does not rely on it to refute the Platonists. For that, he rather refers back to his earlier argument, acceptable to Platonists, that God is a creator through having within him creative principles at the level of *first* actuality, and in his reminder that he has in mind only that being a creator does not require a first actuality to be converted by necessity of nature into a *second* actuality. At most he can be faulted for looking for disanalogies in the shadow analogy. The sun, he says, does not always cast a shadow from a body it is illuminating – not if it is directly overhead.[72] Of course, the Platonists would require only one case for their analogy – say the case of Greece, where the sun is never directly overhead.

(xi) Philoponus, amplifying Aeneas: a shadow is not an effect

Aeneas did hint at an interesting question, missing from Basil and Ambrose, when he alleged that a shadow is *not made* (*poiein*), but merely *accompanies* (*sunakolouthein*) its source.[73] Zacharias had a similar complaint, that a shadow is an *accompaniment alongside* its source (*parakolouthêma*), as if the world existed independently *alongside* God (*parhupostan*).[74]

The two Gazans do not explain why a shadow is an accompaniment, and therefore, as Aeneas says, not made. For this we have to wait until Philoponus, who does explain. A shadow, says Philoponus, is a non-entity (*mê on*), the mere privation or absence of light. This might well seem to be supported by Plotinus, who treated Aristotle's prime matter as a shadow upon a shadow which was not even a being (*on*).[75] Aeneas too had stressed that a shadow is merely the interruption of light by a body.[76] But Philoponus draws the further conclusion that, as a non-being, a shadow is not an effect at all.[77] Could this be taken to mean that it is the point at which the effects of illumination have already stopped? However, Philoponus unintentionally suggests an answer when he points out that blindness is also a privation or absence. Yet surely it would be strange to say that blindness could not be caused, so the point cannot be applied to *all* privations, even if it can be applied to shadow.

[70] Procopius *Commentary on Genesis*, *PG* 87.1, col. 29D. [71] Philoponus *C. Proc.* 78.10–15.
[72] Philoponus *C. Proc.* 16.1–23. [73] Aeneas of Gaza and Colonna 1958: 45.21–3.
[74] Zacharias 1973, lines 529–34, 760–1. [75] Plotinus, *Enneads* VI.3.8 (34–7).
[76] Aeneas of Gaza and Colonna 1958: 46.2–5. [77] Philoponus *C. Proc.* 15.1–16.2.

(xii) Zacharias, Philoponus: sunlight is definitionally too close to the sun to be an effect

There is another potentially interesting argument in Zacharias, that uses some of the same language of mere accompaniment alongside (*par-*, *para-*). But it is presented in too compressed a way, and is clarified only when Philoponus expounds it later.[78] I would translate Zacharias' passage as follows:

> Christian student: Then consider what has been concluded as a whole. If the cosmos is co-eternal with God, and co-eternal things are included among things that are on the same footing, and of things that are on the same footing neither can be the productive cause of the other, then the cosmos cannot have its productive cause from God. If (a) what comes into being is not to be some shadow, nor (b) is the effect to be taken into the definition of the cause and as something that completes (*sumplêrôtikon*) its substance (as with the sun and its radiance), or as having the same substance as the cause (as with God the Father and God the Son), but if rather (c) the cause is something that is productive and rational and exercises will, and produces a being that is displaced from itself (*parêllagmenês*), where in that case, my good sir, will there be room to affirm that God is the productive and substantive cause of the cosmos thanks to its substance being something *distinct* (*hetera*) from God alongside him (*par' auton*), whereas according to *you* people the cosmos is co-eternal with him? Or [where is there room] to say that god and the cosmos are on the same footing from eternity, if we are right in saying that co-eternal things are included among things on the same footing, and of things on the same footing one cannot be the cause of the other?

This is the passage that Zacharias introduces by saying that he wishes to use *also* premises dear and familiar to Ammonius. There are many premises, so it is not clear whether he thinks all of them acceptable to Ammonius. Many of the points made here will now be familiar, but the new point is briefly expressed in (b). Christians and Platonists had each taken a case in which they wanted to argue that the relation of creator and created was too distant to describe a certain crucial relationship. I have already mentioned the Christian case: God the Father does not *create* God the Son, because the relationship of continuous begetting is too close to be a relationship of creating, since it makes Father and Son to be of the same substance. The Platonists agreed with the underlying principle, when,

78 *Zacharias* 1973: 1038–50. I thank Sebastian Gertz for identifying Zacharias' earlier version, which the compression had led me to overlook.

discussing embryology, they insisted that the engendered has a lower status than what engenders it and cannot be of the same substance (*homoousia*).[79] They also described in slightly different language the relation between the sun and the sunlight that fills the transparent sphere on which the sun was thought to be carried around us. The relation was too close for the sun to *create* the sunlight, because, Zacharias says, the sunlight is taken into the *definition* of the sun and *completes its substance.* This gets explained by Philoponus. He refers to the light within the sphere of the sun, evidently the transparent sphere on which the sun was thought to be carried around us. The light within that sphere gives the sun its form and completes its nature (*eidopoion, sumplêrêtikon tês phuseôs*), and the essence and substance (*ti ên einai, ousia*) of the sun accords with that form. The sun is given form and has its being (*to einai*) in accordance with that light. But now the sun and the light (like God the Father and Christ) seem too close to be related as creator and created. Nothing is its *own* creative cause (*dêmiourgikon aition*). Hence the sun is not the creator (*dêmiourgos*) of its own light.[80] This is a bold challenge. It is true that nothing creates itself. But it is less clear that nothing can cause what is definitionally related to itself: murder causes the death that defines it, and it does so without being a cause of itself.

(xiii) Do the Neoplatonists omit the role of the One's will from their account of creation?

The Christian objections to the shadow and light analogies before Philoponus regularly accuse the Platonists of omitting God's goodness and will from the account of creation. But the actual situation is rather different. The Neoplatonists are committed to giving a major role in creation to the *will* of the Creator or demiurge, whether that is their second divinity, the Intellect, or their third, the world soul. For Plato made the Creator's *will* central in his account of creation in the *Timaeus.* The Creator willed nothing bad, but willed that the world should be spared its natural disintegration.[81] What is surprising is that Plotinus, in a passage already quoted, says that the first entity in the chain of created things, the divine Intellect proceeds forth from the One, like brightness, and not by the One's willing.[82] If that were the whole story, the Christians would have been more accurate if they had complained that will appears belatedly in the

[79] Porphyry *To Gaurus On How Embyros are Ensouled* 6.2.42.20 ff.
[80] Philoponus *C. Proc.* 16.24–17.14. [81] Plato *Tim.* 30a, 41a–b. [82] Plotinus *Enn.* v.1.6, 25–39.

Neoplatonist account of creation at the stage of the second divinity, unmotivated by anything in the supreme deity.

But that is not the whole story at all, as I have explained elsewhere, citing Plotinus *Ennead* VI.8, which is called 'On willingness (*to hekousion*) and the will (*thelêma*, here a synonym for *boulêma*) of the One'.[83] The point made is that any description of the One has to use terms that are strictly inapplicable, and which must be hedged with a 'so to speak'. But we have to say that the One is what it wills (*thelein*), or rather casts among beings what it wills. In other words, the One is indescribable, but the best we can say is that, as the Good, it eternally wills other things into being, and they gain their being by willing to partake in the Good.

Proclus also seeks to give a role to *will* in his account of things proceeding from the One.[84] Although the One is indescribable, we need to say that there exists within it goodness, will (*boulêsis*) and providence.[85] These three are analogous to three other things in the One: essence, power (*dunamis*) and activity. Will furnishes the wholes with goods, and power stirs all things to procession.[86] Will depends on goodness, and providence on will, and the universe is generated through the providence, will and goodness of the Father.[87] Proclus here gives a role to what is inexactly called *will* in the One by declaring an analogy between it and what is inexactly called a power in the One that stirs all things to procession. There is no attempt to explain to the unconverted why '*will*' is the closest term we can use for describing the indescribable One. This may be a real gap in Neoplatonism, and the Christians might have had a real chance of pressing such a complaint. But the Christians surveyed here tended instead to accuse the Platonists of omitting reference to the will altogether in their account of creation, even though they discussed in other statements Platonist references to the will of the demiurge.

What is so different about Philoponus?

Philoponus goes beyond the Gazans in many ways, as the arguments sketched here may already indicate. It is important that he almost always attacks his opponents on their own terms, quoting Christian Scripture

83 Sorabji 1983: 317–18, citing Plotinus *Enn.* VI.8.9.44–8, VI.8.13, VI.8.15.1–10, VI.8.18.35–53, VI.8.21.8–19.

84 I am grateful to Dirk Baltzly, who showed me the passages I cite here, and to Carlos Steel and Jan Opsomer, who drew my attention to other passages.

85 Proclus *In Tim.* 1.371.9 ff. Diehl.

86 Proclus *In Tim.* 1.414.27–415.3.

87 Proclus *In Tim.* 1.412.2–7.

only as an addition to the argument, and not as a main consideration, whereas the Gazans often play only to the Christian gallery.

Philoponus is also extraordinarily systematic, opposing Proclus' eighteen arguments for beginningless creation over the 646 pages of the modern edition of *Against Proclus*, with cross-references to his arguments earlier in the book, and with attentive use of the systematic logic of his other main opponent, Aristotle. I have not even mentioned, because it is not mentioned by the Gazans, his use of Aristotle's concept of infinity, accepted by the Neoplatonists, to show that their account of creation allows something impossible in their own terms: that the universe should have finished passing through a more than finite number of past years.[88] But we can see that this is merely the most famous and stunning case of his using his opponents' views in order to refute them out of their own mouths. I think we can also see at least one reason for the contrast with the three Gazans, and to that I shall now turn.

Part II Significance of the commentary tradition in Ammonius' school

Philoponus' skill in answering the Neoplatonist Proclus was partly the fruit of his listening earlier in his career to Proclus' ablest pupil, Ammonius. In the hands of Neoplatonists, commentaries on Aristotle became a source for Neoplatonist views as well as for Aristotle's text. This makes it unsurprising that Philoponus was able to move from editing Ammonius' Neoplatonist lectures on Aristotle to extremely well-informed and dangerous attacks first on Proclus in 529 and later on Aristotle himself. Both attacks related to the eternity of the world. Philoponus' Neoplatonist opponent Simplicius, though outraged, felt obliged to take the attack on Aristotle seriously, though at the time of his reply he claimed not to have brought himself to read the attack on Proclus.[89] Philoponus based his attacks upon years of writing up commentaries on no less than seven of Aristotle's works. At least four commentaries were derived from the lectures of Ammonius, pupil of the even more famous Neoplatonist Proclus, and Philoponus is named as the sole author of another three. Philoponus' switch in 529 to attacking

[88] I have described the infinity arguments in Sorabji 1983: ch. 14. I have added references to better understandings of infinity that may have been achieved by Archimedes (*c.* 287–212/211 BCE), Thabit ibn Qurra (d. 901 CE), possibly al-Haytham (d. *c.* 1040) and Robert Grosseteste (*c.* 1175–1253) in Sorabji 2004b, vol. II: 175–6.

[89] Simplicius *In de cael.* 135.30–136.1. In my introduction to the translation of Simplicius *In phys.* 1.1–5, I express some doubt whether he had read it even at a later date.

his teacher's teacher, Proclus, must have required some very strong motivation, and we have seen one suggested by Edward Watts.[90]

By contrast, Zacharias' 'refutation' of Ammonius, despite a few good arguments, reads more like a cabaret act designed to impress his Christian fellow-students. As such, it would have been rather good – he and Aeneas are talented literati – but not as a refutation of Platonist thought. The contrast must be in part due to another contrast, the difference in Zacharias' own training. Where was he trained? He may indeed have attended Ammonius' elementary classes on Aristotle, but Ammonius will have kept the exposition of Aristotle separate from his own views on the eternity of the world, so Zacharias' fictional confrontation on that subject could not have taken place in that context. For beginners, in any case, Ammonius may well have gone little further than Aristotle's logic. Zacharias' listening is also likely to have been distorted. In his *Life* of an eventual convert to Christianity, Severus, Zacharias tells us that he stayed on in Alexandria a year longer than Severus, to listen to the philosophers and rhetoricians, who were excessively proud of their studies, for a rather idiosyncratic reason – in order to be able to dispute with them.[91] Zacharias may have learnt his actual arguments against Ammonius at the monastery of Enaton, which he also attended outside Alexandria. The monastery could have had more than one motive for resisting Ammonius. Edward Watts has described it as motivated by the condemnation at the Council of Chalcedon of its Christological view that Christ had only one nature, not two, human and divine. This led it to predict the imminent end of the world, and hence to be particularly opposed to Ammonius' teaching about the world's eternity.[92] Whatever the motive, the monastery was clearly interested in opposing Ammonius. Zacharias tells us, again in his *Life of Severus*, that several days of discussion in the monastery at Enaton was enough to complete the conversion of a pagan student of grammar in Alexandria, Paralius, and arm him with arguments to forward not only to his professor of grammar, but to the professors of philosophy, including Ammonius himself.[93] If Zacharias also learnt his arguments against Ammonius at the monastery of Enaton, he too could have learnt them in a brief course. It is also not clear what level of knowledge was possessed by the monastery's teachers, much though Zacharias reveres them. A further possibility,

[90] The desire to take over from the less-experienced pagan philosopher Olympiodorus the imparting to Christians of pagan Greek philosophy.

[91] Zacharias 1903: 46, in the surviving Syriac; for English translation see Zacharias 2008.

[92] Watts 2010: 93, 123–4, 131, 136, 143, 256.

[93] Zacharias 1903: 16.

though it must remain a conjecture, is that their knowledge, or at any rate their students' knowledge, was based on a catalogue of summarised Christian arguments against Platonist philosophy. Either the brevity of the crash course or the brevity of catalogues would help to explain why Zacharias provides a mixture of arguments, a few well worked out, but many naïve and sometimes in a form so compressed that it is an effort to disentangle them. A catalogue of arguments on the subject is exactly what the later Gazan Procopius supplies in his objections to the eternity of the world.

Catalogues of summaries could take any of three forms. There were *florilegia* of doctrines of the Church Fathers from at least as early as the Council of Ephesus in 431 CE on the subject of Church Council debates.[94] There may have been catalogued summaries of doctrines ascribed to pagans, and, perhaps combined with them, summaries of arguments in reply to pagan doctrines. It has been pointed out to me that in 447 after the Council of 431, the Christian Theodoret (d. *c.* 466) had provided both *florilegia* of Christian quotations and an appendix of summarised arguments which he calls a *sullogismos* – argumentation.[95]

David Runia has discussed Christian knowledge of Aristotle and argued that many Christians – far from all, I should stress – relied for their knowledge of pagan thought on collections of summarised pagan doctrines – doxographies – compiled originally by pagans, and subsequently copied from one writer to another. He argues that the Christians Eusebius (d. 339) and Cyril of Alexandria (d. 444) drew on a doxography by Pseudo-Plutarch, and Theodoret drew on Pseudo-Plutarch's source, Aëtius. Many Christians, especially in the earlier second and third centuries, he considers wholly dependent on doxographical sources for their knowledge of Aristotle, and he marks twenty-nine such passages, even though other Christians have been exonerated by himself and by other writing.[96]

I have to say that some Christians possessed a wonderful knowledge of pagan philosophy, and access to Aristotle came from extremely varied directions. To take a superlative example, the Christian Origen (d. 253/4) could not only draw on the knowledge of pagan thought of his predecessors, the first-century CE Jewish thinker Philo of Alexandria and his own Christian predecessor Clement of Alexandria in the second century, but also studied there under Ammonius Sakkas, the teacher of Plotinus.

[94] See Richard 2011: no. 12. [95] I thank Averil Cameron for this information.

[96] Runia 1989, which, however, exonerates Clement of Alexandria from the earlier strictures of Festugière 1932. Hippolytus (d. 235) has been exonerated as regards his knowledge of Aristotle's *Categories* and the Presocratics by Osborne 1987.

His knowledge of Greek thought included Aristotle. I have mentioned above his using a passage in Aristotle's *On Generation and Corruption* 1.5, which has been little discussed even today, in which the perceptible form of an individual organism is compared by Aristotle with an elastic tube. Origen made a characteristically philosophical use of it, in order to explain how people might recover their original bodies on resurrection. He avoided the food-chain problem discussed above by treating Aristotle's tube-like form as calling not for the same *matter* on resurrection, but only for something like a photographic likeness. Origen, though his commentaries were on the Bible, has been hailed as the inventor of the form that commentary on Aristotle eventually took in the commentaries of Ammonius' school.[97]

Nonetheless, other Christians had only summaries available, and this may have been true of Zacharias, whether or not from his monastic teachers, who may have supplied him with his crash course in answering the Neoplatonists. The contrast with Philoponus' detailed understanding of pagan thought provided by prolonged study in Ammonius' school brings out some of the significance of the commentary tradition in that school. I will finish by mentioning another significance. The school was attended by Christian students bilingual in Greek and Syriac. One of them, Sergius of Reshaina (d. 536), it has been argued, acquired his knowledge from personally taking lecture notes in the school, and went on to write a commentary partially in the Ammonian style as regards preface and format, though with Christianised religious aims, on Aristotle's *Categories*, as well as a Syriac adaptation of an earlier commentator's work, that of Alexander of Aphrodisias on the *Principles of the Universe*. This was a very important source – not the only source – for the transmission of the commentary tradition to Arabic writers, since the Syriac was later translated into Arabic. Eventually a Christian educated in a Syriac monastery, Abu Bishr Mattā (d. 940), worked in Baghdad, translating into Arabic some of Aristotle along with Greek commentaries on Aristotle and writing commentaries himself for an increasingly Muslim readership in Baghdad.[98]

[97] Hadot 1987a; Fürst 2011: ch. 1.

[98] The Syriac tradition is discussed in several of the chapters of Lössl and Watt 2011, with a useful overview by John Watt, in ch. 15. That Sergius wrote his commentary on Aristotle's *Categories* on the basis of lecture notes taken directly from Ammonius' lectures is argued (also in Lössl and Watt 2011) by Daniel King in ch. 14: 230.

CHAPTER 5

Gregory of Nyssa on the creation of the world

Anna Marmodoro

Introduction

The Church Fathers hold by and large that God created the world from nothing, by an act of will, at a particular time.[1] This is a radical departure from the Greek paradigm that there is no *ex nihilo* creation; and furthermore, an apparent violation of the causal principle, widely held in antiquity, that 'like causes like' (for brevity from now on, the LCL principle).[2] The general idea that the LCL principle captures is that causation is 'transmission' from *x* to *y* of a condition, say *F*, which *x* has to have already, in order to be able to transmit it to *y*. By passing *F* on to *y*, *x* makes *y* be what *y* is not yet, and *x* is already, namely *F*. That is, *x* causes *y* to be like *x* itself (with respect to *F*). Thus the shorthand reference: like causes like. Aristotle for example expresses the principle this way in his *Physics*: 'it is that which is hot that produces heat, and in general that which produces the form possesses it' (257b9–10). The LCL principle holds not only for cases of alteration, but also of substantial generation; thus Aristotle: 'man begets man' (*Metaph.* 1070b34).

Divine creation appears to be a direct contradiction of the LCL principle. Something non-material, God, causes something material, the world, to come into existence. The Church Fathers, committed to the creation of the world by God, have to address a philosophical dilemma, that is: are they to think that *causation* is different from what they thought it was, or that the *world* is different from what they thought it was? On the

The research leading to this publication was supported by a starting investigator award from the European Research Council (award number 263484). I presented earlier versions of the paper in Oxford and Brussels. I am very grateful for the audiences' helpful feedback on both occasions, and for written comments from my colleagues Jonathan Hill and Brian Prince.

[1] This was at least the prevailing belief after the second century; Tatian is the first person to state it explicitly and Tertullian had to argue for it against Christians who thought otherwise.

[2] For an excellent discussion of the philosophical background to the Church Fathers' discussion of the creation of the universe see Sorabji 1983: ch. 13.

former line of thinking, if the LCL principle does not govern causation, then it is acceptable to think that the world is material even if created by a being with no materiality; on the other hand, if the LCL principle does govern causation, the world has to be somehow immaterial.[3]

The debate on the metaphysics of divine creation was lively in late antiquity. Gregory of Nyssa (*c.* 335–*c.* 395) is a particularly interesting player in this debate. He is acutely aware of the philosophical conundrum the LCL principle generates for the Christian believer in divine creation; he writes, for example, that

> If God is in His nature simple and immaterial, without quantity, or size, or combination, and removed from the idea of circumscription by way of figure, while all matter is apprehended in extension measured by intervals, and does not escape the apprehension of our senses, but becomes known to us in colour, and figure, and bulk and size, and resistance, and the other attributes belonging to it, none of which it is possible to conceive in the Divine nature, – what method is there for the production of matter from the immaterial, or of the nature that has dimensions from that which is unextended? (*De hom. op.* 23; Schaff and Wace 1892: 770)

> You can hear people saying things like this: if God is matterless, where does matter come from? How can quantity come from non-quantity, the visible from the invisible, some thing with limited bulk and size from what lacks magnitude and limits? And so also for the other characteristics seen in matter: how or whence were they produced by one who had nothing of the kind in his own nature? (*Ap. in hex.*, trans. at Sorabji 1983: 290)

In this chapter I investigate Gregory's views on the metaphysics of divine creation, addressing the conundrum he raises in the quotations above (and elsewhere). Is Gregory committed to the view that what the immaterial God causes to come into existence is immaterial, too? Does Gregory allow matter at all in his ontology? If not, is he perhaps the first idealist in the history of philosophy, as has been suggested? But if on the other hand Gregory does admit material objects in his ontology, how does he account for the causality of their creation? It is timely to offer a fresh take on these questions, as they have been addressed with widely diverging answers in the recent philosophical literature. Yet there is much at stake in gaining a full understanding of Gregory's thought. In one camp, as we will see, are

[3] An alternative option is that God be material. Arguably, Tertullian thought so (influenced by Stoicism) and so did the 'anthropomorphites' in the fourth century. The fact that some theologians such as Origen devoted time to arguing for God's immateriality indicates that not everyone accepted it. Overall, however, the notion of God being material was a minority position, not seriously entertained by many philosophical theologians (and certainly not Gregory).

those who argue that Gregory solved the philosophical conundrum of how an immaterial being could create material things by eliminating material things altogether from his ontology. On this interpretation, in full adherence with the LCL principle, Gregory would have held that God creates immaterial things, and the universe contains these only. Some argue that from this it follows that Gregory holds a form of idealism à la Berkeley, or, even more strongly, that in Western thought idealism originated with Gregory. Powerful arguments have, however, been made in the literature against this interpretation, to the effect that not only is idealism absent from ancient philosophy, but that it could not even have arisen at that stage of Western thought.[4]

Making a departure from these received lines of interpretation, I understand Gregory of Nyssa as being concerned to address a metaphysical puzzle of *categorial incompatibility*, of which divine creation is in fact only one instance. To the question of how something immaterial can generate material objects, Gregory's answer is that the immaterial generates intelligible qualities, from which bodies are made up. It is on this basis that Gregory has been interpreted as an early idealist. I argue that Gregory is not an idealist; that his qualities are not thoughts in a mind; and that his matter, which results from the combination of the qualities, is concrete. On my reading, Gregory develops an early account of the metaphysics of *abstract objects*. The qualities are physical aspects of material objects, singled out by abstraction, and having a definition. The definition shows them to be intelligible, but not thoughts in a mind. Being intelligible and immaterial, they can be generated by an immaterial being. Being physical aspects of objects, they can combine to give rise to material body. As I will argue in what follows, the key to Gregory's account is that there is a sense in which physical predicates 'are' true of abstract objects. Being definable abstractions, they are intelligible; being physical, they can constitute matter. This is how Gregory resolves his categorial puzzle of the material resulting from the immaterial.

Does Gregory admit material bodies in his ontology?

As I mentioned, views diverge regarding Gregory's metaphysical commitments: was he an eliminativist about matter, or not? Additionally, did he think of substances as bundles (of instantiated qualities, or of ideas?); or

[4] Richard Sorabji offers a lucid discussion of the arguments, which were put forward by Myles Burnyeat first in 1978, in 1983: 287 and ff.

did he posit a substratum (of what nature?) to which qualities or ideas somehow belong? A brief review of the existing alternative interpretations will enrich our understanding of Gregory's position.

On what I call the eliminativist interpretation, for Gregory the world does not contain matter; so there is no violation of the LCL principle. So, for example, think Arthur Hilary Armstrong (1962), Richard Sorabji (1983 et al.), Johannes Zachhuber (forthcoming) and others. Sorabji has articulated the eliminativist interpretation in a few core theses he attributes to Gregory, most significantly that 'a body, or a piece of matter, is a bundle of thoughts or concepts' (1983: 292–3); hence in this world 'there is nothing corporeal – in the sense of possessing a nature fundamentally different from that of minds – at all'.[5] On this view, there is no metaphysical work left undone that would require positing a substratum; and indeed Gregory does not posit one. From this, Sorabji concludes that Gregory and Berkeley have the same idealist account of reality (although Gregory develops it as an answer to a completely different question from that of Berkeley).[6] Additional, albeit indirect, support for this interpretation comes from Mark Edwards, who draws attention to interesting evidence (for example from Origen's *De principiis*) that he interprets as showing that at the end of the second century after Christ there was an intellectual environment where idealist views were somehow in circulation.[7] A philosophical assessment of this evidence, and particularly Origen's, falls outside the scope of this chapter. Also, the evidence only speaks in favour of the existence of an intellectual environment where idealist ways of thinking might have been circulating. Edwards' considerations about the historical context, as well as Sorabji's proposed idealist interpretation of Gregory, have to be weighed up against the fact that plausibly Gregory (as well as his predecessors and contemporaries) was aware of an important argument by Plato against idealism. In the *Parmenides* (132b–c) Plato argues that if all were thought, objects would be thinking, in the sense that objects would be thought processes; and he deemed this conclusion absurd.[8] It is plausible, I submit,

[5] Hill 2009: 656. [6] For the arguments and a fuller statement of Sorabji's views see his 1983: ch. 13.

[7] Edwards writes: 'Thus, we may be confident that there were Christians at the end of the second century who did not regard matter, or even the creation of matter, as a desideratum for the existence of concrete particulars. This departure from the Greek tradition was a corollary of the doctrine of creation out of nothing and a clear anticipation of a thesis that we commonly associate with Berkeley' (2013: 578).

[8] 'But, Parmenides, maybe each of these Forms is a thought, Socrates said, and properly occurs only in minds. In this way each of them might be one and no longer face the difficulties mentioned just now. . . . And what about this?, said Parmenides, Given your claim that other things partake of forms, won't you necessarily think either that each thing is composed of thoughts, and that all things

that Gregory would have known this argument, directly or indirectly, and would have wanted to avoid its conclusion, presented by Plato as absurd. I take here a conservative approach as it were; unless compelling contrary evidence is produced, I find it plausible that Gregory wanted to avoid a view that Plato had deemed absurd and thus did include matter in his ontology.

A different version of the eliminativist interpretation takes Gregory to hold that objects are constituted of qualities inhering in a substratum – where the latter is assumed to be of mental nature. Thus for example Jonathan Hill writes:

> [Gregory's] strategy [is] to remove matter from the picture altogether and reassign its function to God. On this view, it remains the case that there has to be something 'underlying' the qualities of *physical* objects – but rather than posit some distinct thing called 'matter' to do it, God can do it himself. (Hill 2009: 666)

Thus 'For Gregory, *hyle* no longer denotes some entity or principle distinct from God; it describes one of the things God does' (Hill 2009: 675); Gregory 'uses the term in a functional way, to express God (or his activity) acting in the way matter was traditionally supposed to do' (Hill 2009: 666). On Hill's interpretation, God performs the metaphysical function of holding things together – unifying qualities into the things there are in nature. I take this interpretation to be aimed at showing that Gregory does not require matter in his ontology, because there is something else that can play (at least one of) the metaphysical role(s) that matter is often in antiquity introduced to play, that is to be an ultimate substratum. But even if one granted that God's activity plays the role of being the ultimate substratum in Gregory (which is a line of thinking I do not find in Gregory), Hill's interpretation does not offer an account of matter as physical body. And yet, Gregory ponders about this problem, and ever reports it, as we have seen, as an aporia of his philosophical times: 'How can quantity come from non-quantity, the visible from the invisible, some thing with limited bulk and size from what lacks magnitude and limits? And so also for the other characteristics seen in matter: how or whence were they produced by one who had nothing of the kind in his own nature?' (*Ap. in hex.*, trans. at Sorabji 1983: 290). This remains to be explained in Gregory's system even after positing

think; or that although they are thoughts, they are unthinking? That isn't reasonable either, Parmenides, he said' (*Parm.* 132b–c).

an immaterial ultimate substratum. This is why I find Hill's interpretation ultimately unsatisfactory.

Yet another version of the eliminativist interpretation is by Johannes Zachhuber, who writes that

> What Gregory means . . . must be that matter is not part of the first creation, but only a result of further development. . . . The first creation is that . . . of a kind of intelligible being which is tied to the corporeal world. . . . So, since it must first be in their nature [sc. in the nature of things which have completely existed in the beginning] to develop into corporeal being, and secondly their initial perfection is only a 'potential' one as that of a germ, an imagery popular with many philosophers and resounding in Gregory's account, i.e. they find their full realization only inside the sensible realm. I should thus conclude that what is created fully in the beginning are the principles of corporeal being which, however, are not bodies themselves, but the lowest level of intelligible being. (Zachhuber 2000: 150–1)

Zachhuber concludes:

> In summary then God created, according to Gregory, 'in the beginning' a *plēroma* of *logoi*, of intelligible being, which, however can actualise itself only under the conditions of space and time in order to reach the perfection which is only germinally provided in it. This temporal and spatial development of those forming principles constitute sensible being in the first place. (Zachhuber 2000: 154)

Zachhuber presents every initial being, which Gregory lists, as a *potentiality* that can 'actualise itself only under the conditions of space and time'. But there is a circularity that results from this conception, in view of Gregory's account of the initial beings, which includes space (διάστημα, extension) amongst the 'potential beings' (*Ap. in hex.*, trans. at Migne 1857: 69). The condition of actualisation that is set for the potentialities, namely coming to be in space ('spatial development'), presupposes the actuality of space, which begs the question regarding the actualisation of space in potentiality (which presupposes its own actuality in order to be actualised).

The *non-eliminativist* interpretation, on the other hand, appears more promising in terms of fit with the textual evidence, even if it does commit Gregory to a prima facie violation of the LCL principle. On this interpretation Gregory's world includes material bodies. This view has been defended for example by Darren Hibbs, who argues that for Gregory material objects are combinations of qualities, which are themselves concepts:

> Gregory proposed a solution that explained the creation of material bodies as resulting from a commingling of immaterial, incorporeal qualities that

> are themselves concepts. ... The relevant point is that material bodies are brought into existence by a creative act; they are real items in an inventory of the universe. Gregory's theory is an application of a standard thesis of Platonic metaphysics: the intelligible features of an object are ontologically prior to their material instantiations. (Hibbs 2005: 432)

But although the challenge that Gregory was addressing was how to explain the derivation of a material nature from an immaterial one, on this issue Hibbs tells us that Gregory's solution is silent: 'Exactly how this process [of the creation of material bodies from concepts] is supposed to work, and the status of the concepts after they are brought together in the act of creation, is not made clear by Gregory' (Hibbs 2005: 432). Yet this is where we want the interpretation of Gregory's solution to enrich our understanding. In conclusion, I submit, no one of the existing interpretations of Gregory's views reviewed above as representative of the recent scholarship delivers an account that is philosophically fully satisfactory as well as fitting with the textual evidence.

The challenge of the *immaterial* generating *material* bodies, without violating the LCL principle

In this section I will argue that Gregory's solution to the conundrum of the world's creation is this, in a nutshell: in creating the natural qualities, God *thereby*, without any additional act, created all that was needed for the constitution of material bodies too. But does the creation of the natural qualities respect the LCL principle? Are God and the qualities of the same nature? And if they are, what is that nature, and how could material bodies be constituted by entities of such a nature? In what follows I first offer a reconstruction of the logic of Gregory's argument that accounts for God's creation of the world; I then focus on the metaphysical considerations that Gregory introduces to account for how material bodies come about when immaterial qualities bundle together. In the structured representation of the argument below, the convention I follow is to indent under the conclusion the premises or the sub-arguments pertinent to the support of that conclusion. The premises justifying a conclusion are grouped at the same level of indentation.

C. God created both immaterial qualities and material bodies, with no violation of the LCL principle

1. God created immaterial qualities.
    1.1 God is the source of creation.
    1.2 Creation is governed by the LCL principle.

1.3 God's nature is intelligible.
1.4 The qualities' nature is intelligible.
1.4.1 Qualities are definable.
1.4.2 The definable is intelligible.
1.5 The qualities are immaterial.
1.5.1 The qualities are definable.
1.5.2 What is material is not definable.
1.5.2.1 [Because (presumably) the combination of the qualities in material things is not determinate and expressible in ideas.]
2. Material bodies *are* bundles of immaterial qualities.[9]
2.1 Body dissolves into abstracted immaterial qualities.
2.2 The substratum in material bodies is a logical subject, rather than an additional undefinable entity.[10]

There are many questions that arise from this line of argumentation. In this chapter I focus on what I take to be the most serious philosophical challenge the argument presents us with, that is, that the qualities of *material* bodies are *immaterial*, and nevertheless their combination results in the constitution of material bodies. I aim to show that this is what Gregory is proposing, and that there are both historical and theoretical considerations that can help us make sense of his position.

I should like to begin with a methodological note. Although I believe, with others, that Gregory of Nyssa was mainly influenced by the views of Plato and generally the Neoplatonists, there are many elements of his reasoning that exploit Aristotelian doctrines or arguments.[11] As we shall see this is often explicit in Gregory's content; but even further, in crucial passages Gregory uses philosophical terminology *coined* by Aristotle. Still, Gregory's reasoning is not anti-Platonic. Rather, he uses Aristotelian

[9] The metaphysics underpinning this claim will be discussed shortly.

[10] My premise 2.2. derives from *De hom. op.* 24.212–13: 'For we shall find all matter to be composed of (*sunestanai ek*) qualities (*poiotes*) and if it were stripped bare of these on its own, it could in no way be grasped in definition (*logos*)'. On an alternative way of reading this quotation (particularly the wording 'if it were stripped bare of these [qualities] *on its own*' (my emphasis)), one may want to argue that Gregory posits a substratum underlying material objects (cf. Hill's interpretation above, p. 98). But now this substratum, not being a quality, and qualities being the only items that God creates, would be ungenerated by God. Then who generated it? If this substratum existed, it would generate a clear problem for Gregory's doctrine of divine creation. Rather, I submit, what remains after stripping away the qualities is their logical subject, which has no being of its own.

[11] For a recent discussion of this issue see Rist 2000. The existing evidence to the effect that Gregory voiced negative feelings about Aristotle (see e.g. *C. Eunom.* 1.6) does not preclude or make less plausible that Gregory could make use of Aristotelian views and arguments, in service of his own line of thinking.

arguments to explain further, and make fit for his own purposes, theses that would be thoroughly 'respectable' Neoplatonic and even Platonic positions.[12]

Gregory is clear about the challenge he is facing, which he expresses and frames in metaphysical terminology. He says that

> no one should still be cornered by the question of matter, and how and whence it arose. You can hear people saying things like this: if God is matterless, where does matter come from? How can quantity come from nonquantity, the visible from the invisible, something with limited bulk and size from what lacks magnitude and limits? And so also for the other characteristics seen in matter: how or whence were they produced by one who had nothing of the kind in his own nature? (*Ap. in hex.*, trans. at Sorabji 1983: 290)

How can the immaterial give rise to the material? What is being questioned is not the power of God to generate, but the *metaphysical compatibility* between God's nature and what God generates when matter is created. The problem Gregory poses has the makings of a thoroughly 'recalcitrant', if not hopeless, philosophical puzzle. It is the type of problem that has dominated philosophical thinking for millennia and continues to drive research in present metaphysics. We can think of it as the problem of *categorial incompatibility*. Gregory engages with an instance of it, which is divine creation; in the modern era the problem took the form of the causal gap between the physical and the mental, which still exercises us today. In what follows I engage with the solution that I take Gregory to offer to the version of the puzzle he is concerned with.

How the immaterial generates the material

As we saw, the general problem at stake is the categorial compatibility of the cause and the effect. The LCL principle requires that like generates like – as Aristotle said, man begets man. But the difference between the immaterial and the material is so fundamental that any other likeness between them pales into insignificance. Gregory restates their difference, in order to frame the puzzle vividly and precisely:

> The *corporeal* creation is thought of in terms of properties which have nothing in common with the *divine*. And in particular it produces this

[12] The Aristotelian positions and arguments that Gregory uses are so central to Aristotle's metaphysics and epistemology that they became fundamental tools in the philosophical teaching of succeeding generations.

great difficulty for Reason (*logos*). For one cannot see how the visible comes from the invisible, the solid and resistant from the intangible, the limited from the unlimited, or what is in every way circumscribed by quantitatively conceived proportions from what lacks quantity and magnitude, and so on for everything which we connect with corporeal nature. (*De an. et res.*, trans. at Sorabji 1983: 290, emphasis added)

The puzzle is framed in terms of the violation of the LCL principle. A clarification is in place here. Aristotle held that change is from opposite to opposite, e.g. from being hot the object comes to be cold. This describes the *before and after* conditions in an object – and these two conditions are not related as cause to effect; while the LCL principle is concerned with the relation of the *cause with the effect.* We should not be misled by the terminology of, for example, tangible and intangible, or corporeal and incorporeal, into thinking that Aristotle's principle that opposites come from opposites would apply here too. The corporeal and incorporeal are not two states of a single entity, but the natures of a cause and an effect, which is not what Aristotle is describing in his principles of opposites. Furthermore, Aristotle's dictum does not apply to such distinctions as the corporeal and the incorporeal, but to opposite qualities. Opposite qualities fall under a single genus, e.g. the thermal, for hot and cold; or texture, for rough and soft. But the corporeal and the incorporeal do not share a common genus of any generality. They are not opposites, but the one expresses the privation of the other, designating the two corresponding entities into different categories of being. Here, LCL tells us that generation may occur only between entities that can share the same type of being (that is, that can be like one another).

Thus, there is only one way for Gregory to resolve this predicament: the creator and the created need to be shown to be the *same type* of being. And this is exactly what Gregory undertakes to show. He argues that both God and what God creates are *immaterial* and *intelligible*. By contrast, objects are *material* and *perceptible* but *not intelligible*. Yet, he argues, material objects are constituted by the qualities created by God. I will suggest that the qualities 'are' indeed physical, despite their immateriality and intelligibility.[13]

[13] This is a tacit reference to the special sense in which qualities are true of abstract objects in such Neo-Aristotelian theories of the metaphysics of abstraction as those of Kit Fine (2002), and Edward Zalta (2006).

In order to understand how Gregory resolves the creation puzzle, we can look at its ancestral thought experiment that was famously carried out by Aristotle – stripping qualities off a body:

> When all else is taken away [from a body] evidently nothing but *matter* remains. For of the other elements some are affections, products, and capacities of bodies, while length, breadth, and depth are quantities and not substances. . . . But when length and breadth and depth are taken away we see nothing left except that which is bounded by these. . . . By matter I mean that which in itself is neither a particular thing nor of a certain quantity nor assigned to *any* other of the categories by which *being* is determined. For there is *something* of which each of these is predicated, so that its being is *different* from that of each of the *predicates*. . . . Therefore the *ultimate substratum* is of itself neither a particular thing nor of a particular quantity *nor otherwise positively* characterized; *nor yet* negatively, for negations also will belong to it only by accident. (*Metaph.* Z 3 1029a11–26, trans. Barnes in Aristotle 1984: 1625, emphasis added)

This is a thought experiment which concludes that matter, in itself, has no being, nor privation of being, where the kinds of being and its privation are depicted in an ontology such as that of the classification system of Aristotle's *Categories*. Gregory says, in his turn: 'We shall find all matter to be *composed of qualities* and if it were stripped bare of these on its own, it could in *no way* be grasped in *idea*' (*De hom. op.*, trans. at Sorabji 1983: 291, emphasis added). Just as Aristotelian matter in itself cannot be described by any of the predicates of language (i.e. by any general term), the Gregorian bare matter cannot be described by any idea.

This conclusion is important to Gregory because he wants to show that what is distinctive about matter is that its nature cannot be described by ideas. Although this raises a question about the nature of matter, it does give Gregory what he needs: matter in itself – bare matter – stripped of its qualities, is *not intelligible.* The thought experiment of abstracting away the qualities of a material body shows, both for Aristotle and for Gregory, that what remains does not, in itself, have any type of being. Both Aristotle and Gregory discover, through this thought experiment, not an enigmatic pure potentiality – as some interpreters of Aristotle think, but rather the metaphysical function of subject-hood. This is not a type of being, but a logical role that is played out for every material object. How it can be played out can vary metaphysically, from system to system, and I will not engage with those differences here, but will only note that this interpretation of the ultimate substratum in Gregory, namely of a function to be dispensed with, rather than of an entity (e.g. a bare particular), suits

Gregory's ontology, where there is no space for un-generated un-intelligibles. This is because he will attempt to argue that 'there is an opinion about matter which seems not irrelevant to what we are investigating. It is that matter arises from the intelligible and immaterial' (*De hom. op.*, trans. at Sorabji 1983: 291). So positing a primitive (un-generated) un-intelligible substratum would run counter to the whole endeavour of deriving matter from the intelligible. But how can matter arise from the intelligible immaterial? Gregory tells us how, but this account will need to be supplemented with Aristotelian insights to make it metaphysically 'robust'.

So far we have seen that matter is composed of qualities, and that there is no further being than this in matter. That matter is *composed* of the qualities is shown by Gregory through the stripping-off thought experiment, where all the qualities of matter are abstracted away from body and no being remains. Gregory argues in reverse that 'if we find the absence of these things causes the dissolution of body, we must suppose their *combination* is what generates material nature' (*De hom. op.*, trans. at Sorabji 1983: 291). It is not clear here whether Gregory sees the qualities as being merely compresent or as composing; but either way, they are the sole constituents of material body that possesses them as qualities. This line of reasoning directs our interest to these qualities. What types of entity are they, and whence do they derive? Gregory tells us that

> By his wise and powerful will, being capable of everything, he established for the creation of things all the things through which matter is constituted: light, heavy, dense, rare, soft, resistant, fluid, dry, cold, hot, colour, shape, outline, extension. All of these are in themselves thoughts (*ennoiai*) and bare concepts (*psila noēmata*); none is matter on its own. But when they combine, they turn into matter. (*Ap. in hex.*, trans. at Sorabji 1983: 290)

All the ways in which matter is characterised, namely all the qualities that constitute matter, are said to be in themselves thoughts and bare concepts. This and other statements to this effect have been the basis of interpretations in the literature that attribute a form of idealism to Gregory. But the question arises, how then does the combination of the thoughts turn them into matter? One might think that Gregory is offering a theory of the *supervenience of matter on the mental.* But there is nothing in Gregory or his classical sources that would make such an account plausible; on the contrary, recall Plato's 'damning' comment that we encountered earlier, that if Forms are thoughts and things partake of them, there then follows the absurdity 'that each thing is composed of thoughts, and that all things think; or that although they are thoughts, they are unthinking'

(*Parm.* 132b–c). We would expect Gregory to remark on this type of objection from a prominent predecessor, if he held that objects are made of thoughts. This is why it is more promising to explore a different interpretative direction, which finds foundations and rationale in Platonic and Aristotelian metaphysical positions.

My understanding of the ontological status of the qualities is grounded in this claim of Gregory's, that 'each type of quality is *separated in account from the substratum* [*logôi toû hupokeimenou chôrizetai*], and an idea is an intelligible, not a corporeal way of looking at things' (*De hom. op.*, Sorabji's translation, modified, at 1983: 291). Being *separate in account or definition* (*choriston logôi*) is a technical term that Aristotle introduced (see e.g. *De an.* 413b14–15). It is related to abstraction and indicates that we can individuate by conceptual division what is not physically separable. Thus, a definition or an account of what something is separates it out conceptually from its physical manifestation. For instance, defining the blue of the sea individuates a colour which is manifested in the blue sea. What type of entity exactly is individuated by abstraction can vary; e.g. the colour of this particular sea (that is, a colour trope); or this particular tint of blue (i.e. a specific universal); or as blue (a generic universal). The level of discrimination in individuating a quality within Gregory's ontology will not concern us, as he does not engage with the particularity of the qualities that make up matter. What is significant for our investigation is that separating in account (or definition) allows for the *division by abstraction* of objects into the qualities they are composed of, without requiring their physical division even in principle.

What is it that is being separated in account? The answer is straightforward, namely, what is separated in account is what is described by the account. The account is a definition of a quality and describes *what it is* to be that quality. Thus Gregory says: 'For the explanatory formula envisaged for each of these [qualities] is quite individual according to *what it is*, and has nothing in common with any of the other qualities which we connect with the substratum' (*De hom. op.*, trans. at Sorabji 1983: 291, emphasis added). The account may be given in terms of the species and genera of that quality, or through peculiar characteristics of the quality. For example, Aristotle defines colour as that which can move what is actually transparent (*De an.* 418a31–b1); this is what colour *is*, wherever colour is in objects. The definition of a quality abstracts that quality away from its manifestations in objects; Gregory writes that 'each type of quality is separated in idea from the substratum' (*De hom. op.* 24). What the definition describes is an abstract entity, e.g. colour. Colour is found in nature, not separately as an entity of its own, but embedded with other qualities in objects. So, speaking metaphorically, the definition sets out as if 'in relief' a quality that is embedded in an object.

This is the key to my understanding of Gregory's ontology: each *physical* quality of an object, described by the *definition* of that quality, is an *abstract* entity. The qualities are mixed in the object, but their definitions single each of them out in abstraction. That is, the definition of each of the qualities that belong to the object separates *each quality* by abstraction from the other qualities, all of which in combination compose the object. Thus, the qualities, which are the various aspects of the *physical* constitution of an object, are, *qua* definable, *intelligible* abstract entities. Gregory did *not* think that the abstract intelligible entity is a thought in a mind. It is intelligible in so far as it is fully describable by the definition, through concepts expressed by the general terms of the definition. But what is abstracted and singled out through the definition is a *physical* aspect of an object. That aspect is not separate, distinct in the object, but integrated with the other qualities of the object to which it belongs.

When Gregory talks of *concepts* (*ennoiai* or *noēmata*) he uses concepts impersonally, *not* to refer to the content of a consciousness, but to refer to an *abstract* entity, a quality, which is *definable* and hence *intelligible*. This quality is individuated by abstraction, separated in definition from the object that it characterises, and of which it is *truly predicated*. We read:

> By a process of *mental division* we recognize many things *connected with the substratum* and the idea of each of them [the abstracted qualities] is not *mixed up* with the other things we are considering at the same time.[14] For the ideas of colour and of weight are different, and so again are those of quantity and of tactile quality. Thus softness and two-cubit length and the other things *predicated* [of the substratum] are not conflated with each other, nor with the body, in our idea of them. (*De hom. op.*, trans. at Sorabji 1983: 291, emphasis added)

Following the process of abstraction in reverse with Gregory, one is led *from* the abstract entity to its *embedding* in the object: 'For a thing is not a body if it lacks colour, shape, resistance, extension, weight and the other properties, and each of these properties is not body, but is found to be something else when taken [*by abstraction*] separately. *Conversely*, then,

[14] I take it that it is this 'mixing up' of the abstract qualities in a material object that makes the object undefinable, and hence unintelligible, and merely perceptible. Gregory offers no account of an *interaction* between the qualities, but only a suggestion of the *indeterminacy* of the matter due to the way the qualities come together in an object.

when *these [abstracted] properties combine* they produce material reality' (*De hom. op.*, trans. at Sorabji 1983: 291). What Gregory is describing is a thought experiment of dissolution and re-composition of material objects. He dissolves the objects into their qualities, which he describes with the definitions of the qualities, and then recomposes the objects from these qualities by abstraction in reverse.[15] The *converse* process to abstraction (let us call it 'intraction') is precisely recombining the abstracted, and recomposing the object out of its qualities. Each quality picks out the physical nature of an aspect of the object, which is singled out by abstraction. The combination of these physical natures (qualities) makes up matter – physical matter. Gregory does not think of matter as a concept, an image or a perception of ours; for him matter is as concrete as we believe it is.

Crucial to understanding that it is matter that results from 'intraction' is the sense in which the qualities 'are' physical; for example, the sense in which it is true of the abstracted quality of redness that it 'is' red.[16] Gregory does not himself develop a distinction between the way that an object is red and the way that the abstracted colour of that object 'is' red; but he does assume that for example the quality red is the redness of the colour. At the same time, the quality red is definable and, hence, intelligible. The quality of red is not perceptible for Gregory because it is not concrete, that is, it is not combined with the other physical qualities. So, in so far as qualities 'are' the physical natures they pick out from an object, to that extent the qualities 'are' the constituents of that material object. Gregory thought that the abstracted qualities literally *are* the physical natures of aspects of objects, exactly the natures that the definitions of the qualities specify, and thereby, for Gregory, the combined qualities literally *are* matter; so matter is composed of intelligible physical qualities. Gregory of Nyssa built *material objects out of the abstracted physical qualities* of the material objects.

The challenge Gregory undertook to address is: 'if God is matterless, where does matter come from?' (*Ap. in hex.*, trans. at Sorabji 1983: 290) His answer was that God does not create matter. God creates the abstract

[15] Aristotle's relation of an abstracted entity, separated by definition from an object, is a metaphysically *more complex* account than the one offered above in explaining Gregory's ontology. What gives rise to the further complexity is Aristotle's Homonymy Principle, according to which dividing an object generates entities that are not as such in the object, because division severs and deprives the divided part of the object from its the role and functionality in the whole object. I have not used this principle in my account of Gregory's views because I see no evidence that Gregory employs it in his ontology.

[16] Contemporary metaphysical theories of abstract objects have advanced developments of this idea that abstract entities can be physical.

intelligible qualities which are the physical natures that compose matter. When these qualities are combined with the other qualities that belong to a subject, a material object is constituted. God's nature is non-material, and so are the abstract qualities God creates. These entities are non-material and intelligible, in that they are fully described by definitions. Hence, 'if the conception of these properties is intelligible, and the divine is intelligible in its nature, it is not strange that these intellectual origins for the creation of bodies should arise from an incorporeal nature, with the intelligible nature establishing the intelligible properties, whose combination brings material nature to birth' (*De hom. op.*, trans. at Sorabji 1983: 291). The sense in which these abstract entities 'are' physical is the sense in which the definitions 'are' truly predicated of them. Gregory does not engage with the question of how the definitions of the qualities 'are' true of them, despite their abstractness. How, for example, 'is' the abstracted quality of Socrates' weight a weight (heavy)? Gregory writes:

> But we can say this much on the subject: none of the things we connect with body is on its own a body – not shape, not colour, not weight, not extension, not size, nor any other of the things classed as qualities. Each of these is an idea [i.e. an abstract intelligible entity], but their combination and union with each other turns into a body. (*De an. et res.*, trans. at Sorabji 1983: 290)

Clearly more metaphysical distinctions and theory are needed to fill in the gaps of, on the one hand, the relation between the nature of qualities in abstraction and embedded qualities and on the other, the difference in ontological status of qualities in abstraction and embedded qualities. A metaphysical theory of abstract entities would be needed to offer full explanations of the differences between them. What I have tried to do in this chapter is to explain how Gregory builds the ontology of his system, assuming that such a theory of abstract entities is available, for example from Aristotle's categorial scheme of properties. In so doing I have tried to indicate what metaphysical distinctions would be needed to enable us to follow Gregory's train of thought in claiming that abstract ideas combine into material body.

Conclusions

Gregory makes use of the philosophical heritage of his era with ingenuity, in addressing the metaphysical problem of how an immaterial God could create a material world, given that like causes like. He exploits what he in some sense takes to be a *hybrid entity, delivered to us by our power of abstraction, namely qualities: he derives qualities from material objects, separating them from the objects by abstraction, thereby individuating physical aspects of the material objects as definable, intelligible entities.* His solution, in sum, is that the immaterial God created immaterial qualities of objects, which are physical aspects of objects, and which compose with one another to give rise to material bodies. He needs more metaphysical tools to fill in his account – some of which were developed by Aristotle, but not fully exploited by Gregory. Some further tools of abstract predication have been formulated in recent theories about the metaphysics of abstract objects.[17] Possibly even with these new tools Gregory's account might still be wanting. Thus Gregory has taken us into the very demanding domain of the metaphysics of abstraction, where 'still wanting' may even be a term of praise.

[17] I cannot develop this point further within the scope of the present chapter.

CHAPTER 6

Simplicius on elements and causes in Greek philosophy: critical appraisal or philosophical synthesis?

Han Baltussen

One of Simplicius' contributions on causes in the commentaries, as has been pointed out recently, is that he clarifies the use of 'principle', 'cause' and 'element' in Aristotle, and disagrees with the notion that they can be used interchangeably.[1] His overall exegesis becomes quite distinctive by incorporating many more views from previous exegetes into his textual analysis than one would think necessary. A good example comes at *In physicorum libros* 19.21–20.2, where Alexander is quoted as saying that Aristotle may be referring to axioms (ἀξιώματα) when speaking about general descriptions of principles (τὰ κοινά).[2] Simplicius disagrees: he seems to think that we acquire knowledge of the principles through observation.[3] That the problematic nature of the 'elements' requires further attention is clear from Simplicius' analysis of Aristotle's *Physics* A, the book dedicated to a review of earlier theories on principles.[4]

My aim in this chapter is to examine Simplicius' technique of composition and how it helps structure his evaluative comments. Such an investigation will clarify how his remarkably inclusive selection procedure seeks to draw on whatever sources he thinks useful for his purpose. In past studies some of Simplicius' own views on principles and causes in natural philosophy have been stated with considerable clarity and acumen.[5] In response to Aristotle's text, he will, directly or indirectly, declare his own hand regarding the nature and knowledge of principles and causes. He analyses Aristotle's ideas on elements, matter and their relationship with

[1] See Sorabji 2012: 1; cf. Baltussen 2008, ch. 2 and below p. 114.

[2] Simplicius summarises Aristotle's position as ἀπὸ τῶν κοινῶν καὶ συνθέτων καὶ ἡμῖν σαφεστέρων ἐπὶ τὰς ἀρχὰς τῶν φυσικῶν φησι δεῖν ἀνατρέχειν ὁ Ἀριστοτέλης (19.18–19): 'Aristotle says that we have to work back from the things that are common and composite and clearer to us to the principles of natural things.' For common axioms and demonstration see *An. Post.* 76b14 and cf. *Top.* 155b15.

[3] This passage opens with μήποτε, which normally introduces Simplicius' own opinion (as briefly noted by Konstan 1989: 163, index). A more elaborate case in support of this interpretation is Menn's paper (presented in Trier, October 2012) in Strobel forthcoming.

[4] Cf. Cherniss 1935; McDiarmid 1953.

[5] E.g. Sorabji 1988: 7–22; Hankinson 2002: 6–8.

reference to Aristotle's corpus, to Plato or by applying Neoplatonic ideas.[6] These analyses are often based on his famous essays on place and time (*In phys.* 4), where Simplicius' own views are clearly stated. By contrast it is not so easy to separate out views from his discursive evaluations, and scholars often make assumptions about the relative value of the materials encountered, the different 'sources', so to speak, which he selected and gave a place in his account.

Simplicius' method of reading Aristotle depends on several (unarticulated) assumptions regarding the coherence of Aristotle's thought and the value of his exegetical predecessors. My argument will selectively comment on the manner in which he brings in and evaluates existing exegeses on the text under scrutiny in order to show that his broad sweep of the evidence is both his strength and his weakness. It is a strength because the discussion of principles is thereby enriched and historically rewarding. It is also a weakness because, no matter how hard he tried, he could not achieve one of his overarching objectives: to harmonise the whole tradition as he knew it.[7]

An illuminating passage, which can be found in his essay on time (the so-called 'corollary on time' at *In phys.* 773.8 ff.), illustrates this approach well (lines 9–19):[8]

> I have gone through what he has said from the beginning to the end, showing how it all fits together so far as possible. But since our academic training has not the sole aim of learning Aristotle's opinion concerning the nature of time, but rather of understanding what time is – by which I think that we shall also get a closer grasp of Aristotle's thoughts on time – let us examine (διευκρινησώμεν) that briefly. After that, let us review (ἐπισκεψώμεθα) the opinions of those who have philosophised about time. But since Aristotle at the beginning of his discussion mounted severe attacks (δριμεῶς ἐπιχειρήσαντες) on the reality of time and did not defeat these attacks (τὰ ἐπιχειρήματα μὴ λύσαντος) it would be useful towards our goal to overcome (διάλυσαι) them as far as possible. For otherwise our discussion of time would remain incomplete. (Simplicius 1992: 85)

The passage highlights some typical features of Simplicius' method: a thorough examination of the text ('from the beginning to the end') to demonstrate its coherence, but also a clear emphasis on a philosophical agenda (to learn about time *itself*). The passage also reveals that his philosophical agenda requires at least three things: (1) a review of existing

[6] E.g. *In phys.* 3.10–16.7, 15.4–16.7 Diels; *In de cael.* 1.1–4 Heiberg, with verbal echoes from *An. Post.*
[7] Indications for this ambition at *In phys.* 28.31–29.5, 188.13–16; *In de cael.* 143.16; *In cat.* 2.21.
[8] I discussed this passage in a broader context in Baltussen 2008: 200–1.

theories on time; (2) solving the problems that can be mounted against the reality of time; and (3) the claim that the understanding of time per se will enhance our understanding of Aristotle's views on time.[9]

In the Aristotelian account of principles (*Phys.* 184b22–5), Simplicius appositely sums up Aristotle's position: 'he says that those who investigated what is, were investigating the *principle* of what is' (my emphasis). While this suggests that Simplicius has found an angle to work from, he does not leave it at that. During analysis he brings in many other exegetes. What is the role of these 'fellow scholars' in his analysis? His discussion of the role of elements as causes in Greek cosmogonical theories of causation, in particular Aristotle and Plato, can serve as a good case study for clarifying his method. By considering this topic as it were through his 'Neoplatonist eyes', we can acquire indirect access to his views. As we will see, the approach affected his method in two important ways: first, he was confronted with a number of conflicts in the materials (which he resolved by trying to reconcile them, especially Aristotle with Plato); second, he implicitly shows, based on his wide knowledge of the Greek philosophical tradition, the continuity of the debate on causes.

The reason for this continuity was clear to him: in part it was brought on by the fact that philosophers could not reach full agreement on the nature and role of elements and principles in the creation of the cosmos.[10] Simplicius tried to establish agreement by an extraordinary *tour de force*, even if we reject it as tendentious and forced. But judging his approach is not my focus here. His habit of incorporating long-standing debates is especially important for the question of cosmic creation and its causes – perhaps the most fundamental issue of all. His close familiarity with the preceding works in natural philosophy ranged from the Presocratics to Plato and Aristotle and up to the subsequent commentary tradition.[11] This richness of his sources requires careful unpacking in relation to his own viewpoint.[12] I will move from the debate on cosmic causation to Aristotle

[9] This whole method closely echoes Aristotle's own approach in *Top.* 1.14 and *De cael.* 1.10. Signs that he tries to be comprehensive in sources are his use of phrases like 'all the commentators' (*In phys.* 243.13). The procedure is not always successful: when discussing the question of matter (227.24 ff.) he concludes: 'I do not know how *all those people* who claim to understand matter . . . can be correct' (*In phys.* 232.30–2).

[10] At *In phys.* 36.24–31 he speaks of a 'cavalier reproach of disagreement among the ancients'. See below on 232.30–233.3 and Falcon 2012: 178, 180, 182.

[11] He had access to (parts of?) the works of Parmenides (see esp. *In phys.* 144.25–147.21, where he calls the work 'rare' (σπάνιν, 144.28) and quotes long chunks of the text), Empedocles (*In phys.* 32.5–34.4, 157.25–161.20), Anaxagoras (24.25), Diogenes of Apollonia (25.6–7) and others.

[12] See Wildberg 1987; Sorabji 1987; and Osborne 2009.

and Plato. It must be kept in mind throughout that my account can only be very selective, given the scope and size of the commentaries.

A question of origin(s): explaining cosmic causation

The question of the cosmos's creation (κοσμοποιία, a term first found in Aristotle *Metaph.* 985a19[13]) and its causes became a much-debated and central issue in ancient Greek philosophy.[14] The fundamental questions at the centre of this debate are well known: they concern beginnings and endings, and whether the one presupposes the other. Did the cosmos have a beginning? If so, what could have caused it? Did it have an ending? If so, why? If not, how could it be sustained? In the answers three permutations received support: (1) it had a beginning and an ending (some Presocratics; Christians); (2) it had a beginning but no ending (a literal reading of Plato's *Tim.* as in Aristotle's *De cael.* 280a29–30); and (3) it had no beginning and no ending (Aristotle, and later supported – for different reasons – by Plotinus, Proclus and Simplicius as supra-temporal). Position (1) was formulated in terms of biological processes, making finite existence the prime factor. Position (2) needed a special 'escape clause' for the asymmetrical arrangement, postulating a supernatural power to set things in motion, yet allowing for infinite existence; and position (3) opted for eternity in a very strong sense by claiming that an ending would imply a beginning, while no beginning ought to imply no ending, hence the neat (and symmetrical) solution.[15]

Speaking about 'causes' and 'elements' was an unmistakable legacy from the Presocratic era. The early Greek rationalists had recast 'myths' by looking for the basic constituents of the universe as natural causes.[16] But causal entities became only later described as 'elements' (στοιχεῖα[17]), 'principles' (ἀρχαί) and 'causes' (αἰτία[18]) – a sign of their usefulness and

[13] At *Phys.* 196a22 the word occurs as a possible title of Empedocles' work on nature.

[14] κόσμος is already in Plato, but introduced with some hesitation as equivalent for 'universe', πᾶς οὐρανος at *Tim.* 28b3.

[15] For (1) see esp. Empedocles fr. 57 *DK*; for (2) Plato *Tim.* 28–34. Aristotle opted for (3), see *De cael.* 3.1, 298a24–6 'first heaven and its parts, the moving stars . . . ungenerated and indestructible'.

[16] Famously Democritus was keen on causes (αἰτία, fr. 68B11B–11k *DK*).

[17] Using 'element' for στοιχεῖον is justified, so long as we realise that it is a retrojection of Plato's term (Simplicius *In phys.* 7.13 reports its first occurrence, following Eudemus). It originally referred to a sundial and the components of things which are well organised (e.g. letters of the alphabet).

[18] Usefully summed up in Mann 2011: 30, *stoicheion* is clearly an early formation which refers to a physical component; *archê* is originally 'beginning', but widens its meaning to 'principle', which can be both constituent and cause for a substance.

lasting importance, even if descriptive styles differed.[19] A case in point is Empedocles, whose four elements were called 'roots' (ῥιζώματα, fr. 31B6), a biological metaphor to signify the foundational role of elements; but he introduced an innovation by adding the 'powers' Love (attraction) and Hate (repulsion) in order to resolve the problem of inertness of the elements: Love and Hate would make elements move, change and interact. Simplicius' contemporary Priscian recognised the two powers by calling them the 'kinetic principles, love and hate' (κινήτικαι ἀρχαί), and linking their agency with Plato's *Timaeus* on the cognitive soul functions (*In de an.* 27.34–7; *Tim.* 48b–c).[20] Yet the three terms still needed clarification, since some physical elements are not principles.[21]

In his commentary on the *Physics* Simplicius regards the Presocratics as respectable participants in the debate, largely because Aristotle did so (Plato's importance was a given).[22] Simplicius improved on the schematic *dihairesis* of *archai* in Aristotle by offering a more complete version of the possible options for basic elements.[23] Again, his main aim is comprehensive reporting rather than mere summative compression. While Aristotle took some pleasure in pointing out their failings, Simplicius was more constructive, consulting Empedocles independently, although the (in our eyes crucial) testimonia on the elements are not found in his commentaries.[24] Whether these specific choices of elements were satisfactory or not (one may ask whether in all theories they are primitives), their attempts at explanation carved out new pathways of understanding the world, thus determining the future direction, vocabulary and methodology in natural philosophy.[25] Both Plato and Aristotle had carefully examined existing views on the nature and genesis of the cosmos (and its causes). Both ended up accepting the four (Empedoclean) elements (*Gen. et corr.* 330b1–2; cf. *Tim.*). Aristotle pays attention to these in the context of discussing the four qualities (hot, dry, cold, wet) and their transmutation: 'for all those who

19 On continuities in ancient physics see Solmsen 1960: chs. 1, 4–9 and 22.
20 A vast literature on this issue exists: see Martin and Primavesi 1999 with earlier literature.
21 For an exploration of the terms and their use in Aristotle see Mann 2011 with earlier literature.
22 His verdict: 'the ancient wisdom remains irrefutable' (*In phys.* 77.11). McDiarmid (1953: 89) agrees that Aristotle influences Simplicius: 'The sentences with which he introduces each group of philosophers are clearly patterned on similar sentences in Physics.' But his analysis does not go much further.
23 See esp. Mansfeld 1989 for this aspect of his analysis.
24 See esp. 31B6, 7, 12 and 14 *DK*. Cf. Mansfeld and Primavesi 2012: 446–9, texts 49–51 (elements); 126–150 (cosmology); crucial information on elements (e.g. text 49) not in Simplicius.
25 *Natural* causes *within* this world, physical entities as fundamental constituents of all bodies (Thales: water; Diogenes: air; Anaximenes: aether; Heraclitus: fire; Anaxagoras: infinite; Empedocles: all four) leading to a simplified theory by way of a reductionist model.

make out that the simple bodies are elements make them either one, or two [Parmenides] or three ('as Plato does in the divisions') or four [Empedocles]' (*Gen. et corr.* 330b7–9, 19). While Aristotle clearly used this survey to demonstrate that there are only *four* simple bodies (cf. 330b31), even if he offered little proof beyond giving the obvious examples, Simplicius claimed that he did deduce them by argument in the *Physics*.

Simplicius defended Plato's 'likely account' of the *Timaeus*, which includes the genesis of elements (a form of 'geometrical atomism') against Aristotle's criticisms (rejects the divine Craftsman, the creation story as a 'probable account', the elements as geometrical[26]). Plato's cosmogony speaks of several causes for the world's genesis, but his use of the terms for principles is not clear-cut. On a broad scale, he postulates two – necessity and intelligence – as causes of the world as we know it.[27] But halfway through the dialogue the Forms and the Receptacle also come into play, since the former provide copies of themselves for the instantiations of particulars, the latter being the space within which this occurs.[28] Simplicius claims that he defined three main causes and three subsidiary ones (as did Proclus: see p. 124 below).[29]

Theophrastus famously defended his teacher's views on the eternity of the world, against the Stoic Zeno (fr. 184 FHSG = Philo *De aetern. mundi* 23.117–27.149), even though he himself disagreed on details, probably rejecting the 'fifth element' attributed to the cosmos as a whole (fr. 161A FHSG).[30] But for Platonists the notion of causality was different from that of Aristotle.[31] The preoccupation with the question of the cosmic creation was still a point of fierce debate among the Platonists, starting with Plotinus (*Enn.* III.2.1.38–45; cf. VI.5.8.20–2), continuing under Proclus, who wrote an enormous commentary on the *Timaeus* (extant only up to *Tim.* 35), and reached a climax with Simplicius' polemical attacks on the Christian Neoplatonist Philoponus.[32]

[26] Plato's core text for the *eikos logos* is *Tim.* 29d. Aristotle's objections are formulated at *De cael.* 3.3; *Phys.* 191b35 ff. (Cherniss 1944: 84–96); *Metaph.* 1.6; *De cael.* 3.1, 298b33 ff.; *De gen. et corr.* 1.2, 315b30 ff.

[27] Cornford 1956: 159; Silverman 1992: 96.

[28] Silverman 1992: 92 ff., disagreeing with Zeyl about the status of the particulars (pp. 94–5).

[29] Sharples 1995: 67–73.

[30] We also know from Simplicius that Theophrastus considered the question of how many causes there are: see Sharples 1998: 87–94 and n. 52 below on Xenarchus.

[31] For a concise account of causality among the late Platonists see Helmig and Steel 2012.

[32] On this polemic see esp. Sorabji 1987; Hoffmann 1987a.

Tracing the elements and defining matter

One important issue that Simplicius tries to resolve is how Plato and Aristotle can be read as agreeing on the qualities of the elements and their relationship to matter. Aristotle (and Theophrastus) establish that the causes of the physical world must be physical themselves by reviewing earlier theories, as well as rigorous reasoning (Aristotle *Phys.* 1.7; Aristotle *Metaph.* 1.3; Theophrastus *Phys.* 1 = Simplicius *In phys.* 9.7–11 = fr. 144B FHSG). Simplicius seems in agreement when he adds that it is convention to assume principles, since 'neither the geometer nor the doctor demonstrates his own principles (τὰς ἑαυτοῦ ἀρχὰς ἀποδείκνυσιν), but they use them as given'.[33] When Simplicius makes an effort to bring Plato and Aristotle together on this point later on, his harmonising approach cannot get around the question (*In phys.* 230.35 ff.) how Aristotle's elements, which have properties (hot/cold, wet/dry), can be aligned with Plato's notion of matter (a kind of 'substrate'), which has *no* qualities – except for being 'all-receiving' (*Tim.* 51a7: ἀνόρατον εἶδος τι καὶ ἄμορφον, πανδεχές . . .). He therefore tries to argue that Plato and Aristotle say basically the same thing (*In phys.* 232.6–7): 'But how can these things be harmonised with Aristotle and Plato, who think that matter is a substratum for contrariety? Or does what other people say reduce it to the last body?' Simplicius next infers that extension is important for Aristotle in both heavenly bodies and sublunary ones. This conclusion he bases on *Physics* 209b6–9 ('in so far as place is thought to be the extension of a magnitude it will be thought to be matter'), ending in Neoplatonic style by taking matter as the ultimate deviation of the One, that is, furthest from the cause of all things.

The intricacies of Plato's elemental cosmogony are too complex to consider here in full. The material elements of the world as they appear in Plato *Timaeus* 53b–c pose considerable interpretative problems, some of which were already recognised in antiquity.[34] When Plato (or his narrator

[33] Porphyry considered the question of principles a matter of metaphysics (*In phys.* 9.11–12; cf. n. 56 below). Cornford (1956: 45) comments on *Tim.* 31b–32c: 'Empedocles has taken the four elements as given fact [but cf. *In phys.* 1186.5–14]; Plato deduces the need for four primary and simple bodies by an argument.'

[34] Thus Epicurus criticised Plato's definition of the elements (*On Nature* 14, cols. 33–9: see Schmid 1936: 22 ff.; Baltussen 2003: 57–62). For Aristotle see note 54. Among the Middle Platonists the problem of pre-cosmic 'order' exercised the minds of several thinkers. Plutarch commented on the role of the elements, paraphrasing *Tim.* 30a3–5 and 52a2–7 at *De an. procr.* 1016D, but *quoted* accurately 53a8–b5 at 1016E (see Deuse 1983: 236). Other pre-cosmic narratives occur in *DL* 3.76, Alcinous and Calcidius (Deuse 1983: 239 ff., who also notes that matter is described as active, not passive (240 n. 14)).

Timaeus) is engaged in explaining how the physical world and its constituents 'come into being', it is part of a wider analysis of the genesis of particulars in our world and, because Plato does not think of them as ontologically stable, how we should take care not to suggest stability through our language, which may deceptively suggest fixed characteristics.[35] Thus this section of the *Timaeus* is of immense importance: by offering a two-tiered approach to the problem of coming-to-be (components, descriptors), it becomes programmatic in Timaeus' narrative of the creation.[36] Not only does it sketch a way to view the physical process, but it also suggests why it is so difficult to speak about a *process*. Thus Plato shows full awareness of the difficulty that the phenomenal world poses to clear linguistic expression: what is in flux cannot be expressed satisfactorily. After all, words tend to suggest stability, yet when phenomena keep changing they problematise that presumed stability: they are not physical objects in any traditional sense of the word.[37] When he outlines the process of coming-to-be in temporal terms, he postulates a 'before', during which things had a pre-cosmic nature up until the point that they actually materialise in 'physical' form. Plato's phrase used for this pre-cosmic state is peculiar: ἴχνη, often translated 'traces'. One must presume that they are 'proto-elements', before they are given their final form.[38] Simplicius quotes it without comment (*In phys.* 228.6). Whether any cause is involved here remains unclear (does Necessity have a role here? The 'inklings' do seem *predisposed* to become a particular thing).[39]

The more fundamental problem Plato is dealing with is of course metaphysical: if he does not want to commit to permanent, stable particulars, he has to construct the phenomena in a way that expresses their ephemeral nature. At *In physicorum* 227.23 ff. Simplicius goes out of his way to investigate the claim that 'both according to Aristotle and according to Plato the very first matter is qualityless body (τὸ ἄποιον σῶμα)'. He then lists the views on matter, only to reject the 'dominant

35 For a detailed reconstruction of Timaean particulars see Silverman 1992 and note 37 below.

36 It also happens to be placed at a crucial juncture in the dialogue, almost exactly halfway, as Cornford (1956: 35) and Silverman (1992: 96) emphasise.

37 Silverman (1992: 88–9). On the particulars and elements see also Cherniss 1935; Vlastos 1939–40; Vlastos 1967; Morrow 1968; Keyser 1995.

38 The metaphor of 'trace' is problematic: when a footprint leaves a trace, it is the *result* of the presence of the foot; if they are meant to represent a 'foreshadowing' of the 'real thing' (something fiery, watery etc.), 'trace' is inappropriate. Is Plato groping for a notion of potentiality? If so, his terminology falls short.

39 Silverman (1992: 87) announces that he does not think that matter 'is a primitive, unanalysable factor in Plato's ontology'. For a detailed discussion of the various options, see his section II, 95–9.

notion of matter' (matter is a three-dimensional body without qualities).[40] He quotes *Timaeus* 52d and 53b, refuting at least part of the claim that first matter is qualityless body (with an appeal to Plotinus), since Plato made the planes and triangles elements of body, hence prior to matter. At *In de caelo* Simplicius deals with this difficult issue in an anachronistic way.[41] For instance, at 59.24–91.20 Heiberg he considers the constitution of the heaven. Since Plato allocated to the whole universe the geometric shape of the dodecahedron (*Tim.* 55c), Simplicius aims to bring in closer alignment Aristotle's view that heaven is composed of a simple element, which also has its own type of motion.[42]

Causes, cosmos and the natural world

In his commentary on Aristotle's *De caelo* (considered earlier than the *In phys.*) Simplicius' interest in first principles is evident from the first lines. He speaks about the elements when he wants to establish the purpose (σκόπος) of the treatise.[43] Simplicius claims that Alexander of Aphrodisias understood the work to be about 'the world as a whole' (περὶ κόσμου, *In de cael.* 1.2 Heiberg; cf. 1.10–12 where the scope is expanded, 'the five bodies of the heaven and the four bodies below the moon'). By contrast, Simplicius posits that the work is about the simple bodies (ἅπλα σώματα) as principles, an interpretation for which he draws on the first century BCE Peripatetic Adrastus (*In phys.* 4.11–15 and 6.4).[44] This interpretation also explains why he reads the *Physics* as being about natural principles in books 1–3, and about motion in books 4–8.[45] Moreover, Simplicius reveals that concerning the meaning of 'comes to be' he has taken extra care to 'set out things at length in the desire to articulate *my own thought*' (*In de cael.* 102.15–16; my emphasis) on the basic contraries and elements as principles.[46]

[40] He ends (233.2–3) by saying that the discussion has dragged on for so long because it is the dominant view about matter (διὰ τὴν κρατοῦσαν περὶ τῆς ὕλης ἔννοιαν) – yet he does not like it! (οὐκ ἀρεστὴν ἐμοί). See Sorabji 1988: 9–10; Sorabji 2012, 4–5; and Mueller 2012: n. 100.

[41] It is possible that he had read Proclus' commentary on this part of the *Tim.*, but that part of the commentary is not extant, so it is difficult to verify.

[42] Later called *aether*, 'fiery air' (trans. Mueller 2011: 25–6). Cf. Solmsen 1960: 255.

[43] A standard point of the school's reading practices (see Mansfeld 1994), the 'aim' (*skopos*) was introduced by Iamblichus.

[44] He repeats this position (without mentioning Adrastus) at the start of book 4 (520.7–16).

[45] A different division of books is found at *In de cael.* 226.19–20 (four-four) and *In phys.* 802.8–11, attributing it to Porphyry (the same division also in Philoponus, *In phys.* 2.16–21).

[46] *In de cael.* 912.3 ff. Here he refers back to the *Physics* where this was demonstrated. The whole chapter has a striking 'syllogistic' form, that is, Simplicius reconstructs Aristotle's argument in terms of major and minor premises.

Simplicius is also keen to show how the fundamental discrepancy between Plato (on a literal reading: 'the universe comes to be') and Aristotle ('it does not') can be defused and that they are not contradicting each other (*In de cael.* 91.30–2). Simplicius decides to take the universe's coming-to-be in the *Timaeus* as a continuous process of ontological progression, starting out from the One, the source of all being. Thus it is not a literal creation in time, but an ontological dependence relation of beings from Being.

When we move to his commentary on Aristotle's *Physics*, we find further relevant passages. At the start he offers his views on principles (*In phys.* 3.13 ff.), in which we see the first signs of his harmonising tendency: by describing the principles as crucial to physics, and listing both main causes and auxiliary ones (on which more below); at 3.18–19 he mentions Plato as having 'added the paradigmatic (τὸ παραδειγματικόν) to the [main] causes and the instrumental (τὸ ὀργανικόν) to the auxiliary causes'. His use of 'added' (προστίθησι) shows how natural it is for Simplicius to slip in ideas on the issue as complementary to those of others (presumably including Aristotle's). Simplicius also acknowledges that Aristotle had used the term 'principle' for different purposes (e.g. 216.11–15 'cause' and 'principle' are used for matter; 216.15: 'the discussion is seeking elemental principles'; 216.30–6 it should be understood that it is one thing to seek principles and elements of natural things from which as primary ingredients, they are per se and not in an accidental sense, and another to seek principles of change (paraphrase of *Phys.* 190b17–20)).

Because Simplicius examines Aristotle's notions of elements and cause in the context of the latter's own review of others, it becomes clear how the tradition in which Simplicius has placed himself brings with it a certain method. Because Simplicius' interest in causes does not just come from the text, his own notion of cause is not a simple borrowing from the classical thinkers. The idea of causality and agency in the Neoplatonist universe includes notions of form and transcendence.

Simplicius' discussion of elemental principles is thus both historical (in the sense of taking on board previous material) and philosophical. It should be viewed partly in the light of Presocratic ideas (especially *In phys.* 21.22–22.8, 22.22 ff.) as well as Plato's creation account in the *Timaeus*.[47] He is a fierce defender of the imperishable heaven in Plato (*In de cael.* 106.1–108.13 Heiberg). Interestingly, at *In de caelo* 202.31–203.1 he shows

[47] E.g. *In phys.* 7.6–7, 31.24–8, 35.28–36.1, 254.28–31; *In de cael.* 87.3–11, 154.6–156.24 Heiberg. Cf. Hankinson 2002: 7–8.

his awareness of the power of principles, when discussing the proof that elements are finite. After going over several claims regarding the (in)finite nature of the elements (Anaximander: infinite; Leucippus and Democritus: finite) he agrees with Alexander that they must be finite. But the final comment demonstrates how important this conclusion is, because it is given a more general significance: 'Thus he elegantly shows that the slightest oversight in the case of the postulation of the smallest magnitude becomes responsible (αἴτιον) for the greatest errors on account of the very great power (μεγίστην δύναμιν) it has as a principle (ὡς ἀρχήν)'.[48] The point is clear: argument may be about a small thing, but it is not trivial, and Alexander's argument is central for making the case. This variety of arguments from authority is an important characteristic of the *Physics* commentary. Alexander plays a major role (he is referred to about 700 times), and so do Theophrastus (especially in the first thirty pages or so), Eudemus and Porphyry. Simplicius is using Alexander's commentary on Aristotle's *Physics* (extant in fragments only[49]) on almost every page.

The dual nature of his discussion of principles (historical and philosophical) in *In Physicorum* is worth pausing over. How does a scholar-exegete of the early sixth century CE manage to organise his materials in order to achieve his timely and rich quotation style? I will focus on some specific aspects typical for his analysis of principles.[50] The question of organisation is particularly relevant. We get a glimpse of his working practices when we notice that he regularly places certain judgements of earlier authors closely together.[51] Because he manages to collect comments on specific points from a range of earlier exegetes we may hypothesise two obvious possibilities for how he did this: either he had a phenomenal memory to recall such relevant points at the right time or he had a set of preparatory notes that were organised by themes alongside the Aristotelian text.[52] Two passages may illustrate the latter procedure.

At *In Physicorum* 9.7–10.25 the sources referenced in close proximity are Theophrastus, Alexander and Porphyry. The discussion takes as proven that there are three basic elements (above p. 116). Three causes fits the

48 Translation from Hankinson 2004: 19.

49 See Rashed 2011 for fragments of Alexander's *In phys.* books 4–8, found in the margins of a Paris manuscript.

50 I have discussed broad features of Simplicius' methodology in Baltussen 2008.

51 On clusters see also Baltussen 2008: 91, 122, 167.

52 A third possibility is that the earlier commentaries had such notes already in place, but the range of his sources spanning across seven centuries (e.g. Theophrastus, Eudemus, Boethus, Xenarchus, Alexander, Plotinus, Porphyry, Iamblichus, Proclus, Damascius) clearly does not allow just *one* source.

passage in the discussion of the role of the elements in Timaeus' creation story (*Tim.* 48e–49a). Simplicius attributes three causes to Plato, but Theophrastus is quoted as saying that Plato stipulated two principles (*In phys.* 26.11–13).[53] Simplicius questions this claim and draws on Theophrastus and Porphyry to develop the argument. The former must have based his understanding of the causes in the *Timaeus* on a very cursory reading (does he ignore the third cause or not know of it?), while 'Porphyry says that it does not even belong to the natural scientist to investigate whether there are principles of natural things, but to one who has risen higher' (*In phys.* 9.11–12).[54] Theophrastus thus interprets the issue as reported up to *Timaeus* 48e, while Porphyry suggests that the question of principles belongs to metaphysics. Of course Aristotle would not disagree,[55] given that he discussed them also in his *Metaphysics*. But the two points of view are not incompatible: the natural scientist deals with natural things and assumes certain things as given (ὁ γὰρ φυσικὸς ὡς δεδομέναις χρῆται, 9.12), while the first philosopher tackles the deeper causes and nature of things.

A second passage that illustrates the method of clustering and which seems to result in a synthesising approach negotiating an outcome based on all the relevant material can be found at *In Physicorum* 11.15 ff. Here Simplicius mentions Alexander, Eudemus and Porphyry in close proximity and goes up against their views on principles as causes. Not long after examining Aristotle's survey of predecessors, he criticises Alexander and Eudemus for misunderstanding what 'principles' are (*In phys.* 10.15 ff.):

> It is worth our while investigating what a principle (ἀρχή), a cause (αἴτιον) and an element (στοιχεῖον) are. For (16) Alexander says that these differ from each other in that the efficient cause – that from which motion begins – is especially called a 'principle' (ἀρχή), while that for-the-sake-of-which (οὗ ἕνεκα = final cause) and the form (εἶδος), which is the final cause in natural things, are especially called 'cause' (αἴτιον), and the substrate (τὸ ἐνυπάρχον) as matter is especially called 'element' (στοιχεῖον).

Here Simplicius clearly takes his lead from Alexander in assessing the importance of principles in physics (*In phys.* 10.16–17).[56] Note, however,

[53] On counting principles see esp. Sharples 1995, and below note 67.

[54] Contrast Aristotle *Gen. et corr.* 1.3, 317b14, who refers back to *Phys.* 1.6–9. [55] See note 54.

[56] Elsewhere Alexander is quoted as saying that there are two types of principle (*In phys.* 197.30–198.1): 'Some are pre-existent and ungenerated and numerically everlasting, like the creative agency and matter. The others are those in accordance with which generation and perishing [occur]' (trans. Share and Atkinson in Baltussen et al. 2012: 37). Simplicius attempts to show that his view is incoherent.

that he presents the Aristotelian commentator as following Eudemus in his interpretation. To support this he quotes from Eudemus' own *Physics*, whose account of principles, causes and elements is taken as closely resembling that of Aristotle:[57]

> While 'cause' is said in four ways, 'element' is said in the sense of matter; for the elements seem to be present in (ἐνυπάρχειν) a thing, like the letters (τὰ γράμματα) in words. . . . Thus they call the principle and the element *causes* (αἰτίας), but the for-the-sake-of-which (τοῦ ἕνεκα) does not admit the account of an element, for it is not present in what it causes, e.g. health in walking, nor does health seem to be a *principle* of walking, but rather a *cause*. And the for-the-sake-of-which and the form seem to be very close and often the same. So for this reason the for-the-sake-of-which seems to be a cause.

It is obvious from this procedure (the importance of definition and the correction of existing definitions) how much the continuous effort to synthesise materials guides Simplicius in his analysis and reinforces the interconnections *within* the Peripatetic tradition. But it gets complicated when he seeks to formulate a coherent view. In this particular case, Eudemus' view introduces the notion of 'present in' as a condition for a cause to be a principle. Both authors suggest some kind of hierarchy among the causal factors, relating each to a different aspect of causality. Porphyry is found guilty of giving too narrow an interpretation of 'principle' and 'cause', when he claims that Aristotle uses cause and principle in different ways (*In phys.* 10.25 ff. 'from which', 'out of which' etc.). In fact, Simplicius disagrees and goes on to refute all three, responding to each of them individually (11.16 ff.), with some interesting comments:[58]

> 'Cause' and 'principle' are not said in the same number of ways (ἰσαχῶς); rather every cause is also a principle, but the starting-point of a thing like a journey or an action would not be called a cause (οὐκ ἂν λέγοιτο αἴτιον). Also, 'principle' is not even prior in thought to 'cause', if the cause must exist prior to the effect (ἀποτέλεσμα), and by contrast the principle – either as a part that precedes or as an element – exists together with the effect. (*In phys.* 11.23–9)

[57] Though extant in fragments only, Eudemus' work is known to us in outline on the basis of Simplicius' references to individual books. See Sharples 2002; Baltussen 2002; Baltussen 2006.

[58] To be fair, he could have taken into account Aristotle's comments at *Metaph.* 1070b23, 1088b27 where *archê* is given a different scope from *stoikheion* (noted in LSJ, s.v. 'στοιχεῖον').

When he continues to offer his own view, it consists of an interesting mix of Platonic and Aristotelian ideas (evidenced by terminology), declaring that Aristotle probably divided the common term 'principle' into other causes such as the final and efficient cause, but also into 'auxiliary causes' (συναίτια), a term which, although in Plato (*Tim.* 46d1; cf. 76d6, *Gorg.* 519b2), is used in the way other late Platonists did (Proclus *In Tim.* 1.3.1, 1.17.15, 1.261.15 etc.; Damascius *In Phlb.* 113.7). All in all, one gets the impression that Simplicius is drawing on a well-organised set of notes, from which he can take the relevant material when he needs it. His own view, clearly Neoplatonic, is used as a 'corrective' of views in Aristotle and others: after the claim of a division in Aristotle into (main) causes and added causes, he further rephrases the principles within the broader framework of the Plotinian universe, in which a distinction is made between proximate causes/principles and first causes 'in the primary and strict sense' (αἱ πρώτως καὶ κυρίως ἀρχαί, *In phys.* 11.33–4).

In his commentary on the *Physics* Simplicius does not just follow the main order of *Physics* A, but also interweaves his own preoccupations and concerns, selecting from other commentators and writers who responded to Aristotle's work. These regularly interrupt the direct textual exegesis. He carefully follows Aristotle's account of principles, compares and aligns it with Plato's, critically evaluating and rejecting Alexander's idea of principles as elements. When his own opinion emerges he tends to align Plato's three causal factors with those of Aristotle (e.g. when he reviews Aristotle's justification for the substrate, 199.6–26, esp. 6–8: we need a 'third principle, the substratum for the contraries, namely matter'). Intriguingly, Simplicius' arguments use terminology that echoes Platonic and Empedoclean notions (impermanence; theory of the primal forces love and strife, 199.15–20):

> That the contraries on their own produce nothing is clear from the fact that a thing that acts, acts on something that sustains (ὑπομένον) the effect, as when black acts on body, but the contraries, if on their own, act on nothing. <The contraries> do not sustain each other (ὑπομένει) and white does nothing to black or love to strife. Love does not collect together strife (οὐ γὰρ τοῦ νείκους ἐστὶν ἡ φιλία συναγωγός[59]), but the underlying elements (ἀλλὰ τῶν ὑποκειμένων στοιχείων). A thing that acts, acts on

[59] The term συναγώγος (cf. *Tim.* 31c) is not normally used in connection with Empedocles' notion of love and strife (more common are *diakrinei* and *sunkrinei*; but cf. *Protag.* 322c; Aristotle uses συνάγει at *Gen. et corr.* 330b19–20 (= A36 *DK*)). I note that συναγώγος is used again at 631.38 and 640.11 and very common in Proclus for 'unifying and ordering force', e.g. *In Tim.* 1.192.20; cf. *In remp.* 2.78.14; *Theol. Plat.* 6.14.15. It also seems a favourite term in Simplicius' commentary on

something that holds its ground (ὑπομένον), a contrary does not sustain (ὑπομένει) a contrary.[60]

All through the passage (and before) the emphasis is on physical elements/principles (much like the start of the discussion of principles from 179.20 onwards, esp. 182.9 ff.). The language not only reinforces the notion of (im)permanence, present in both Plato and Aristotle, but also shows how the discussion becomes coloured by examples from the Presocratic era. In other words, the newly articulated position on principles is the result of a comprehensive synthesis, and thus a kind of composite of many positions, *in so far as* Simplicius finds them acceptable.

Simplicius also seeks to evoke a broader context for the principles. In his *In de caelo* he launches into a Platonist discussion of the grander metaphysical theory of the world, in which the principles are placed within a new framework, after he had started by evaluating Alexander's interpretation of the purpose of the work (*In de cael.* 1.1–6). As in the *Physics* commentary, Simplicius mentions other scholar-exegetes (*In de cael.* 20.1 ff.), including Ptolemy (his lost *On the Elements*[61] and optics), Plotinus and Xenarchus. Each of these thinkers formulated points in response to Aristotle's ideas, in particular the link between an element and its natural motion. Such a link raised problems: the number of simple motions (two: up and down) was not easily compatible with the number of elements (four). In making use of the post-Aristotelian works, Simplicius creates a richer discussion. Their combined authority offers a way out of the issue (they are presented as a group at *In de cael.* 20.10 and 20.21) and with additional support from Aristotle himself (*De cael.* 310a33–4; *Meteo.* 341b18–22). Their reading suggests that motion in a straight line only attaches to the elements 'when they are still coming to be and are in an unnatural region, when they have not yet taken on their natural one' (*In de cael.* 20.12–15). Strikingly, Simplicius goes on to rely primarily on Xenarchus' interpretation (*In de cael.* 20.31, referring forward to 21.33–33.10), while he answers a possible further objection

Epictetus' handbook with reference to the high-order analysis of nature (e.g. *In Epict.* 83.41 'gathering of similar things').

60 Translation Share and Atkinson 2012 39, with minor adjustments. Four different occurrences of *hupomenein* appear in Share and Atkinson as infinitive each time, which is slightly misleading. They translate *hupomenei* as 'face up to' and 'holds its ground'. I prefer (the less personified) 'sustain' which also keeps the sense of the prefix *hupo-* intact.

61 Mueller 2012: 129 n. 49 suggests that the work is perhaps used at 37.33–4.

with a position foreshadowing a Platonist solution ('different from natural motion because it belongs to something better which dominates it', 21.22–3).

The discussion of Xenarchus' objections takes up considerable space, as mentioned earlier (*In de cael.* 21.33–33.10), but since this section has been well analysed recently, it need not detain us for long.[62] What Simplicius extracts from the discussion of Xenarchus' objections (ἀπορίαι) is of interest for its correction of Aristotle's view of how the element fire moves once it reaches its natural place; for us it helps to 'appreciate how unusual Aristotle's conception of the natural world was in antiquity'.[63] Alexander's role in this context becomes clear when he is brought in to 'dissolve(s) these objections of Xenarchus on these matters' (24.20). Others make an appearance as well: on the natural place of elements (*In phys.* 639.13–640.12) Simplicius takes care to quote from Iamblichus to show that he was in agreement with Theophrastus, that place is not a substance but something which is 'predicated in relation to the order and position of bodies, according to their natures and powers, equally in the case of animals and plants, and generally, of things composed of different elements' (639.13–21).[64] Iamblichus is then presented as claiming that body and place are closely connected and created together.

These examples strongly suggest that Simplicius' overall method in these polemical and discursive sections of his commentary is both critical appraisal *and* philosophical synthesis. It resembles that of Aristotle in the sense that he takes care to review previous views on, and interpretations of, the natural world. In the case of Xenarchus we have a good example of the aporetic method as advocated by Aristotle – a method of evaluating puzzles, characterised as the demonstrations of the difficulty of its contrary position (αἱ γὰρ τῶν ἐναντίων ἀποδείξεις ἀπορίαι περὶ τῶν ἐναντίων εἰσίν, *De cael.* 1.10, 279b5–6).[65] But while Aristotle undertook such reviews partly out of a conviction that all men can contribute to the truth, Simplicius has a more ideological agenda, in trying to unite the pagan front against the Christianised world around him. It is this objective, I submit, that underlies much of his elaborations, polemical excursions and analyses.

[62] See Falcon 2012: 57–96: T3–T7. [63] Falcon 2012: 168; cf. also 25 ff., 36–40 and 176–83.
[64] Baltzly et al. 2007: 53 n. 50.
[65] The method is well clarified by Quandt 1983 as typical for Aristotle's expository practice.

Epilogue

I have argued that Simplicius' approach to the discussion of physical principles can be further illuminated by emphasising how he reaches his positions by way of a distinctive and unusual working method. He not only offers a rich interpretative appraisal of Aristotle (and Plato) on the topic, but also illustrates how the various exegetical points of view create new and critical angles on the material. Both aspects arise from a blend of historical data and philosophical considerations, leading to an amalgamated representation of Greek philosophical notions of causes and elements. This is not a trivial point: his own interpretation is closely intertwined with the method of gathering and organising his materials.[66]

Thus the direction of Simplicius' examination of principles becomes determined by the different materials he consults.[67] Although he has his own agenda, his treatment is also coloured by the analysis of Presocratic ideas and Plato's creation account in the *Timaeus*, and influenced in more decisive ways by Alexander's exegesis. His analysis weighs up and sometimes combines different points of view, but remains focused on clarifying and reconciling Aristotle and Plato where possible – the result of his rather persistent harmonising strategy. This even includes elaborate discussion of views which he disagrees with (e.g. Philoponus) or which he thinks will complement what Aristotle has reported. We have seen that, in the evaluation of the principles of creation, he follows the lead of Aristotle, Alexander and Plotinus to *construct the theoretical perspective*, and reviews questions raised in Xenarchus, Eudorus, Ptolemy, Moderatus and Iamblichus to *evaluate potential problems*. The many examples in which he then tries to realign the outcomes with Aristotle's own ideas (found in the work under discussion or elsewhere in the corpus) demonstrates how he intends to maintain the presumed coherence of Aristotelian thought. To an attentive modern reader it is abundantly clear that the final result is a further development and reinterpretation of Aristotle's views on nature.

[66] This is perhaps the place to note an omission: the work of his teacher Damascius *On Principles* does not occur in this discussion of principles, at least not explicitly. Damascius' name occurs eighteen times (*In phys.* 4) only. Did Simplicius not have access to the work?

[67] Aristotle considered his critical assessment of Plato and the Presocratics an improvement on all his predecessors: *Gen. et corr.* 335b ff., esp. b6–8, 'the third principle [formal cause] must be presented as well – the cause vaguely dreamed of by *all our predecessors*, definitely stated by none' (trans. Barnes in Aristotle 1984: 1.549, my emphasis). Cf. Mann (2011: 47), who highlights the comment in *Metaph.* 984b15 ff. where Anaxagoras is 'said to be the "sober one" among the Presocratics for having introduced *nous* as a cause'.

In addition, this rich and well-resourced reading of Aristotle is also intended to show that the exegetical tradition itself should be viewed as both relevant and coherent. Thus Simplicius' perspective is always a *retrospective* arising from an attitude based on the considered view that the tradition matters and assists in creating a richer and more robust understanding of Aristotle – admittedly an Aristotle of the sixth century CE.

PART II

The Origins of Human Agency

CHAPTER 7

Divine and human freedom: Plotinus' new understanding of creative agency

Kevin Corrigan

In late antiquity Plotinus develops three new important models of creation, production or creativity,[1] at least partially in critical dialogue with the Gnostics, models that deserve to be better known than they are – certainly they deserve to be as well known as the theory of 'emanation' generally associated with Neoplatonism.[2] These are, first, a model of creative contemplation or insight (developed in treatise 30 in the chronological order, or III.8); second, a model of non-deliberative demiurgic production (developed in treatise 38, or VI.7); and, third, a model of agency and divine self-causality (developed in treatise 39, or VI.8). These models are not different from Plotinus' theory of hypostatic causality set forth in the earlier works (and developed, as A. C. Lloyd has argued, in relation to Aristotle's physical model of causality: 1990: 98–101), but they are nonetheless radical developments for their, or even for our own, times that represent, in significant measure, interpretations of Aristotelian thought. It is part of my argument that what is at stake throughout these important middle works is, first, the internal nature of creative production and, second, only gradually the problem of what divine and human agency really means. While significant threads of Plotinus' account are worked out in treatise 30 (III.8) (and treatises 31 and 32 (V.8 and V.5)), and 38 (VI.7)), the real problem of agency is only fully tackled in treatise 39, with quite revolutionary consequences for the subsequent history of thought.

Let us start with Plato and Aristotle. In the account of the making of the World's body and soul in Plato's *Timaeus* (and also in the myth of the *Statesman*), God, the demiurge (together with the lesser gods), is represented as doing or making many things: the demiurge looks to the

[1] Against 'creation' in Plotinus see O'Brien 2012: 72–6; in favour, Zimmerman 2013. For 'creative' power see Wilberding 2006: 102, who points to Alcinous *Didask.* 15.2.2–3; Ammonius Sakkas (Photius *Bibliotheca* 461b8–9); Origen, *De princ.* 3.6.6 and *C. Cels.* 5.23.22.

[2] For a good account see Wallis 1995: 61–9.

paradigm of the Eternal Living Creature, deliberates, plans, desires, frames, and so on, not unlike the God of the Bible (*Tim.* 28a–37d; *Plt.* 271d–274e). This is why Plato can seem to the later tradition none other than 'Moses speaking Greek'.[3] And in that tradition the demiurgic function comes sometimes to be divided (as in the *Chaldaean Oracles* or Numenius[4]) between a contemplative and a doing or making aspect, or even acquires an evil function, like the irrational world soul in Atticus and Plutarch of Chaeronea (Proclus *In Tim.* 1.381.26 ff.), or becomes simply ignorant and blundering like the Valentinian demiurge.[5] But throughout the Platonic tradition a divine or semi-divine force does or makes many things. In other words, God makes, moves, intervenes, withdraws – whether for good or ill.

By contrast, in Aristotle's *Metaphysics* and elsewhere, God or *Nous* moves, not by doing or making, but by being unmoved, that is, as a final cause or 'as being loved'.[6] For Aristotle God's thought is the originative source of motion only as the object of desire, real or apparent; and this divine 'thinking of thinking' is activity or energy in the purest sense (*Metaph.* 12.9, 1075b34), that is, contemplation or pure consciousness (as A. C. Lloyd translates the word *θεωρία*: 1990: 182–3), a life that we sometimes participate in, but rarely and fleetingly (*Metaph.* 12.7). God moves, but does or makes nothing. On the side of divine activity, one may think that the Unmoved Mover is hard to associate with any positive creativity, love or desire: and on the side of human reality, the contemplative life, which is the highest life for Aristotle, seems solitary and hard to associate with the ideal of the practical life,[7] in which the human being enjoys community intrinsically and needs good friends for his or her best self-development.[8]

Moreover, this apparent lack of connection between contemplation and action or making is exacerbated by several other features of late ancient thought. First, the Peripatetic view (of Alexander of Aphrodisias) that divine providence only extends as far as the movements of the heavenly bodies and the maintenance of sublunary species, but not as far as *sublunary individuals*.[9] God's creative and sustaining activity does not reach or

[3] Numenius fr. 8, des Places numbering, in Petty 2012.
[4] *Chaldean Oracles*, the First God as thinking alone: frs. 37, 39, 40; the Second God as divided: frs. 5, 33, 35, 37 Majercik; Numenius, fr. 16 Petty.
[5] See, for example, *Trip. Tract.* 104.25 ff. For overview see Turner 2001.
[6] Aristotle *Metaph.* 12.7, 1072b3; cf. Alcinous *Didask.* 10.2.5–9.
[7] For assessment see Joachim 1970: 241–3, and cf. 284–97; Lear 1988: 309–20; Irwin 1988: 347–72.
[8] See esp. *Eth. Nic.* 8–9; Joachim 1970: 241–61.
[9] In the Arabic *De providentia*, 1.1–9.2, trans. Ruland in Sharples 1982.

touch our lives at all. Second, according to many Gnostic texts all sorts of contemplation or revelations are possible – but only for the Gnostic enlightened few (cf. II.9.18.36 (treatise 33)) and only with what appear to be properly registered guides. Furthermore, demiurgic activity tends to occur fairly low down the spiritual hierarchy, and it is generally represented by anthropomorphic figures full of flaws and passions, if they are not outright evil themselves (e.g. *Trip. Tract.* 74.19–84.37).

Third, even major Middle Platonic thinkers, such as Numenius, seem to have no clear idea of what divine activity might be. For instance, Numenius says in one fragment that the 'first God' is 'idle of all works', while the demiurgic God rules 'as he goes through the heavens'.[10] The demiurge 'directs', 'steers' and 'takes judgment from contemplation [of the First], but is impelled by desire' (fr. 18);[11] and in another striking fragment, contemplation seems to alternate with demiurgic activity: 'for the second, being double, self-makes his own idea and the cosmos, being demiurge, then wholly contemplative' (fr. 16).[12] In other words, in one of the principal pre-Plotinian figures (whom Plotinus was accused of plagiarising: Porphyry *Vit. Plot.* 17.1–7) there is no coherent account of divine activity or agency. Does the demiurge deliberate after contemplation and draw motive power or impulse from desire? And after making both himself and the cosmos, does he return to contemplation? In such a scenario there seems no intrinsic or internal relation between desire, making and contemplation. Platonic and Aristotelian elements seem thrown at random into the mixing bowl. And if we put this into the context of Peripatetic thought (that appears to reject divine agency in the lives of individuals) and of Gnostic thought (that tends to present an externalised view of contemplation in privileged visions presented as interpretations of ancient texts (of the Bible, Plato, Aristotle etc.),[13] privileged because access is open only to the few by contrast with the oblivion of ordinary humans), then we can see with some clarity the pressing issues of divine and human agency with which Plotinus had to wrestle. An independent treatment of the question of providence has to wait until III.2 (treatise 47) and III.3 (treatise 48), but

[10] Numenius fr. 12.12–14: τὸν μὲν πρῶτον θεὸν ἀργὸν εἶναι ἔργων . . ., τὸν δημιουργικὸν δὲ θεὸν ἡγεμονεῖν δι' οὐρανοῦ ἰόντα.

[11] Numenius fr. 18.10–14: τὴν ἁρμονίαν δ' ἰθύνει, ταῖς ἰδέαις οἰακίζων, βλέπει τε ἀντὶ τοῦ οὐρανοῦ εἰς τὸν ἄνω θεὸν προσαγόμενον αὐτοῦ τὰ ὄμματα λαμβάνει τε τὸ μὲν κριτικὸν ἀπὸ τῆς θεωρίας, τὸ δ' ὁρμητικὸν ἀπὸ τῆς ἐφέσεως.

[12] Numenius fr. 16.10–12: Ὁ γὰρ δεύτερος διττὸς ὢν αὐτοποιεῖ τήν τε ἰδέαν ἑαυτοῦ καὶ τὸν κόσμον, δημιουργὸς ὤν, ἔπειτα θεωρητικὸς ὅλως.

[13] On Sethian Gnostic thought generally see Turner 2001.

Plotinus develops a revolutionary new understanding of the relation between contemplation and making in III.8, a profound critique of anthropocentric rationality as the paradigm for thinking about demiurgic activity in VI.7, and gradually comes to tackle the deeper problem of divine and human free agency in VI.8.[14] The scenario I shall sketch here, then, will include some key features of III.8, VI.7 and VI.8, contrasting in each case divine and human activity with questions of agency.

Ennead III.8 (30) argues for the view that contemplation or philosophic wisdom, far from being private or external to the world, as the Gnostics appear to hold (Cf. II.9.18.35–6 (treatise 33)), is the fundamental form of all natural making and, indeed, of all life. Everything – even plant life – is either contemplation (so that even nature's life, which Plotinus quaintly represents as a silent contemplation constantly giving rise to bodily forms (III.8.4.3–10), is a form of living intelligibility) or thought (*νόησις*), no matter how lowly; and forms of living thought (plant life, making, action, sensation, imagination, and intellectual activity itself) become more unified the more they 'hasten' to the intimate unity-in-duality of intellect, where thinking and object of thought are one (cf. III.8.8.1–8). So, as Plotinus concludes the first part of his argument, everything is either contemplation (in the sense that it contains its intelligibility within itself, as does intellect)[15] or a product or consequence of contemplation (in the sense that if you unpacked the intelligibility in anything whatsoever, it would lead you to everything else in the universe or to a more comprehensive view of reality as a whole) or, finally, a substitute for contemplation (in the sense that action and production are ways of coming to see or understand a reality that is at first too densely compacted for us to grasp it altogether: III.8.7.1 ff. (treatise 30)). Contemplation or insight, therefore, is the primary creative force in both the spiritual and physical worlds.

Plotinus is aware of the paradoxical – even revolutionary – nature of his project, as he attempts to uncover the contemplative reality of everything from plants to the divine. In contrast to the Gnostic elitist relation between a hierophant and favoured initiate, Plotinus' method is dialogically more inclusive and radically democratic, starting in fact from the principle of all-inclusive play:[16] 'Well, as this arises among ourselves (πρὸς ἡμᾶς) there will be no risk of playing with our own things. Are we now contemplating as we play? Yes, *we and all who play* (ἡμεῖς καὶ πάντες ὅσοι παίζουσι) are doing this or at any rate this is what they desire as they play'

[14] For contemporary treatment of this problem see Tracy 1994.
[15] III.8.7.1 ff. (treatise 30); 8 *passim*; cf. VI.7 (38) 1–7. [16] As Socrates advocates in *Rep.* 7, 537a–c.

(III.8.1.8–12 (treatise 30)). This democratic emphasis also runs through the next two works of the *Grossschrift*, V.8 (treatise 31) and V.5 (treatise 32). In V.8.1 the central question posed is how can *anyone* contemplate intelligible beauty and its cause from the here and now of historical existence; in other words, the goal of the inquiry is not to privilege names, individuals or groups but to show the Beautiful and the Good to *anyone*, and this is a motif that reaches its culmination in the next treatise V.5.12.34–5 (treatise 32): 'The Good is gentle and kindly and gracious and present to *anyone* when *anyone* wants.'

At the same time, Plotinus interprets Aristotle's heritage in a new way. Divine contemplation or thought is the fullest reality that extends to, and moves, everything – not simply as a final cause, but as an internal formal cause. A. H. Armstrong has observed that in III.8.7 (treatise 38) Plotinus goes far beyond Aristotle, establishing a new universal sense of *θεωρία* that takes its starting point from *Nicomachean Ethics* 10.6 and 7, but ends up abolishing 'Aristotle's distinction between *praktike* and *theoretike episteme* or *dianoia*[17] and [making] the whole life, not only of man but of the universe, philosophy in Aristotle's sense' (Armstrong 1966–88, III: 382–3 n. 1). In my view, however, Plotinus does not really abolish Aristotle's distinction between *praktike* and *theoretike episteme*; nor does he create, as John Deck has suggested with some plausibility, a new form of formal-efficient causality in developing the notion that *θεωρία* is productive (Deck 1967: 107–8). I suggest that he is aware of another line of thinking in Aristotle, since Aristotle does say that the right functioning of the two rational parts of the human being is contemplative in both directions – in that of particular variable things and in that of things that cannot be otherwise (*Eth. Nic.* 1139a5–6). It is also true that while *sophia* and *nous*, as contemplative in the highest sense, do not literally 'make' anything (*Eth. Nic.* 10.8, 1178b20–1), Aristotle does say a little later in *Nicomachean Ethics* book 6 that they do make or produce happiness in a different way, not externally but 'as health makes health', that is, not as an efficient or motive cause, but rather as a formal cause.[18] This is, in fact, integral to Plotinus' argument from the outset, namely, that desire or final causality operates as a formal cause throughout all of nature internally; and this is why he cites Aristotle at the conclusion of the first part of his argument: 'for all other things (apart from the first principle) desire this if the goal for them all is their originative principle' (III.8.7.1–15 (treatise 30);

[17] Cf. *Eth. Nic.* 1, 1095a5, 6, 1139a21–b4, 10, 1179a35ff.

[18] On this see Gauthier and Jolif 1970, II: 542–7.

Eth. Nic. 6, 1143b10). In other words, he emphasises Aristotle's own dictum that *nous* is both *arche* and *telos*.

As A. C. Lloyd has shown, Plotinus adapts Aristotle's model of physical causation to non-physical causation (Lloyd 1990: 99; also Rutten 1956); but Plotinus here also adapts this model to the internal workings of physical causation in so far as these are activities (and not simply qualities, for instance).[19] Just as teaching and learning involve two different subjects, but constitute a single activity (*energeia*) from different perspectives (Aristotle *Phys.* 8, 255a33–b5 and 3, 202a13–21), so also what is an action or an external production from one viewpoint is a manifestation of the real, and from another an *energeia* or piece of living insight. They are not two distinct activities, separate from one another (though they may become distinct and they can be viewed as such),[20] but a single activity seen from two different points of view. And since the real is not a patchwork of pieces, but a whole expression or participation in the life of God, my making of something can be 'mine' from one viewpoint, and a window into reality or divine thought from another; in moral action, for instance, to the degree that I get something 'right', that action embodies contemplation or insight. Energy or contemplation, therefore, is a formal activity that internally makes my action or production possible. Plotinus puts this succinctly in treatise 39, VI.8.6.19–22: 'In practical actions, self-determination and what depends on us are not referred to practice or outward activity, but to the inner activity which is the thought and contemplation proper to its best functioning.'[21] Contemplation, as creation or co-creation, then, is what really *makes* the physical world at the heart of all forms of action and production. The inner activity of action is its thought and contemplation.

So, Plotinus appears to hold the view that nothing in the universe is entirely private, or unconnected with anything else, and that this is because all desires – even apparently blind reproductive impulses – reflect, however dimly or unconsciously, a developmental desire to manifest the living, totality-in-one of intellect. As life, this contemplation is already unrestricted. 'Contemplation and vision have no limits', Plotinus states at III.8.5.29–30: 'and that's why they are everywhere.' And in the case of

[19] On the distinction between activities and qualities see *Ennead* II.6 (treatise 17), and on the relation of intelligibility to *logoi* see VI.2.21.32–51 (treatise 43).

[20] Aristotle *Phys.* 202b7–8; cf. Plotinus VI.8.6, VI.8.19–22 (treatise 39); cf. the argument of IV.4.28 (treatise 28) culminating in IV.4.28.69–72; for the two-act theory see Rutten 1956; Lloyd 1990: 98–101.

[21] For further treatment of VI.8.6.19–22, see note 33 below.

intellect desiring the One, Plotinus applies his theory of creative contemplation to the whole of reality in a striking formulation: intellect 'is always desiring and always attaining' (III.8.11.22–4 (treatise 30): καὶ ἐφιέμενος αεὶ καὶ αεὶ τυγχάνων). Intellect too therefore cannot be conceived as a static, fixed essence; its real nature is dynamic – to be drawn out of itself incessantly into itself and the Good.[22]

Is there, then, any place for individual agents in Plotinus' dynamic model of contemplative creation? Agency, at first sight, seems to disappear into desire and contemplation. Yes, we, the readers and interlocutors, actively trace the lines of creation internally through ourselves and other physical things into the intensifying unity of subject and object in intellect (III.8.1–7 (treatise 30)). Implicitly, the deepening range of unity as we ascend is where organising subjectivity may appear, and so Plotinus traces a series of the 'ones' of everything back to the One itself (III.8.10 (treatise 30)). But though the One is 'fountainhead', and 'root' of a mighty tree, 'why would it have to see or be active at all?' (III.8.11.9 (treatise 30)) Plotinus' answer is effectively that the One is 'the power for all things', a power that has 'given the trace of itself on intellect to intellect to have by seeing'. Subject-agency, then, in III.8 (treatise 30) is undiminished giving at its fullest, the highest form of unrestricted vision that is productive contemplation for all things. But it is not clear how or where precisely in creative contemplation qualities such as freedom, will or choice fit into such a picture.

In *Ennead* VI.7 (treatise 38) (commentary in Hadot 1988; Fronterotta 2007) Plotinus develops another new model of divine causality (one already prefigured in V.8 (treatise 32) and V.8.4–7 (treatise 31)), according to which the making of the cosmos occurs neither in time nor by deliberative thought, but spontaneously and intelligibly in such a way that a single enfolded totality becomes unfolded in time and space so that what is, in reality, all together, when unfolded, can be experienced sequentially as a 'this after this' (VI.7.1.54–8 (treatise 38)). For Plotinus, one major problem bequeathed from Plato is the need to overcome the apparent necessity for deliberation or planning valorised by the (mythical) representation of the demiurge deliberating or taking thought. A second problem is the apparent absurdity of including wild animals in the intelligible paradigm of the *Timaeus*. What then does divine demiurgic agency entail?

[22] This is one source, I suggest, of Gregory of Nyssa's doctrine of *epektasis*, according to which the soul is continually drawn out of herself into God (see Daniélou 1944: 309–26).

To counter the first problem, Plotinus argues that Plato cannot mean that the demiurge actually reasons or deliberates because such a representation of rationality in any form of creation is a *defect*, not an achievement. Reason is necessary to work things out after the fact; but understanding grasps reality all at once without the need for deliberation (cf. VI.7.1–11 (treatise 38) and V.8.6–7 (treatise 31)). If divine activity cannot be in any sense defective or incomplete (that is, demiurgic creation cannot be an Aristotelian incomplete motion, as Numenius had appeared to claim),[23] then it must be 'whole and entire' and 'in anything of those things that belong to the divine everything must inhere' so that we can unfold it later in temporal succession as a 'this after this' (1.45–55). The total 'all-togetherness' of intellect means that it contains its cause in itself (1.57–8) as the implicate or enfolded interconnectedness of everything. Even in the unfolded or explicate physical world, Plotinus argues, we can sometimes see this interconnectedness in the simultaneous unity of cause and fact in knowledge and perception (as in Aristotle's example of an eclipse, where cause and fact are identical; *An. Post.* 93a14–b20; *Metaph.* 8, 1044b9–15). A similar dynamic grasp of cause and fact can also be glimpsed in the parts of natural organisms, Plotinus argues; the eye of a living being, for example, is organically related to all the other parts so that the causal interrelation of the parts makes each a cause in respect of all the rest (VI.7.2.18 (treatise 38); cf. VI.8.14.18–29 (treatise 39)).

In other words, the all-togetherness of divine activity and its outpouring in the making of the world without deliberation or rationality is an integral part of our present experience of ordinary things and events in the physical world. And Plotinus then goes on to extend this analysis to all living creatures in order to counter the second problem – how can *nous* embrace the goods of all things? – and this results in a highly sophisticated, new view of intelligence:

> But someone will say, 'I grant the valuable living animals, but ... how could the cheap and irrational ones [be there in the Complete Living Creature]?' ... Now, there ... intelligence (*to noein*) is different in man and in the other living creatures, and reasoning (*to logizesthai*) is also different, for there are present somehow also in the other living creatures many works of deliberate thought (*polla dianoias erga*). Why then are they not equally rational? And why are human beings not equally so in comparison to each other? But one must consider that the many lives, which are like movements, and the many thoughts [are] different ... in brilliance and

[23] Numenius fr. 12: the demiurgic God rules 'as he goes through the heavens'.

> clarity. . . . For just as any particular life does not cease to be life, so neither does an intellect of a particular kind cease to be intellect . . . since the intellect appropriate to any particular living being does not on the other hand cease to be the intellect of all, of man also, for instance, granted that each part, whichever one you take, is actually all things, but perhaps in different ways. For it is actually one thing, but has the power to be all; but we apprehend in each what it actually is; and what it actually is, is the last, so that the last of this particular intellect is horse . . . as the powers unfold they always leave something behind . . . and as they go out they lose something . . . and in losing different things different ones find and add on something else because of the need of the living being . . . nails . . . claws and fangs. (translation adapted from Armstrong 1966–88)

Here intelligence, far from being a single paradigm of rationality, as it tends to be in the modern world, is more like a variegated continuum of different intensities of organised life, that allows for a sort of natural selectivity, of which we see only the last manifestation. There is a kind of geological depth to each species that prevents us from recognising that each is, in fact, a holographic representation of a much larger intelligible organism which manifests design or purpose without a designer or deliberative agent. All animals have reason or implicit rationality in such different ways that the barriers are porous. Even human beings are not all equally rational. In fact, we are more 'life-kinds' than separated rigidly into different human and other-animal species (as Stephen Clark has observed).[24] And Plotinus in subsequent chapters extends this holistic, demiurgic understanding even to the elements that possess different relations to soul and intellect (VI.7.10–11 (treatise 38)). In other words, Plotinus eliminates deliberation, planning, deliberate design and divided movement from divine production, and instead provides a reasonable case for seeing specific times and spaces as single strands or separated viewpoints of what is ultimately the complete activity of a divine being eternally creating and sustaining the world.

But what does this mean for agency? There is a paradoxical result to Plotinus' arguments in VI.7.1–13 (treatise 38). Plotinus starts VI.7 (treatise 38) by getting rid of the only agent readily available – Plato's demiurge – and instead works towards a larger contextual model in which will and freedom do not really appear at all. Certainly, if one reads carefully, one may discern in chapters 4–5 a view of what the human active subject, as a complex compound, may be: Plotinus asks what it is that '*makes* this

[24] Clark 2011: 52: 'Evolutionary change is no great surprise for either Platonists or Aristotelians.'

human being, a *logos* not separate, but indwelling in the compound' (4.28–30; for *logos* see Corrigan 2004: 112–16). And in the next chapter he defines 'the human being here' as a complex compound that requires substantial agency derived from the presence of soul: 'Soul in a specific kind of *logos*,' he replies, 'the *logos* being a specific kind of activity' (that is, an organic body), 'and the activity being unable to exist without the acting subject' (VI.7.5.2–5 (treatise 38)). So Plotinus wants to define the agent-subject that makes, transforms and also emerges from the complex unity that is the human being. Or again, later, in VI.7.7 (treatise 38), he sees the confluence of different agent forces, as individual souls articulate the outline-traces made for them by the world soul and as they become those particular traces in shaping themselves (VI.7.7.8–17 (treatise 38)). Or still later, in chapter 15, there is a remarkable subject-agent picture of intellect as 'an all-face thing shining with luminous faces' (VI.7.15.26 (treatise 38)). And, finally, the question of the Good's present agency is posed with some urgency: granted that the Good has made all things, what does the Good *make now*, Plotinus asks at the end of VI.7.23 (treatise 38): 'Now as well it is keeping those things in being and making the thinking things think and the living things live, inspiring thought, inspiring life and, if something cannot live, existence.' Certainly, the idea of agency seems to be coming more to the forefront as this treatise progresses – and, arguably, treatise 38 is Plotinus' greatest work. Yet it is not too difficult (according to the scenario I am sketching out here) to imagine Christians or Jews (perhaps a Gnostic known to have frequented Plotinus' school or simply a visitor) asking him the following questions with increasing urgency: 'Yes, your One is good, but surely It cannot do anything other than make by spontaneous reflex, indistinguishable from necessity or blind chance? In addition, if the One does not need anything, it must be indifferent, as the Epicureans suppose? In fact, it is neither a believable subject nor a meaningful agent. By contrast, the biblical God is a real creative agent, intervening and caring for his people and, in the New Testament, sending his only Son to die on His people's behalf. Your One and your Demiurge cannot even compete. Besides, even if we grant that God does not deliberate, how can we have a God who doesn't even *will anything*? A will-less God will be a chance event, completely incompatible with a biblical, Gnostic or Peripatetic notion, and, in fact, even with Plato's notion, of the primacy of freedom ("virtue is without a master").[25] How do you answer?'

[25] *Rep.* 617e3; Epicurus in Diogenes Laertius 10.133.9; Plutarch *Mor.* 740d2; Alcinous *Didask.* 27.179.10 ff.; see also Romano 1999.

Something like this is how I interpret the 'audacious argument . . . from a different way of thinking' or simply 'from the other side of the question' that Plotinus mentions in VI.8.7 (treatise 39), and goes on to attack, an argument to the effect that the Good acts by chance, being neither master nor cause of itself, and therefore has no freedom or autonomy. Six major candidates for this audacious argument have been proposed: a materialist, perhaps an Epicurean; a Gnostic (Bréhier 1924–38, III: 121 notice; Cilento 1973: 108);[26] a pure thought-experiment (*Gedankenexperiment*) initiated by Plotinus himself (Harder, Beutler and Theiler 1960–7, IVb: 372; Leroux 1996); a Christian (Armstrong 1982); a Peripatetic (Aristotelian school) (Lavaud 2007: 182–4 notice and notes); or simply Plotinus in dialectical argument with himself (O'Meara 1992: 345–9). Any 'visitor' might have proposed this, but it emerges dialectically through Plotinus taking up the other side of the question.[27]

Ennead VI.8 (treatise 39), I believe, is Plotinus' answer to such critical questions both from himself[28] and, perhaps, from someone external to the school, but familiar with Platonic and Peripatetic thought. His answer is of major importance for the following reasons: first, it develops an entirely new way of thinking about God – Plotinus is the first to argue, in the precise way he does, that will comes before intellect and to posit the One as pure freedom, a self-establishing cause of Itself[29] – and, second, it articulates a creative idea of agency and radical freedom by showing how such terms as desire, will, self-dependence and freedom in the human ethical sphere can be genuinely applied to Intellect and the One in a non-homonymous manner that nevertheless preserves the radical inability of all metaphysical language to express anything about God or gods.[30]

Ennead VI.8 (treatise 39), then, examines the nature and source of real agency in human beings, Intellect and, ultimately, the One. Its immediate and larger context is Alexander of Aphrodisias' denial that the notion of 'what depends on us' (τὸ ἐφ' ἡμῖν) cannot be applied to the gods since their

[26] More recently Narbonne 2007. For overview, Leroux 1990: 104–23; Lavaud 2007: 266n.105–267n.106.

[27] Both views, a dialectical Plotinus and a 'visitor', are fully compatible. For this dialectical contrast between correct thinking and persuasive argument, see VI.7.40.2–5 (treatise 38); V.3.6.1–18 (treatise 49); Beierwaltes 1991: 202–5; O'Meara 1992.

[28] For Plotinus' earlier thinking about freedom in IV.8 (treatise 6), IV.3 (treatise 27), IV.4 (treatise 28) see Charrue 2013; for commentary on VI.8 (treatise 39), Westra 1990; Leroux 1990: 104–22; Lavaud 2007 (with detailed notes).

[29] On the vexed question of the discovery and history of the will see Dihle 1982; for other references, Horn 2007: 153, nn. 1–2.

[30] For two other approaches to this question, by contrast with Leroux's view of a new positive theology in Plotinus (1990: 13, 38–9, 104–5), cf. O'Meara 1992 and Lavaud 2007: 189–94.

being what they are is necessary and, therefore, cannot be said to be in their power (*De fato* 32.204.12–15, 32.204.22–5 Bruns). To determine whether the notion of 'what depends on us' can be applied to the gods and, above all, to the One, we first have to determine 'whether anything happens to depend on us' (VI.8.1.15–16).[31] In ordinary life, Plotinus notes, we find ourselves besieged by opposing compulsions and passions, and we think that we as subjects are nothing. We hope that if we could escape from these compulsions, what we wish might be free (VI.8.1.22–33). In other words, we hope with Aristotle that an action depends on us if it originates in the agent. And certainly, what is 'voluntary' is what we do without compulsion and with knowledge; and 'what depends on us' is what we are masters of, or competent in, doing. However, unlike Aristotle, Plotinus argues, we need more than knowledge of the circumstances, but a 'general' knowledge – otherwise, like Oedipus, we will kill Laius in supposed freedom but in ignorance of the real circumstances or of what we ought to do. Not to know what we should know cannot be truly voluntary (VI.8.1.33–44). To what then should we attribute 'what depends on us' – to desire, passion, lust, or rational calculus with correct desire, or imagination? But if we are 'led by' any one of these, how can this be self-determining? We must then trace freedom to 'will' residing in correct reasoning with knowledge (*episteme*), since correct opinion on its own is not necessarily self-determining. Nor can we put self-determination in perception or knowledge (*gnosis*) since each simply perceives or knows: 'something else leads to action', Plotinus argues (VI.8.2.30 (treatise 39)). But if we attribute freedom simply to intellect, this will not be freedom 'in action', since everything involving action is mixed and, here, self-determination is not pure (VI.8.2.35–7). So Plotinus argues that we do not have to go as far as Intellect proper, since we can realistically trace self-determination to 'intellectual activity' and grant that 'the premises derived from there are truly free and that the desires awakened from thinking are not involuntary' (VI.8.3.22–6), for desire of the Good does not lead either Intellect or us outside ourselves, but rather makes us enter our proper

[31] Plotinus uses several terms to express the notion of what we might call will: *τὸ ἐφ' ἡμῖν* (what depends on us), *ἑκούσιον/ἀκούσιον* (voluntary/involuntary), *τὸ αὐτεξούσιον* (self-determination), *ἐξουσία* (freedom), *βούλησις* (wish, will), *θέλησις* (will), *τὸ ἐλεύθερον* (freedom), *προαίρεσις* (choice). While *βούλησις* and *θέλησις* seem synonymous, *ἑκούσιον* and *τὸ αὐτεξούσιον* are different, since an action can be voluntary but not truly self-determined. Freedom is therefore closer to self-determination than to what is voluntary. In Aristotle, προαίρεσις or choice very often refers to the choice of means to an end (and is discursive), whereas *βούλησις* is the will to the end, and is 'inseparable from *νόησις*' (Joachim 1970: 104–5). For *τὸ ἐφ' ἡμῖν* see Frede 2007; Eliasson 2008.

nature; and this is true for soul too, he argues 'when it is active according to intellect and does things according to virtue'. Should we then grant self-determination to the agent, but not to the accomplishment of the action, Plotinus asks (VI.8.5.1–8);[32] however, if we do this, we risk putting freedom outside action, whereas what we actually experience is the transforming, dynamic power of virtue actually *doing* things; and this gives us freedom, so that virtue is a question of decision, that is, of will and choice: being good is in our power 'if we will and choose it' (VI.8.5.30–2).

So Plotinus concludes the first part of his argument about human free will in chapter 6 as follows:

> So that *in actions* self-determination and what depends on us *should not be referred to acting or to the outward activity but to the inner activity, that is, the thought and contemplation of virtue itself.*[33] ... So it is still clearer that the immaterial is free, and it is to this that what depends on us is to be referred and the will that has the mastery and is self-dependent, even if something directs by necessity to what is outside. All therefore that comes from this will and because of it depends on us, *both external and in itself.* (VI.8.6.19–30)

It has been suggested that the above passage refers to 'acts of will that do not result in action'.[34] But this is not so. Plotinus specifies that he means *in actions* and that we refer *everything* self-dependent and in our power, *both external and internal,* to free will rooted in the immaterial. What is real in action is the inner activity, *νόησις* and *θεωρία*, as we saw in III.8 (treatise 30) above. And only the free subject truly acts or makes in the physical world, since his or her *praxis* manifests externally the inner energy that forms one reality with it; and the inner activity is the action's formal and final cause, '*the thought and contemplation of its healthy functioning*'. Free agency, therefore, is a fully real, historical fact, grounded in each subject's firm orientation to Intellect and the Good through which it wills, orders and makes the world to be good.

The contemplative intellect, 'the primary one', that Plotinus goes on to mention immediately, possesses a still higher degree of freedom, in that 'its work in no way depends on another ... it is all turned to itself ... and it rests in the Good ... living according to its will' (VI.8.6.32–6).[35] Intellect is

[32] On the Stoic background here and throughout see Horn 2007: 160; Lavaud 2007: 259–60 nn. 67–9.

[33] That is, of its best functioning in action, though not 'referred to' action, 6.19–22: τὸ ἐν ταῖς πράξεσιν αὐτεξούσιον καὶ τὸ ἐφ' ἡμῖν οὐκ εἰς τὸ πράττειν ἀνάγεσθαι οὐδ' εἰς τὴν ἔξω, ἀλλ' εἰς τὴν ἐντὸς ἐνέργειαν καὶ νόησιν καὶ θεωρίαν αὐτῆς τῆς ἀρετῆς.

[34] Dillon and Gerson 2004: 166n.; but see Leroux 1990: 279 line 29.

[35] Cf. Aristotle *De an.* 3.9, 432b27–9; cf. Lavaud 2007: 264–5 n. 95.

not bound by the necessity of its own nature, then, but is identical with its nature, has its cause in itself, and is 'causative substance' (VI.8.14.37) beyond chance and randomness,[36] a free complete substance that without deliberation 'gives the why and the being together as a whole' to soul and physical things (VI.8.14.32–3). Intellect, like soul, is grounded in the Good, but its freedom – like that of the Good – is the pure spontaneous creativity of its will that makes and sustains the world in itself and through us. Only of Intellect in VI.8 does Plotinus use *προαίρεσις*. 'Everything here is as it would have been if the *free choice* of its maker had willed it' (VI.8.17.2–4). The significance of this, I think, is that while his notion of freedom is not decisionist, namely, freedom as a choice between alternatives, Plotinus does not eliminate the meaning of *προαίρεσις* in our choosing of the Good. The demiurge does not deliberate, as Plotinus had argued in VI.7, but it establishes the paradigm of what deliberative free choice should be.

In the case of the One's freedom, Ernst Benz in 1932 saw Plotinus as the precursor of later voluntarism, and the first to overcome the tradition of Greek intellectualism (Benz 1932: 301). Christoph Horn has recently argued, to the contrary, that Plotinus' account is intellectualist in the broad sense 'that the divine will is strictly dependent on God's insight', that it never 'transgresses the Corpus Platonicum', but that it does develop a new dynamic, spontaneous view of Divine Free Will (Horn 2007: 155 ff.). The terms of this debate are anachronistic and, therefore, misleading, but there is some truth in both accounts. Certainly, Plotinus builds into his treatment of the One's freedom some striking features of Gnostic thought, such as the characterisation of the One as love and the self-production of the First Principle, as in the Valentinian *Tripartite Tractate* and other texts,[37] as well as earlier Platonism.[38] But his emphasis on the dynamic primacy of Divine Will from different perspectives in chapters 13–21 is striking for its daring quality.[39] And even though he notes at many points

[36] Cf. VI.8.14.31–42.

[37] E.g. the aloneness of the Good – *Ennead* VI.8.7.38–9, VI.8.9.10–13; *Trip. Tract.* (NH I) 51.74; the image of root and tree – VI.8.15.33–6; *Trip. Tract.* 51.17–19 and 74.10–13; the characterisation of the One as love – VI.8.16.34, VI.8.19.13; *Trip. Tract.* 55.22; and the all-important self-production of the First Principle – VI.8.54, VI.8.21.17; *Trip. Tract.* 56.1–4. See Lavaud 2007: 183–4. For Gnostic and Hermetic self-generation see *Trip. Tract.* NH I.5.56.1–6; *Gospel of the Egyptians* NH IV.2.79.5–6; *Three Steles of Seth* NH VII.5.124.25–9, VII.5.126.1–7; *Zostrianos* NH VIII.1.20.4–14, VIII.1.74.20–4, VIII.1.124.17–19; *Allogenes* NH XI.3.56.10–15, XI.3.65.22–7; *Corpus Hermeticum* IV.10.17; *Eugnostos the Blessed* NH III.3.75.2–8.

[38] Numenius fr. 16, for example, or Alcinous *Didask.* 10.3.15–18: The good 'by his own will . . . has filled all things with himself, rousing up the soul of the world,' or the *Chaldean Oracles* 37.2.4, 81.4, 107.4.

[39] Cf. Plotinus' caveat at VI.8.13.4–5 (treatise 39).

that he is not speaking properly, he does not reduce the One's agency to that of either a divine theatrical character or an anthropomorphic subject.

What is remarkable is the extent to which Plotinus, after denying things of the One in chapters 7–12, is prepared for the sake of 'persuasion' to develop modes of expression in chapters 13–21 that appear to contradict his entire metaphysics: The One makes itself (VI.8.7.53, VI.8.13.55); it is exactly what it wishes to be (VI.8.9.44–5); the One is ruler of himself (VI.8.7.13, VI.8.7.35, VI.8.9.44 etc.); it didn't 'happen to be' by chance or randomness; it is not by necessity the sort of thing it is, but because being what it is is the best (VI.8.10.25–6); its will is identical with its existence (VI.8.13.55–6); it is cause of itself, exists by itself, and for itself (VI.8.14.41–2); it brought itself into being (VI.8.16.15); it is as it woke itself up to be (VI.8.16.33); it is entire will (VI.8.21.14).

In these later chapters, then, Plotinus weaves together several perspectives on the One's free will, some of which are worth sketching briefly here: First, the Divine will is 'choosable': 'the nature of the Good is more worthy of choice for itself' (VI.8.13.15–20) – clearly this does not involve a decision between alternatives, but such a will constitutes the paradigm for all affirmative *decisive* power.[40] Second, Divine will is pure agency and the source of subjectivity in everything else: 'If then by This each thing itself makes itself, it becomes clear . . . that That is primarily the kind of thing it is by its own agency, by which the other things are able to be by their own agency' (VI.8.13.24–7). Third, Divine will is desirable and desire, not only the goal of all appetition, but love and self-loving: 'And that same self is lovable, love, and love of himself' (VI.8.15.1). An active love brings the One into new focus against the background of Christian thought, in particular, but it also poses the question for later (Neoplatonic) thought how all created things and their deepest aspirations can be meaningfully included in that love.[41] Fourth, Divine will is everywhere and nowhere, bringing itself into subsistence and 'giving to the rest to be, to lie alongside him in the everywhere' (VI.8.16.1–16). Here Existence is prior to essence, in the most eminent mode possible, and providentially present to all individual beings, against the Peripatetic view that providence does not extend to sublunary individuals. Consequently, the Divine will is supremely causal: 'he is the father of reason, cause and causative substance'; it is also self-creative, as the Gnostics and others (e.g. Numenius) had supposed: 'he as it were makes himself and is not as he chanced to be, but as he wills'

[40] Cf. Aristotle *Metaph.* 12, 1072a26, 1072a35; cf. Alexander (against the Stoics) *De fato* 196.25 Bruns.
[41] See, for example, Iamblichus *De myst.* 5, 26.237.6–239.10.

(VI.8.16.21–2). In other words, the One constitutes causality, not by transferring chance or randomness into himself, but – with Aristotle – by rendering chance or spontaneity as transparent as they can be in the light of regular causality in the world.[42] Divine will is therefore also a self-related and a structuring power: 'such an inclination of himself to himself ... makes him be what he is' (VI.8.16.24–6); 'making and self are concurrent' (VI.8.20.26–7); 'His waking transcends substance and intellect ... these are from him and not from another' (VI.8.16.33–7). Finally, the Divine will is autonomous, unique and primordially spontaneous as expressing the One's purest independence. It is reliant purely on itself.

Here then is a Divine Will that is eminently choosable, desirable, self-productive, supremely causative, a self-related, structuring agency, operative everywhere and present to anything, no matter how lowly. This is a Will that goes beyond anything in the earlier tradition. The Divine Will is, in fact, not dependent on insight or intellect, but '*primarily* will' (VI.8.21.16). And it is not solitary or detached: 'holding himself together has to be understood ... as meaning that everything else is held together by this' (VI.8.21.19–21). The will of the One is the comprehensive *connectivity* of everything (*Metaph.* 12, 1074b2–3; Leroux 1990: 400) and it is a radically simple, universal experience, 'something like waking up' (VI.8.16.32; *Metaph.* 12, 1072b17 and 9, 1048b1–5; Leroux 1990: 364–5). It is striking in the last sentence of VI.8 (treatise 39), then, that Plotinus should return to the democratic viewpoint that connects subjects in their upward connective movement: 'for *even you can touch upon something* about which it is not possible any more to speak or apprehend; but it is something that lies above, this alone truly free, because it is not enslaved to itself, but only itself and really itself, while everything else is itself and something else'. The phrase 'for it is possible for even you to touch upon something' is a *Symposium*-montage, first, of Diotima's words to Socrates at the beginning of the 'greater mysteries' ('pay attention and even you can learn')[43] and, second, of her final words about 'touching upon the beautiful' at the end of those mysteries, the ladder of ascent to the Beautiful (*Rep.* 212a2–5). The last sentence of VI.8, then, is addressed to anyone who can touch, through the transformed identity of Socrates, the One as ultimate

[42] Cf. *Phys.* 198a9–10.

[43] *Symp.* 209d5–210a1. In Socrates, all interlocutors, *agathoi* and *phauloi*, can be included, from Aristodemus and Apollodorus to impossible people like Thrasymachus in the *Republic*, who is definitely another 'audacious' fellow 'from elsewhere'.

guarantor of agency. In addition, the 'something' (*tinos*) that lies above is ineffable, but also remarkably positive and concrete, a reflexive memory, perhaps, of V.5 (treatise 32) 12: 'The Good is gentle, kindly and gracious, and accessible to anyone whenever anyone wishes.' The *free will* of the One – 'only itself' – is open to the *free wish* of anyone – 'itself and something else' – by virtue of that which wills or wishes in each subject.

In short, while there is a complex tradition about the freedom of the virtuous will from Plato onwards (see Romano 1999), and while Gnostic models of Divine willed self-production pre-date Plotinus, the three models I have outlined here – models of creative contemplation, non-deliberative demiurgic production and, finally, agency and divine self-causality – provide entirely new ways of thinking about the spontaneous non-rational causality of both divine and physical subjects, and ultimately show how divine activity and human action can cooperate and yet retain their independence in a single activity. In VI.8 (treatise 39), finally, Plotinus first determines the reference and function of human agency and then applies this daringly to Intellect and the One, weaving together several convergent perspectives of the One's free will that lead to the positing of a radical Divine free will prior to being and intellect, a will that is the ground of all subjectivity, from the Divine Intellect to the most ordinary subject. Despite his repeated warnings that he is not speaking properly about the One (in VI.8.13–21 (treatise 39)), it is striking that Plotinus should finish the treatise without a full retraction of his positive statements, and this perhaps intimates his concern to show dialectically (VI.8.7–12 and VI.8.13–21 (treatise 39)) how Gnostic and other views might be incorporated into his own thought by an examination of the meaning of agency without reducing the Divine to a 'sound and light' action sequence. We have also seen throughout how the interpretation and development of Aristotle in relation to fundamental Platonic problems figure prominently at every point of Plotinus' argument.

Whether or not this is the beginning of a new voluntarist tradition, it is certainly a ground-breaking development in the history of thought, whose consequences are profound, as can be seen in Werner Beierwaltes' claim that Plotinus is the predecessor of Spinoza's later concept of God as *Causa sui* as well as the founder of the modern idea of radical freedom.[44] These consequences are already to be found in Marius Victorinus[45] and also, I suggest, in later Christian Trinitarian thought. In subsequent

[44] Beierwaltes 2001a; cf. Hadot 1971; Narbonne 1993.
[45] *Adv. Arium* 1.55.19–21; cf. 1.52.28–30; cf. Lavaud 2007: 195–6.

Neoplatonism, Iamblichus (*c.* 245–325 CE) rejects Plotinus' model of the Good's self-causation; only the term 'Unbegotten' can appropriately be applied to the Godhead. Iamblichus' view was transmitted to Eunomius, Bishop of Cyzicus (d. *c.* 393 CE), who argued that God is only 'Unbegotten', an absolutely simple Being into whose nature no act of generation can be introduced without blasphemous duality. Against Eunomius, the great Cappadocian Fathers – Basil of Caesarea, Gregory Nazianzus and Gregory of Nyssa – developed their own Trinitarian thought, and we can find in Gregory of Nyssa a hidden 'Trinitarian' model already articulated by Plotinus (Corrigan 2008). Whatever the case, the later chapters of VI.8 (treatise 39) provide for subsequent thinkers a Trinitarian model of the purest Divine Unity. In chapter 20, for instance, Plotinus argues that although we use such terms incorrectly of the One, the (1) activity, (2) substance and (3) perfection of (1) making, (2) selfhood and (3) eternal generation must be identical in the Good:

> Now certainly an activity not enslaved to substance is purely free and in this way he himself is himself from himself ... if he is rightly said to hold himself together he is both himself and the bringer of himself into being, granted that what he by his nature holds together is what from the beginning he has made to be ... but now if he was what he is before eternity existed, this 'he had made' must be understood to mean that making and self are concurrent (*σύνδρομον*); for the being is one with the making and the, so to speak, eternal generation. (VI.8.20 (treatise 39))

If one puts this passage alongside Iamblichus' and Eunomius' claim that *Agennetos* is a more appropriate term for God, then we are in an intellectual climate, I suggest, close to the emergence of Cappadocian Trinitarian theology and, indeed, the word 'concurrent' (*σύνδρομον*: VI.8.13.29, 20.26 (treatise 39)) from the passage above is used of the Trinity in Gregory of Nyssa.[46]

Finally, Werner Beierwaltes' view that through VI.8 (treatise 39) Plotinus is the founder of the modern idea of radical freedom finds strong confirmation in the great American Neoplatonic thinker Ralph Waldo Emerson. Plotinus' view of radical free will in the Good as self-production and self-reliance, and as *the* model for understanding all freedom, together with Plato's description of virtue as 'without a master' – this view is at the root of Emerson's notion of self-reliance, as in the following passage that clearly derives from VI.8 (treatise 39): 'This is the ultimate fact which we so

[46] See *Ap. in hex.* 6.9.1; *Orat. Catech.* 2.33, 8.150; *C. Eunom.* 1.1.396.2, 1.1.440.3, 1.1.662.7, 2.1.227.3, 3.7.22.9. For later Fathers see Lampe 1961, s.v.

quickly reach on this, as on every topic, the resolution of all into the ever-blessed ONE. *Self-existence is the attribute of the Supreme Cause*, and it constitutes the measure of good by the degree in which it enters into all lower forms. All things real are so by so much virtue as they contain' (Emerson 1979: 40). Self-existence/self-reliance is the birth of freedom in the One that extends to all subsequent forms of existence – and this self-reliance has come, strangely enough, to characterise the birth and continuing legacy of a New World.

CHAPTER 8

Consciousness and agency in Plotinus

D. M. Hutchinson

An agent is one who acts, or more precisely, one who initiates one's own actions. This concept of agency had been treated with sophistication long before Plotinus wrote the *Enneads*. In dialogues such as the *Phaedrus* and the *Laws* Plato develops a view according to which soul is essentially a self-mover and is the source of motion to all things that move and change in the cosmos, and in the *Republic* and the *Philebus* he develops a view according to which soul is the source of motion to one's actions.[1] In treatises such as the *Eudemian Ethics*, *Magna Moralia*, *Nicomachean Ethics* and *Physics* Aristotle develops a view according to which human beings are principles of action and stand in efficient causal relations towards their voluntary actions.[2] In the extant fragments and select treatises the Stoics develop a view according to which adult human beings initiate their action by rationally assenting to their impulsive impressions, which is 'up to them'.[3]

Plato, Aristotle and the Stoics influence Plotinus' theory of agency each in their own respective ways. However, Plotinus' theory of agency is unique in that it requires a degree of inwardness that Plato, Aristotle and even the Stoics lacked due to the fact that he locates the three principles of reality – One, Intellect and Soul – also in us.[4] He writes, 'Just as in nature there are these three of which we have spoken, so we ought to think they are present also in ourselves. I do not mean in [ourselves as] beings of the sense world – for these three are separate [from the things of sense] but in

[1] See *Rep.* 437b1–444a1; *Phaedr.* 245c5–e7; *Philb.* 34c7–d3; and *Laws* 892a–896a.

[2] See *Eth. Eud.* 2.6, 1222b15–1223a20; *MM*, 1.10, 1187a30–1.11, 1187b8; *Eth. Nic.* 3.1, 1110a15–18 and 5.8, 1135a17–28; and *Phys.* 2.3, 194b30–33, 8.5, 256a3–13 and 257a27–30. See Meyer 2011: chs. 4 and 6.

[3] See Diogenes Laertius 7.49–51 (*SVF* 2.52, 55, 61; *LS* 39A); Origen *De princ.* 3.1.2–3 (*SVF* 2.988, part; *LS* 53A); Stobaeus *Ekl.* 2.86,17–87,6 (*SVF* 3.169, part; *LS* 53Q); Plutarch *Stoic. rep.* 1037F (*SVF* 3.177, part; *LS* 53S); and Alexander of Aphrodisias *De fato* 13.182.6–19.

[4] I am not claiming that the Stoics lacked inwardness, but rather that Plotinus has a richer notion of inwardness than the Stoics. On the development of inwardness in Stoicism, particularly Epictetus, see Kahn 1988; Long 1991; and more recently Remes 2008.

[ourselves as] beings outside of the realm of sense-perception' (v.1.10.5–8).[5] For Plotinus the intelligible realm is not a supra-cosmic place from which one is separated during one's life and to which one hopes to return in the afterlife. Rather, since it is also located in us the intelligible realm is 'the self at its deepest level' (Hadot 1989: 25) and can be reached by turning inwards and ascending upwards (III.4.3.21–4, IV.8.1.1–11).

This theoretical doctrine has profound implications for practical agency. In order to be the sole causal source of our own actions Plotinus holds that we must derive the premises of our actions from Intellect. For which reason, the key to understanding his theory of agency is to understand his theory of consciousness, since being an agent involves being conscious in a very particular sense. In what follows I will show that awareness plays an essential role in Plotinus' theory of agency since in order to progress towards self-sufficiency and self-determination one must, first, be an embodied subject with the minimal level of unity that is required for impulse-directed movements and, second, one must become cognisant of one's intellect and establish right reason as the guiding force in one's embodied life. Without awareness neither of these would be possible since it plays the role of unifying the qualified body in the sensible world, and enabling the soul to turn inwards, ascend upwards, and recognise what is its own in the intelligible world.[6]

Consciousness

Extended in space and divisible into parts, bodies are by their very nature subject to being scattered (IV.2.1.12–17). Due to the fact that the parts that constitute a body differ from one another and from the whole of which they collectively are parts, bodies possess a limited degree of unity. They are one by continuity, i.e., they are unified to the extent that their parts are contiguous with one another (IV.2.1.60–3). The human body is no different. So how does something that is essentially extended, divisible, and subject to dispersion acquire the unity that is required for impulse-directed

[5] Unless otherwise specified, all translations are taken from A. H. Armstrong's Loeb Classical Library Edition of the *Enneads*.

[6] It is important to note that although awareness is required for us to become agents and establish right reason as the guiding force in our embodied lives, conscious awareness is not required for all of our embodied actions after having placed reason in charge and acting from reason. In III.8 (treatise 30) *On Nature and Contemplation and the One* Plotinus treats nature's contemplative activity as the paradigm case for action, which produces without consciously reflecting on its activity, and holds that we strive to contemplate and act on the model of nature's productive actions. See Wildberg 2006 and Wilberding 2008a.

movements?[7] The answer lies with soul, and in particular with two psychic capacities whose role it is to unify the qualified body at the level of nature.[8]

Plotinus holds that animals and embodied human beings share a type of consciousness, which enables them to unify their qualified bodies and function as structured and coherent wholes. The qualified body (*to toionde sōma*) is the body, which has been informed by the lower soul. However, in the case of human beings the qualified body is not who 'we' really are. Plotinus uses the first-personal and reflexive pronoun, 'we', to refer both to discursive reasoning and the qualified body when discussing the embodied person. However, properly speaking 'we' *are* discursive reasoning; the qualified body merely *belongs* to 'us', in the way that a tool belongs to a craftsman. Nonetheless, 'we' can sink to the level of the qualified body by adopting a way of life that involves acting on the basis of non-rational sources of motivation such as appetite or passion, since the dominant pursuits of our life-activities dictate the level of self at which we live and act.[9]

This type of consciousness is awareness (*sunaisthēsis*). It differs from sense-perception (*aisthēsis*) by being directed towards one's own internal parts and activities.[10] Awareness is the most general form of consciousness in Plotinus, occurring at all levels of the human self and in each of the hypostases. However, at the level of the qualified body it functions as a form of proprioception or bodily self-awareness that enables living beings to recognise that the parts and activities that constitute themselves are their own, and to activate specific bodily parts over others in order to accomplish specific tasks, such as standing on our feet in order to walk or extending our hands in order to grab.

Closely related to awareness is the notion of sympathy (*sumpatheia*). Plotinus borrows this notion from the Stoics,[11] but significantly modifies it. He agrees with the Stoics that soul unifies the body and provides stability and coherence; however, he disagrees that soul sustains the body by means of the tensile movement of air and fire (or anything physical,

[7] I am borrowing this term from Susanne Bobzien, who defines impulse-directed movements as 'movements that are the result of an externally induced impression in the mind, which prompts the impulse, which in turn brings about the movement – provided no hindrances interfere'. See Bobzien 2001: 379.

[8] I have argued for this at length in Hutchinson 2012. I here summarise the main results of my argument. For further evidence and argumentation please refer to the earlier article.

[9] See III.4.2.3–24, IV.4.45.40–8 and VI.7.6.18.

[10] See V.8.11.32, III.4.4.11, IV.4.24.21–2 and V.3.2.4–5.

[11] See Nemesius *De nat. hom.* 81.6–10 (*SVF* 2.790; *LS* 45C); Alexander of Aphrodisias *De mixtione* 216.14–218.6 (*SVF* 2.473; *LS* 48C) and Hierocles *Elements* 4.4–22 and 4.32–53.

for that matter) and that sympathy is a shared affection between soul and body. The basis for his disagreement is his view that soul is incorporeal and, therefore, non-spatial, indivisible and impassible. Interestingly, Plotinus assigns to sympathy the role of unifying the qualified body, but he does so without specifying a physical mechanism such as tensile movement. This may appear to be a shortcoming in Plotinus' view, but it is entirely consistent with his 'top-down' approach towards natural phenomena. In so far as souls are unitary substances that occupy a higher level in the hierarchical ordering of reality and organise bodies by delivering formative principles into matter, the unity, structure and coherence that bodies exhibit comes from a higher level.

Accordingly, Plotinus maintains that soul is present to the body as a whole in all the parts of the body, and that this unique mode of presence places all the parts in a community of common feeling with one another.[12] It is this sympathetic relation that living beings have towards themselves that enables them to function as unities despite being composed of a multitude of bodily parts. Thus, awareness and sympathy work together to unify the qualified body that is attached to us and enable the human being to engage in impulse-directed movements. Sympathy provides a unified subject in which awareness can occur; awareness provides the recognition that the bodily parts and activities that constitute this subject are one's own or belong to oneself.

However, in order to achieve higher levels of unity and to progress towards self-sufficiency (*autarkeia*) and self-determination (*autexousios*) one must turn inwards and ascend upwards in order to direct one's attention away from the concerns of the lower soul and towards the good of the higher soul (I.1.3.23–7).[13] Self-sufficiency and self-determination are the normative ideals towards which our embodied efforts are oriented.[14] Their accomplishment involves living according to the noblest part of us, intellect, acting on premises derived from the noblest principle, Intellect, and fulfilling one's nature as a moral and intellectual being by voluntarily striving towards the Good. Realising these ideals enables us to engage in the activity that constitutes our essence without anything external or beyond our control impeding our activity. Moreover, by identifying

[12] The three main passages on which I base my interpretation of sympathy in are V.2.1.48–53, IV.7.3.31–6 and VI.4.9.36–7.

[13] Plotinus distinguishes between a lower soul and a higher soul. The former is the form present in the form–matter composite that organises the body and provides it with life capacities; the latter is that which makes the form in the form-matter composite and is separate from it. See IV.3.20.38–9.

[14] See III.2.10.19–21, VI.8.3.18–26, VI.8.4.12–17 and IV.4.18.21–2.

oneself with one's intellect and shaping oneself towards this end one becomes one's true self. As Pauliina Remes notes, 'Ultimately there remains one major responsibility, that of self-improvement of the embodied self, that is, realizing aspects of the ideal and autonomous self at the level of ordinary human existence' (2007: 209).

Now that we have investigated the role that awareness plays at the level of nature, let us turn to the role awareness plays at the level of Intellect. Plotinus treats Intellect paradigmatically, characterising it as the primary thinker and its unique mode of self-thinking as primary thinking, whose being and activity we aspire towards.[15] Intellect is the primary thinker because it is the primary form of life and activity, which process directly from the One and takes part in creating and sustaining the universe through providing essences to all things, including individual intellects (III.8.8.26–9 and 41–3, IV.7.9.11–15, V.4.2.22–6; Gerson 1994: 57; Horn 2012: 215, 217). Intellect's self-thinking is primary because it is a self-contained activity in which the subject is directed towards itself as its own object, which is the simplest form of activity beneath the One (V.3.7.18–20, V.6.1.1–14). For this reason, Intellect is the principle of human intellects, and insofar as discursive reasoning is the intellect unfolded on a lower level, what Plotinus calls the dividing intellect (V.9.8.21–3, VI.5.2.1–7), it serves as the principle of discursive reasoning as well.[16]

Awareness plays a crucial role in the generation of Intellect from the One. During the procession Intellect generates the world of Forms by halting and looking back towards its source and, upon doing so, begins to think. Its thinking is directed towards itself since it generates the world of Forms internally and it shares the same actuality with the Forms. Moreover, Intellect becomes aware of itself and its contents the moment it constitutes itself as a definite entity (VI.7.16.19–21, VI.7.35.31–3). Important for my purposes, awareness also plays a crucial role in Intellect's eternal act of self-thinking. Plotinus writes at V.3.13.12–22:

> For in general thinking (*to noein*) seems to be an awareness (*sunaisthēsis*) of the whole when many parts come together into the same [subject]. This occurs when something thinks itself, which in fact is thinking in the primary sense. Each one is just itself, and seeks nothing. However, if thought will be of what is external it will be deficient and not be thinking in the primary sense. That which is entirely simple and truly self-sufficient

[15] See II.9.1.47–52, V.3.6.1–3 and V.6.1.1–14.

[16] See V.1.11.1–8, V.3.3.7–13 and V.3.4.1–24. Cf. Emilsson 2007: 207–13.

needs nothing. That which is secondarily self-sufficient, that which needs itself, this is what needs to think itself; and that which is deficient in relation to itself produces self-sufficiency by being a whole, with a sufficiency deriving from all its parts, being with itself and inclining towards itself. Since indeed awareness (*sunaisthēsis*) is a perception of something that is many: even the name bears witness to this. (translation mine)

In this passage Plotinus distinguishes between that which is truly self-sufficient and that which is secondarily self-sufficient. The One is truly self-sufficient because it does not stand in need of being completed by anything external to itself and its activity is not impeded by anything outside itself beyond its control, whereas Intellect is secondarily self-sufficient because although it is a purely self-directed and self-contained activity, it nevertheless stands in need of being completed by the One.[17] Moreover, Plotinus claims that in order to be secondarily self-sufficient Intellect must collect its parts (i.e. itself as a subject, act of thinking and the world of Forms with which it is identical as object) through the activity of self-thinking, which he characterises as an 'awareness of the whole when many parts come together into the same [subject]'. Even though Intellect is the paradigm form of life and activity, it stands in need of being aware of its parts and activities because it is a multiplicity (VI.7.41.17–29). Awareness is playing the same role it does at the level of nature, namely enabling a living being composed of a multitude of parts to unify itself into a structured whole. Intellect is no exception (VI.7.41.17–29).

To return to our intellects, when we ascend to the intelligible world and recognise that our true self is an intellect we become aware of our kinship with intelligible Being and adopt the mode of intellection and self-awareness appropriate to Intellect.[18] This involves thinking the Forms directly and all-at-once, sharing in the actuality of Intellect, Being and the Forms, and becoming aware of our identity with Intellect. Plotinus writes:

> When it [i.e. the soul] is purely and simply in the intelligible world it has itself too the characteristic of unchangeability. For it is really all the things it is: since when it is in that region, it must come to unity with Intellect, by the fact that it has turned to it, for when it is turned, it has nothing between, but comes to Intellect and accords itself to it, and by that accord is united to it without being destroyed, but both of them are one and also two. When therefore it is in this state it could not change but would be

[17] See III.8.11.12–19, V.1.6.42–3, V.3.13.18–22, V.3.17.6–14, VI.7.2.40–3, VI.7.9.45–10.10.

[18] See IV.7.10.30–7, III.4.3.22–3 and VI.5.7.

> unalterably disposed to intellection while at the same time having an awareness of itself (*sunaisthēsin autēs*) as having become one and the same thing with its intelligible object. (IV.4.2.23–33)

It is in virtue of this heightened self-awareness and recognition that our intellect is identical to the intelligible Beings in Intellect that we are capable of determining ourselves. As we will see in the next section, becoming a self-determining agent involves identifying oneself with one's higher soul, acting on premises derived from Intellect, and establishing right reason as the guiding force in one's daily life. Given the fact that Intellect is also in us, the process by which we achieve this autonomy is essentially an inward process that requires a unique mode of consciousness.

Agency

Plotinus develops the concept of agency throughout the *Enneads*. The most detailed analysis of this concept occurs in III.8 (treatise 30) *On Nature and Contemplation and the One* and VI.8 (treatise 39) *On Free Will and the Will of the One*. However, the set-up for these treatises occurs in the natural-philosophical treatises devoted to providence and destiny, namely II.3 (treatise 52) *On Whether the Stars are Causes*, III.1 (treatise 3) *On Destiny* and III.2–3 (treatises 47–8) *On Providence (i–ii)*. The central problem that Plotinus faces in the latter set of treatises is the compatibility of autonomous agency with universal causal determinism, since he holds that autonomous agency consists in being the causal source of one's actions but that destiny appears to fully determine the sensible world in which human beings live and act. Importantly, the Stoics shape Plotinus' understanding of both universal causal determinism and autonomous agency. By the second century CE Stoic compatibilism was a lively topic of debate amongst philosophers of the era (Bobzien 2001: 358), and Plotinus was eager to stake his position in this debate.

Plotinus attempts to solve the central problem by introducing his view that the higher soul is a principle (*archē*), which is capable of initiating its own actions. Moreover, he sets up his conception of autonomous agency, which he fully develops in III.8 (treatise 30) and VI. 8 (treatise 39), by distinguishing between actions based on non-rational sources of motivation deriving from the lower soul and actions based on rational sources of motivation deriving from the higher soul, and establishing that the former actions are subject to destiny but the latter are outside the causation of the physical universe (*kosmikēs aitias exō*). It is in virtue of the lower soul that human beings are rooted in the sensible world and experience non-rational appetites and passions, whereas it is in virtue of the higher soul that human beings retain a foothold in the intelligible world, occupying a unique place in the intelligible realm and embodying a unique perspective on the intelligible Beings (IV.7.10.8–11, IV.7.10.30–8; Kalligas 1997: 223–6 and 2000: 25).

Plotinus follows the moral psychology of Plato and Aristotle in holding that embodied human beings have both non-rational and rational motivations for action (I.1.7.18–25). Moreover, he appears to follow Aristotle in dividing desire (*horexis*) into rational desire (*boulēsis*) and non-rational desires (*epithumia* and *thumos*), with rational desires being of what is good and non-rational desires being of what appears good.[19] Uniquely, however, Plotinus locates appetite (*epithumia*) and passion (*thumos*) in the lower soul[20] and willing (*boulēsis*) in the higher soul.[21] The former desires are concerned with goods of the soul–body compound, but the latter desire is concerned with the good of the soul or intellect. Importantly, each of these sources of desire is a motivation for action. Consistent with the entire Greek philosophical tradition, Plotinus viewed reason as a motivating force for action (Cooper 2012: 11–16).

Let us now turn to the treatises on providence and destiny in order to understand the set-up to Plotinus' conception of autonomous agency. Plotinus identifies Intellect with the Platonic demiurge since it is ultimately responsible for the generation and organisation of the sensible world, being prior in nature to the sensible world and the model on which the sensible world is based. However, Plotinus often includes the activities of Soul and the world soul within the demiurgic activity since they are needed to carry out the activities of Intellect, on a lower level, by delivering formative principles into matter and establishing the ordering in the All.[22] Unlike the creation story of the *Timaeus* (at least on the literal interpretation) the All does not come into being as the result of rational planning (see III.2.14.1–5, V.8.7). Rather, it is an eternal image of Intellect that emerges due to the principle of plenitude (III.2.1.20–7). As an image, the All possesses the features found in Intellect but in a dimmer, less pure form due to its diffusion into space and time and proximity to matter

[19] See *De an.* 413b23–24, 414b1–2, 432b5–7, 433a17–25; *MA* 700b21–9.

[20] See I.1.5.27–8, I.1.4.6–10, I.1.5.22–9, IV.4.18.9–21, IV.8.8.17–25, IV.4.8 and IV.4.28.

[21] See I.4.6.9–19, I.6.7.1–6, IV.8.4.1–3, VI.8.6.37–8 and VI.1.21.16–23.

[22] Compare V.1.8.1–14 with IV.4.10.1–5, IV.3.10.10–42, IV.3.12.31–5 and II.3.17.15–18. Gary Gurtler also notes this. See Gurtler 2002: 100.

(v.8.7.17–18). For this reason, Plotinus tells us, 'providence for the All is its being according to Intellect' (iii.2.1.22–3).

What he means by this is that providence sees to it that the sensible world reflects the beauty of the intelligible realm by producing a universe that is in sympathy with itself, by equipping living things with the parts and capacities they need to flourish, by establishing an inescapable system of karmic justice whereby wrongdoers are punished for their actions in this life or the next, by natural phenomena operating according to causes, and individual causes taking place in a network of causes that are oriented towards the good of the whole. Although providence reaches the sublunary world and even extends towards individuals (see iii.2.6.23–6, iii.2.13.18–20), it is actually destiny that carries out the activities of providence in the sensible world (iii.3.5.15–17).

Soul directs the All according to a rational order (*logos*) by establishing that all events in the realm of becoming happen according to causes, that individual causes are woven together into a network, and that the network is ordered to the good of the whole. The precise nature of Soul's direction is brought out clearly in chapter 16 of his late treatise, ii.3 (treatise 52) *On Whether the Stars are Causes.*[23] Herein he develops his own view through critically engaging with three Stoic-inspired determinist positions. The first holds that Soul creates the natural kinds and lets the interweaving and succession of consequences that follow from their interaction with each other occur without playing an additional causal role (lines 6–13); the second holds that Soul creates the natural kinds but effectively causes the interweaving and succession of all consequences that follow (lines 13–15); the third is a middle path between the first two, holding that Soul creates the natural kinds and knows the interweaving and succession of consequences that follow, through its possession of formative principles (*logoi*), but that it is not the efficient cause of everything that ensues. Crucial to the third option is the idea that

> The forming principles certainly exist, but not as causing (*poiountōn*) but as knowing (*eidotōn*) – or rather the soul which contains the generative rational principles knows the consequences which come from all its works; when the same things come together, the same circumstances arise, then it is altogether appropriate that the same results should follow. Soul takes over or foresees these antecedent conditions and taking account of them accomplishes what follows and links up the chain of consequences, bringing

[23] I am thankful to Pavlos Kalligas for sharing his commentary on this passage with me. See Kalligas 2014.

antecedents and consequents into complete connection, and again linking to the antecedents the causes which precede them in order, as far as it can in the existing circumstances. (II.3.16.18–27)

In other words, Soul causes individual things to come into existence and furnishes them with capacities associated with their nature (antecedents); individual things act on each other in ways associated with their natures from their own impulses (consequences); and Soul, in virtue of containing the formative principles derived from Intellect, foresees these events and links the consequences with the antecedents and places them into a harmonious network (IV.4.39.6–18).

Plotinus finds the first view unappealing since he holds that the providential ordering extends to particulars in the realm of becoming and, as such, Soul could not be indifferent to the consequences that follow. He finds the second view unappealing since it entails that Soul is the efficient causal source of everything that happens. This view, perhaps belonging to a Stoic or a Middle Platonist, he argues against in his early treatise III.1 (treatise 3) *On Destiny*, on the grounds that a single animating principle that permeates and sustains the cosmos leaves no room for animate beings to act from their own impulses or for human beings to have actions which are their own (III.1.4.21–30, III.1.7.13–24).[24] However, he does find the third option appealing since it entails that Soul knows the consequences that follow but does not cause them, which leaves open the possibility that animate parts of the whole contribute to the good of the whole from their own impulses (II.3.13.11–13).

Plotinus is committed to preserving the idea that although animate beings are subject to destiny they are not restricted to reacting mechanically or automatically to external stimuli. Instead, they are conscious of the effect external stimuli have on their bodies and respond in ways that give rise to bodily movements through appetites, desires and, in the case of human beings, voluntary actions through reasoned thoughts, that is, thoughts about what reasons there are for acting and whether the reasons for acting are good reasons. Of course, the Stoics are also committed to preserving this. The late Stoic theory that Alexander of Aphrodisias reports and criticises in his *De fato*, perhaps belonging to Philopator, holds that

[24] Plotinus discusses this view in chapters 2, 4, and 7–10. Emile Bréhier and A. A. Armstrong identify this view as belonging to a Stoic or a Stoicising Middle Platonist. See Bréhier 1925: 4–5, 10, 14 and Armstrong 1967: 6–7, 18–19. More recently, Paul Kalligas has strengthened the case that it belongs to a Middle Platonist by showing the similarities it has with Atticus' theory, according to which Destiny is treated as a substance identified with the cosmic Soul. See Kalligas 2014: Introduction to III.1 (treatise 3) and Commentary on III.1.4 (treatise 3).

although our actions are necessitated and are brought about *by* fate, nonetheless they occur *through* impulse and assent, and therefore are 'up to us'.[25] However, Plotinus holds that a corporeal soul embedded in an inescapable network of corporeal causes is not capable of being the causal source of its own actions. Only an incorporeal soul capable of living according to a higher code of laws (*nomothesia*) can achieve this (IV.3.15.11–25).

Although Plotinus does not subscribe to (his interpretation of) Stoic determinism due to the restrictions he thinks it places on agency, it is worth pointing out how close his view comes to universal causal determinism. For he holds the general causal principle that nothing happens in the sensible world without a cause: 'as for things which come into being, or which always really exist but do not always act in the same way, we must say that all always have a cause for coming to be; nothing uncaused can be admitted' (III.1.1.14–17). The second clause refers to individual souls, who always exist but whose activities change, and the lines that follow make it clear that this principle applies not just to coming into being but to human action in the realm of becoming (lines 16–24). And he appears to hold the further specified causal principle that ensures regularity or uniformity between types of causes and types of effects: 'when the same things come together, the same circumstances arise, then it is altogether appropriate that the same results should follow' (II.3.16.21–3).

Although the appearance of the specified causal principle occurs in a polemical passage, the remaining lines of the chapter, and related comments he makes elsewhere, suggest that he endorses the third view. However, he endorses this view with one crucial qualification: it does not apply to the higher soul or, more precisely, the embodied human being who identifies with the higher soul. Together with his view, in accordance with the Myth of Er, that each individual soul chooses its lot based on its previous life and this choice determines the position in which it is born and the role it plays in the cosmic drama, suggest that the sensible world in which the human being lives is a determined world. However, it is in virtue of the higher soul that we, like actors in a play, can decide whether or not to play our assigned role well or badly.[26]

In virtue of being informed by the lower soul the embodied human being is subject to destiny in the sensible world. The reason for this is that

[25] See Alexander of Aphrodisias *De fato* 13.181.12–182.20 and 34.205.25–206.3. Susanne Bobzien provides a convincing argument that this view belongs to Philopator. See Bobzien 2001: 367–70.

[26] See II.3.15.1–13, III.2.7.15–28, III.2.17.24–90 and III.4.5. Cf. *Rep*. 617c–618b.

the soul acquires a vehicle in the heavens in the course of its descent from the intelligible world, which transports it downward through the celestial regions until it reaches earthly bodies. Upon acquiring the soul-vehicle and descending through lower regions of the cosmos, the lower soul acquires the capacity to undergo affections and act on the basis of non-rational impulses and thus becomes a part belonging to the powers of the whole.[27] Consequently, Plotinus is willing to grant that 'more remote causes' such as the stars do play a limited causal role in our lives, since they contribute to our bodily constitution, bring about changes in our temperaments, and can even foretell our fortunes. However, neither the position of the stars nor the motion of the planets is responsible for our characters or ways of life. These are *our* responsibility.[28] But what does it mean for our actions to be subject to destiny?

As I mentioned above, Plotinus holds that all living beings act on the basis of their own impulses (*hormai*), but that human beings also deliberate about whether or not it is good to act from certain impulses and decide whether or not to carry their impulses through to action on the basis of their deliberations (III.1.1.14–24, III.1.7.13–25). In the case of human beings, impulses can be rational and come from the higher soul or they can be non-rational and come from the lower soul. Actions that are subject to destiny are those committed on the basis of impulses deriving from the lower soul, and can be based either on the soul–body composite or even on a mixture of the soul and the soul–body composite. Take the following example:

1. Eating in order to indulge a craving for some particular food
2. Eating in order to indulge a craving for some particular food but knowing that maintaining a healthy body requires providing it with sustenance

In the case of impulses deriving solely from the composite (case 1), the source of the impulse lies in some pre-existing circumstance in the external world. In the case of impulses deriving from the mixture of soul and the soul–body composite (case 2) the source of the impulse still lies in a pre-existing circumstance in the external world but it also involves making a choice (*prohairesis*) to act on the impulse, which requires reason. The difference between case 1 and case 2 is that in case 1 I am compelled to act and not in charge of my action, whereas in case 2 I am still compelled to act but I exercise some authority over my action (III.1.8.11–18). However,

27 See II.3.9.7–31, II.3.10.4–8, II.3.15.13–15, III.1.8.4–21 and IV.3.15.1–12.
28 See III.2.4.37–47, III.2.7.15–28, III.2.10, III.3.3.1–17, III.3.3.35–7 and III.3.4.5–8.

in both cases the efficient cause of my action is ultimately the desired food item even though case 2 involves the intermediation of reasoning. Thus, in both cases I am living under destiny since my actions can be traced back to external causes that fully or partially determine them. As we will see below, Plotinus holds that actions that involve reason and the non-rational affections – 'mixed actions' – remain subject to destiny because they are not truly in our power.[29] While engaged in mixed actions we are agents to the extent that we are *a* causal source of our actions, but we are not autonomous agents engaged in self-determination since we are the not the *sole* causal source of our actions.

In virtue of possessing the higher soul the embodied human being can avoid living under destiny. Plotinus writes, 'now when the soul is without body it is absolutely in charge of itself and free (*kuriōtatē te autēs kai eleuthera*), and outside the causation of the physical universe (*kosmikēs aitias exō*); but when it is brought into body it is no longer in all ways in charge, as it forms part of an order with other things' (III.1.8.9–12). The higher soul is outside the realm of becoming, and consequently outside the causation of the physical universe. When disembodied it is in charge of its actions and free, and therefore purely self-determining. However, embodiment subjects it to natural necessity and forces it to lose complete authority over its actions, which can result in error or vice (I.1.9.5–16). Although the phrase 'without body' in this passage refers to an actual separation of soul from body, the ensuing chapters in which Plotinus develops his view make it clear that human beings can achieve this state while embodied through establishing right reason in charge of its impulses (see III.1.9.5–17, III.1.10.4–15). Moreover, related passages make it clear that we achieve this state of being 'without body' by turning inwards, ascending upwards, and identifying with our higher soul or intellect (II.3.15.15–18, II.3.9.24–32). Thus, when the embodied human being acts on the basis of rational impulses it is fully in charge of its actions, free, and outside the causation of the physical universe. It is no longer a part that belongs to the powers coming from the whole, but an autonomous agent that belongs to itself.

The basis for this conception of autonomous agency is that the higher soul, and not just Soul or the world soul, is a principle (*archē*). After claiming that embodied human beings can rise above the powers coming from the whole and preserve the ancient part[30] of the soul, Plotinus writes:

[29] See III.1.9.1–2, IV.4.43.19–22, IV.4.44.5–7 and VI.8.2.36–7. Cf. O'Meara 2003: 133.

[30] Plotinus uses the adjective *archaios* in reference to the higher soul at II.3.15.16–17, II.3.8.13–15, IV.7.9.28–30. Cf. Plato *Rep.* 547b6–7.

> For we must not think of the soul as of such a kind that the nature which it has is just whatever affection it receives from outside, and that alone of all things it has no nature of its own; but it, far before anything else, since it has the status of a principle, must have many powers of its own for its natural activities. It is certainly not possible for it, since it is a substance, not to possess along with its being desires and actions and the tendency toward its own good. (II.3.15.17–23)

His view of the soul as a self-subsisting and self-moving entity, with a permanent residence in the intelligible world, leads Plotinus to regard the individual soul as a principle in its own right. By this he means that soul has a nature of its own, has the capacities to engage in its own activities, and is the efficient causal source of its own voluntary actions. Plotinus reiterates this point several times in the treatises on destiny and providence, but it is not until the next treatise in the chronological order after II.3 that he explains which capacities he means (see I.1.5–13 (treatise 53)). Briefly, Plotinus has in mind rational desire for the good and the capacities of imagination and discursive reasoning in virtue of which we pursue the good of the soul, namely virtue. It is owing to the possession of these capacities that embodied human beings can deliberate over whether or not there are reasons to act and can cause their voluntary actions through decision.

It is important to note that standing outside the causation of the physical universe does not entail that our actions are causeless or that we are not involved in the ordering of the All. Plotinus denies the indeterminist view that actions could occur without causes, on the grounds that acting without causes would render us more compelled than acting on the basis of antecedent causes that determine our actions, since we would be carried around by movements that are uncaused, unwilled and, as a result, would not belong to ourselves (III.1.1.16–24; cf. III.1.8.2). Moreover, he claims that providence and the ordering of the whole includes us 'as the persons we are', by which he means as the embodied human being who is capable of acting freely and being in charge of his actions (III.2.10.16–20, III.3.3.1–4, III.3.4.6–8).[31] What it does entail is that we cause our own actions and we contribute to the interweaving of causes as co-authors of the providential ordering (see also Dillon 1996a: 330; Leroux 1996: 310–11). In other words, we are no longer dragged around passively by acting on the basis of impulses stemming from the soul–body composite; rather, we contribute actively to the ordering by

[31] Laura Westra also notes this. See Westra 2002: 132, 135.

acting on the basis of impulses stemming from the higher soul in the intelligible world (II.3.13.18–32, IV.3.15.11–25).

Once we identify with our higher soul and initiate our own activity we not only pursue our own good but, in doing so, we also contribute to the interweaving of causes that promote the good of the whole since both are oriented towards the Good (IV.4.35.33–5). In keeping with the analogy between the cosmos and a drama alluded to above, we no longer merely play the roles assigned to us by the playwright but we help write the script. As John Dillon notes:

> If Plotinus is not a Stoic determinist, it is only, I think, because of a daring conception of his which sees the highest element in us, the 'undescended' intellect, as in fact the autonomous component of the hypostasis Intellect, and thus in its own right (since every intellect in Intellect is in a way coextensive with the whole) a guiding principle of the universe. (1996a: 330)

In the early treatise on destiny and the late treatises on providence Plotinus does not explain what he means by being in charge of one's actions or being free.[32] His primary concern in III.1 (treatise 3) is to show that destiny does not exclude human agency, and his primary concern in III.2–3 (treatises 47–8) is to show that providence is not responsible for evil and that human beings are morally responsible for their actions. The discussion of these issues occurs in the opening chapters of VI.8 (treatise 39) *On Free Will and the Will of the One*, which I turn to next.

Plotinus departs from Plato and Aristotle by holding that the virtuous life does not consist in moderating the appetites and passions but in completely detaching oneself from them, and attending to them only when necessary and without experiencing their emotional excitement. For the appetites and passions produce an involuntary impulse (*to aproaireton*) that is compulsory and leads us away from the Good (VI.8.4.15–17, I.2.5.13–22). We are led away from the Good because, when an external object moves the appetitive or passionate powers an impression (*phantasia*)[33] is produced in the imagination, which informs the soul of the experience the body is undergoing and demands that we should follow along with the

[32] Due to space constraints, I will discuss the meaning and significance of being in charge of our actions. For a discussion of freedom see Kevin Corrigan's chapter in this volume (Chapter 7).

[33] Plotinus uses the terms *phantasia* to refer both to the imagination and to images produced in the imagination. In the case of the latter usage, he also uses *phantasia* interchangeably with the Stoic terms *tupos* and *tupôsis*. Plotinus follows the Stoics in maintaining impressions (*phantasiai*) reveal themselves and their cause (*LS* 39B; *SVF* 2.54) but disagrees that they are affections of the soul and have extension. See III.6.1.7–12 and IV.3.26.30–3. Hence *phantasia* can be translated either as image or impression.

impression and obtain the desired object. When this occurs false opinions concerning what should be pursued or avoided and what is good or bad are produced, resulting in 'us' falling into a state of perplexity and becoming increasingly ignorant of the Good (IV.4.17.12–20). Plotinus thinks the imagination stores impressions in an emotionally laden way and when the impressions 'come like a perception and announce and inform us of the experience' they do so in an emotion-triggering way that has the effect of making demands on the soul (IV.3.32.3–7). Thus, the Plotinian virtuous agent aims to satisfy the needs of the body without sharing in the emotional excitement of the lower impulses since, in doing so, he runs the risk of evaluating the pleasures that result from satisfying the lower impulses as good (I.8.4.8–13) and self-identifying with the soul–body compound (I.4.4.13–18, IV.4.18.16–19).

This is the background Plotinus has in mind when he asks in VI.8.2 how our actions can be said to be up to us if impression and non-rational desires compel us to act? Were we merely to react automatically to our non-rational impulses we would be no different from children, animals or madmen who are carried wherever their impulses lead with nothing under their authority (III.1.7.13–25, VI.8.2.5–9). However, following the Stoics Plotinus holds that in between impression and impulse there occurs a rational assent that governs our response to non-rational impulses and ensures us ownership of our actions.[34] Moreover, this assent of reason constitutes the motivating psychological impulse that impels us to act and carries us through to action (VI.8.2.30–7).[35] For which reason, acting on the basis of a rational impulse is a *willing*. Plotinus employs the Stoic term for assent (*sunkatathesis*) only once,[36] probably because of its association with the Stoic physicalist theory of soul. However, it is clear from his discussion of the relationship between the will and non-rational impressions in VI.8 that he holds this view.[37]

We are now in a position to see why, in order to be the *sole* causal source of our own actions, we must act from reason and derive the premises of our actions from Intellect. Let us begin with the notion of the 'up to us' (*to eph' hēmin*). Plotinus reasons that actions that are up to us are not those which are enslaved to impulses that carry us in whichever direction they lead or

[34] See Plutarch *Stoic. rep.* 1057a (*SVF* 3.177, part; *LS* 53S).
[35] See Stobaeus *Ekl.* 2.86.17–87.6 (*SVF* 3.169, part; *LS* 53Q).
[36] See I.8.14.4–5. Cf. similar usage of *epineuô* at IV.4.43.7.
[37] Michael Frede and John Cooper have argued convincingly that Plotinus is heavily influenced by the Stoic theory of adult human agency. See Frede 2011: 57–9, 62–3 and 125–52 and Cooper 2012: 363–81.

those constrained by external circumstances that prevent us from accomplishing what we wish to accomplish. Rather, actions that are up to us are those whose efficient cause is our will and those the accomplishment of which depend solely on us willing them. He writes: 'What is up to us is enslaved to the will (*tē boulēsei*) and would occur or not depending upon whether [or not] we willed it, for everything is voluntary (*hekousion*) that is done without force but with knowledge, whereas what is "up to us" is, in addition, what we are in charge (*kurioi*) of doing' (VI.8.1.31–4). In order for an action to depend solely on willing it must be more than voluntary in the Aristotelian sense, however. It must also be one that we are in charge of doing or have authority over doing, which, as we will see, requires us to act solely from reason.

Plotinus' characterisation of the voluntary as an action that is 'done without force but with knowledge' is an obvious reference to Aristotle, who defines voluntary actions as those which are not forced but originate in the agent and are performed with knowledge of the particulars of a given situation (*Eth. Nic.* 1111a21–4). However, the Aristotelian notion of voluntariness is insufficient for Plotinus since one could act voluntarily without being in charge of what one is doing. For example, if while walking in front of my favourite café and noticing a flaky croissant I suddenly experience an appetitive desire to eat the croissant, and I act on this appetitive desire, I am doing so voluntarily. No one is forcing me to do it, and I am doing it with full knowledge of the particulars of the situation. However, the premises (*protaseis*) on which my practical action is based originate in the soul–body compound since the motivating psychological impulse is coming from the non-rational desire.

By 'premises for action' Plotinus probably has in mind the kind of reasoning employed in the Aristotelian practical syllogism in which the universal premise identifies some good or apparent good (e.g. flaky croissants should be eaten), the particular premise spots the good to be achieved in some present situation (e.g. this is a flaky croissant) and the conclusion results in action (e.g. eat this).[38] Even though I am using reason in conjunction with appetite to enter the café and eat, it is ultimately the non-rational desire that serves as the efficient cause and sets the reasoning in motion. Or in Plotinus' terms, 'reason does not produce the impulse [to act] but the non-rational also has an origin in the premises derived from the affection' (IV.4.44.5–7). It is only when reason serves as the efficient cause and sets the desire in motion that I am fully in charge of my actions.

[38] See *Eth. Nic.* 1147a25–32; *De an.* 434a17–22; and *MA* 701a7–701b1.

The notion of reason Plotinus has in mind is not just any instance of reasoning or calculation, but rather right reason (*orthos logos*) that belongs to the understanding (*epistēmē*). Acting on the basis of a right belief, without knowing why one's belief is right, would not count as truly self-determining since the basis for the right belief could be chance or the imagination, which are beyond our control (VI.8.3.2–8). Hence the rational assent that constitutes the motivating psychological impulse must be right (*orthos*) and the agent must know why what he is doing is right, which requires understanding (*epistēmē*). But what makes a rational assent right?

Plotinus does not employ the phrase 'right reason' often. However, his usage of the phrase and related variants suggest that the *orthos logos* is inborn, belongs to the purest and most untroubled part of us, is oriented towards the Good, and that when we act according to it we are free and active, but that it becomes weakened and fettered when the appetitive and passionate parts of the soul are in control and impel the person to action.[39] Taken with VI.8.3.2–8, this suggests that we know why what we are doing is right and act on right reason, as opposed to right belief, due to the fact that our intellects are activated and in touch with the Forms in Intellect.[40] This is consistent with Plotinus' view that practical wisdom (*phronēsis*), the intellectual virtue responsible for deliberating well concerning what is good or bad for a human being, derives its principles from dialectic and is a superior form of reasoning concerned with grasping the universal.[41]

This usage of right reason highlights a crucial feature of the Plotinian notion of voluntariness: voluntary actions are those that are naturally inclined towards the Good (VI.8.4.13–20). Hence, for Plotinus voluntary actions are not simply those that stem from within and with knowledge of the particulars, but also with knowledge of the universal (VI.8.1.36–45). Plotinus inserts orientation towards the Good into the notion of voluntariness due to the close relation he sees between voluntariness

39 See III.1.9.5–17, III.1.10.4–15, III.5.7.31–9, IV.4.17.21–4 and IV.4.35.32–4.

40 Although *orthos logos* is commonly associated with Aristotle (*Eth. Nic.* 1107a1–4, 1138b18–25) and the Stoics (Diogenes Laertius 7.88 and Stobaeus *Ekl.* 5b10, 11ae), Plotinus may have employed this notion in connection with a passage from Plato's *Phaedo* to highlight the inborn connection we have with the intelligible world. While introducing the theory of recollection Plato writes, 'there is one excellent argument, said Cebes, namely that when men are interrogated in the right manner, they always give the right answer of their own accord, and they could not do this if they did not possess the knowledge (*epistēmē*) and the right explanation (*orthos logos*) inside them' (73b7–10). For Plato and Plotinus human beings have the right reason in virtue of their connection with the Forms. I am thankful to Charles H. Kahn for bringing this passage to my attention.

41 See I.3.6.6–14. Cf. I.2.1.17, I.2.3.15, I.2.6.13–15, I.2.7.7, I.2.7.14–31, I.8.9.2–3 and I.8.15.7–9.

and self-determination. The more our actions are inclined towards the Good the more they derive from within ourselves, and the less they depend on external factors that are beyond our control and that constrain us from acting the way we rationally desire to act. For which reason, voluntary actions belong only to those who are self-determining. Plotinus writes, 'We will designate those as self-determining who, owing to the activities of Intellect, are free from the affections of the body. Referring "up to us" to the most noble principle, the activity of Intellect, we will designate as really free the premises that come from there and claim that desires that arise from thinking are not involuntary (*ouk akousious*)' (VI.8.3.20–4).

The desire that arises from thinking is voluntary because it originates in the best part of us, intellect, and is concerned with truths contained in the noblest principle, Intellect. Following Lloyd Gerson, I take the 'thinking' to be concerned with universal truths found in Intellect and the 'premises that come from there' to refer to the universal premises in the kind of reasoning employed in a practical syllogism (Gerson 1994: 161). Taken thusly, Plotinus is saying that the self-determining agent is one whose universal premise of a practical syllogism contains a universal truth that such and such is good, who recognises that the universal truth that such and such is good is also good for oneself, and desires to act with this conformity of our good with the Good in mind. Unlike the above case, the premises on which this action is based originate within our intellect, and the motivating psychological impulse comes from rational desire. With reason serving as the efficient cause and setting desire in motion, we are fully in charge of our actions.

Deriving our premises from Intellect requires more than self-awareness, however. Importantly, it also requires turning inwards and ascending upwards through practising the virtues. Briefly, Plotinus holds that there are three levels of virtue: the civic, the purificatory and the intellectual. The civic virtues are those that impose limit and measure on our non-rational desires of appetite and passion and abolish false opinions arising in the compound (I.2.2.14–20); the purificatory virtues are those that separate the soul from the body by stripping away everything alien to it, thereby enabling it to act independently of the non-rational desires and opinions arising in the compound (I.2.2.11–23); and the intellectual virtues are those possessed by a soul, which, upon being purified from its involvement with the body, realises its nature as an intellect and fully absorbs itself in contemplation of the Forms (I.2.6.7–27; cf I.6.6.1–21). It is only when we have realised our nature as intellects, through purification, that we can derive our premises from Intellect and establish reason as the guiding force in our embodied lives. Crucial to my emphasis on the role of consciousness, this process of purification involves turning inwards, ascending upwards, and consciously shaping ourselves towards our ideal selves (I.6.9.7–16).

This spiritualisation of the virtues has profound implications for Plotinus' concept of autonomous agency. We are self-determining, and therefore in charge of our actions and free, only when we are operating at the level of our intellects. This can be seen from Plotinus' discussion of the civic virtues. These virtues are those that result from habit and training as opposed to thought (I.I.10.12–14), and correspond roughly to what Plato and Aristotle refer to as the moral virtues. When we act in circumstances that require performance of these virtues, for Plotinus, we are not truly free or in charge of our actions. In order to be brave there must be a war; in order to be just there must be injustice; in order to be liberal there must be poverty. In each of these circumstances the virtuous person's actions are constrained by circumstances that he himself would not choose. A truly virtuous person would prefer rather that there not be wars, injustice or poverty in the first place, just as a physician would prefer that her patients not be sick and in need of medical treatment (VI.8.5.7–20). To be truly free and in charge of our actions would involve us acting in circumstances that we ourselves choose and that we would always choose under ideal circumstances, since these choices are oriented towards what is best, namely the Good. This can only occur when we are operating at the level of intellect and we 'leave behind in actuality' the civic virtues and engage in the intellectual virtues. However, should the circumstances arise we can perform the civic virtues guided by right reason since we retain them potentially (I.6.9.11–31).[42]

With this framework of virtue in mind, we can see why Plotinus holds that self-determination is not achieved when 'we' engage in practical reasoning or practical actions but when our intellects contemplate the Forms. He writes 'So, also in actions, that which is self-determining and "up to us" is referred neither to the acting nor to what is external but to the activity of the interior, that is, thinking or the contemplation of virtue itself' (VI.8.6.20–2). Because practical action constrains us from acting the way our intellect rationally desires to act, Plotinus attributes the 'up to us' to the intellect detached from actions, engaged in the contemplation of virtue (VI.8.5.23–37). Due to being external to the soul and other-directed,

[42] On the question whether we perform virtuous actions via deliberation or automatically see Wilberding 2008a.

practical reasoning and action split our attention in multiple directions, rendering us multiple and in need of objects outside ourselves to flourish (in the mistaken way that the soul–body compound identifies). It is only when we have purified ourselves of the non-rational desires and opinions stemming from the compound and contemplate the Forms that our activity is entirely self-directed, our attention is purely focused in the singular direction of what is best, and we are completely self-sufficient. At this level, our will and our intellect coincide and what we know to be good we will to accomplish (see Rist 1967: 136–7).

Paradoxically, it turns out that in order to be the causal source of our actions we must not engage in actions associated with the civic virtues but rather engage in activities associated with the intellectual virtues. In what sense then is this a theory of agency? Like much of Plotinus' philosophy his notion of agency is worked out at the ideal level. However, that should not distract us from seeing the effects of the ideal on the mundane realities of daily life. What Plotinus has shown, I think, is that even in a world governed by destiny human beings can initiate their own activity, in varying degrees, depending upon the extent to which they are free and in charge of their actions. However, in order to do this we must strive to attain the freedom and authority belonging to the best part of ourselves, namely our intellects. Although most of our actions are 'mixed' due to the demands of the non-rational desires serving the needs of the body, we nonetheless should strive to establish reason as the guiding force in our lives. Doing so results in us achieving self-determination and self-sufficiency and, to the best of our abilities while embodied, approximating the One.

Conclusion

I have argued that in order to engage in impulse-directed movements awareness is required to unify the qualified body. Moreover, to become an agent capable of mixed actions or an autonomous agent capable of self-determination and self-sufficiency, awareness is required to turn inwards, ascend upwards, and establish right reason in charge of our embodied lives. The Plotinian theory of agency requires us not only to 'look inwards' and correctly use our impressions along Stoic lines, but also to 'ascend upwards' and identify with our higher soul and derive our premises for action from our knowledge of the Forms. I hope to have shown that accomplishing this feat requires a degree of inwardness and a type of consciousness unattested in the earlier Greek philosophical tradition.

CHAPTER 9

Neoplatonists on the causes of vegetative life

James Wilberding

In the Neoplatonism of late antiquity there was an exciting and revolutionary development in the understanding of the aetiology involved in the generation of living things, and here it will be argued that this extended all the way to the Neoplatonic understanding of the causes of vegetative life. In a way, this should come as no surprise. Hippocratics, Aristotle and Galen had all viewed the processes involved in the generation of plants as analogous to those in the generation of embryos.[1] In fact, the embryo was commonly held to have the life-status of a plant, with the mother taking on the role of the earth, at least at the earliest stages of its generation.[2] As a result, these thinkers saw the same causal models that govern the generation of embryos at work in the generation of plants. Indeed, Galen even advises those who wish to investigate the formation of embryos to begin by looking into the generation of plants. The above-mentioned analogy is certainly part of the motivation behind Galen's counsel, but equally important is that plants are simpler, in terms of both their physiology and their psychology, and thus more perspicuous objects of study. This is what gives us 'hope to discover among the plants [biological] administration in its pure and unadulterated form'.[3] What *is* surprising is the conception of vegetative generation and life that results for Neoplatonists. For, as I shall show here, they ultimately concluded that the vegetative souls of individual plants are not self-sufficient. That is to say, the dependence of individual plants on the earth, in terms of both their generation

[1] Within the Hippocratic corpus see esp. *Genit.* 9.3 (51.2–12 Joly = 7.482 Littré) and *Nat. Puer.* 22–7 (68.19–77.7 Joly = 7.514–28 Littré); Aristotle, e.g. *De gen. animal.* 731b5–8, 740a24–7; Galen, e.g. *De sem.* 94.24–96.21 De Lacy (4.544–6 Kühn). And see Lonie 1981: 211–16.

[2] See references in previous note and Aristotle *De gen. animal.* 736b12–13, 740b8–10, 778b32–5; *SVF* 2.806; Galen *De sem.* 178.15 De Lacy (4.625 Kühn), 194.7–8 De Lacy (4.639 Kühn); Galen *De foet. form.* 66.24–6 Nickel (4.663–4 Kühn), 70.12–13 Nickel (4.667 Kühn); [Plutarch] *Mor.* 907c–d.

[3] See Galen *De foet. form.* 68.9–23 Nickel (4.665–6 Kühn).

and their preservation, extends beyond mere nutritive needs into the psychological domain of their life activities.

In order to see how they arrived at this surprising conclusion, it will be necessary to begin with a brief sketch of Neoplatonic embryological theory, such as it can be found across a wide range of core Neoplatonic authors and texts.[4] This theory may be encapsulated into four theses. (i) First, all Neoplatonists are one-seed theorists: there is no female seed. In this the Neoplatonists were in full agreement with Peripatetic embryology and in opposition to the two-seed theories advanced by the Hippocratics and Galen, though this opposition remains only implicit, as they never even acknowledge the possibility of a female seed.[5] (ii) Second, Neoplatonists universally understand the seed to be a collection of form-principles (*logoi*) corresponding to individual parts of the father (and by extension of the offspring).[6] Since these form-principles are immaterial, they are wholly present in every part of the seed, allowing the seed to be completely homoiomerous.[7] Finally, (iii) these seminal

[4] I defend the claims in the following two paragraphs at greater length in other work currently in progress.

[5] Although the later Platonists do not acknowledge this to be a contentious issue, their commitment to the one-seed theory may be inferred from concise descriptions of conception and insemination that refer to the mother as receiving a seed (in the singular) and herself providing matter. See, e.g., Ammonius *In Porph. isag.* 105.1–3 Busse; Asclepius *In meta.* 345.30–2 Hayduck (cf. *In meta.* 57.36–58,1 Hayduck, 397.16–17 Hayduck, 448.4–5 Hayduck); Damascius *In Parm.* 2.42.9–15 Westerink (2.94.13–16 Ruelle); Porphyry *AG* 2.4 (35.24–25 Kalbfleisch), 3.1 (36.16–17 Kalbfleisch), 10.1 (46.14 Kalbfleisch), 10.5 (47.16–26 Kalbfleisch); Porphyry *De abst.* 4.20 (263.14–16 Nauck); Proclus *In Tim.* 1.300.1–13 Diehl; Proclus *In Parm.* 792.3–15 Steel (cf. Proclus *In remp.* 2.33.16 Kroll, 2.35.24–5 Kroll); Simplicius consistently discusses conception in terms of a single seed and menses (*In de cael.* 101.23–6 Heiberg, 110.5–8 Heiberg, 127.2–3 Heiberg; *In cat.* 244.2–3 Kalbfleisch; *In phys.* 219.29–32 Diels (citing Aristotle), 248.23–249.5 Diels, 313.7–9 Diels, 362.6–7 Diels, 391.25–7 Diels; Syrianus *In meta.* 97.21–4 Kroll. Plotinus *Enn.* v.7 and Philoponus *In GC* 295.24–7 Vitelli appear to be *prima facie* exceptions to this rule, but they can be accounted for (see above, note 4).

[6] For seeds consisting of *logoi* see Plotinus esp. v.9.6.10–24 (see also ii.6.1.10–12, iv.3.10.10–13, iv.9.3.16–18, v.1.5.11–13, v.3.8.4–9); Asclepius *In meta.* 408.8–9 Hayduck; Iamblichus *In Nic. arith.* 82.1–5 Pistelli and Klein; [Iamblichus] *Theol. arith.* 16.4–6 De Falco and Klein (cf. 21.17–19 De Falco and Klein); Michael Psellus *Opusc. psych. theol. daemon.* 32.15 ff. O'Meara; Olympiodorus *In Alc.* 109.24–110.2 Westerink; Philoponus *In phys.* 93.2–5 Vitelli, 320.1–2 Vitelli (and cf. 247.22–9 Vitelli); Simplicius *In cat.* 210.9–10 Kalbfleisch, 306.23–4 Kalbfleisch; and the references in the next two notes.

[7] See Asclepius *In meta.* 38.6 Hayduck and 202.25–6 Hayduck; Damascius *De princ.* 3.55.6–10 Westerink (1.252.1–4 Ruelle) and 3.91.18–23 Westerink (1.274.24–7 Ruelle); Iamblichus *In Nic. arith.* 82.1–5 Pistelli and Klein (cf. 81.23–4); Olympiodorus *In Alc.* 1 109.24–110.1 Westerink; Olympiodorus *In Phaed.* 13.2.27–32 Westerink (p. 169); Philoponus *In de an.* 13.30–4 Hayduck (cf. 238.9–12 Hayduck); Plotinus ii.6.1.10–12, iii.2.2.18–23, iv.7.5.42–8, iv.8.6.7–10, v.9.6.10–13; Porphyry *AG* 7.2 (43.23–4 Kalbfleisch); Porphyry *Sent.* 37.33–9 Brisson et al.; Proclus *In Tim.* 1.396.10–26 Diehl (reporting Porphyry's view with approval = Porphyry *In Tim.* fr. 51 (38.30–39.10) Sodano) and 2.47.22–8 Diehl; Proclus *In Parm.* 754.10–13 Steel, 792.7–9 Steel; Proclus *De dec. dub.* 8.30–41 Isaac; Philoponus *In de an.* 13.26–35 Hayduck; Simplicius *In phys.* 382.15–21 Diels.

form-principles are in a state of potentiality,[8] and (iv) they must be led to a state of actuality by an external cause that possesses these same principles in actuality. This cause is generally identified with the nature of the mother, who is additionally responsible for supplying the matter in the form of menses.[9]

It is these final two theses that establish the Neoplatonic theory as an exciting new development in ancient embryology. On Aristotle's one-seed theory, by contrast, the male seed serves as the formal and efficient cause of embryological development, requiring only matter from the female. Aristotle establishes the self-sufficiency of the male seed as an efficient cause by attributing *actual* motion to it.[10] And even on Galen's two-seed theory, where one might have expected the female to be granted greater causal efficacy in the embryological process, the male seed remains the sole efficient cause, with the female seed more or less demoted to serving as nourishment for the male seed.[11] What is revolutionary, therefore, in the Neoplatonic account of embryology is its placing the female on equal footing with the male in terms of their causal contributions in embryology. This new conceptualisation of the respective contributions of the male and female should be seen as resulting from the application of the Neoplatonic metaphysical framework of procession and reversion to embryology. Within this framework the creation of an offspring consists of two distinct causal moments. In the first moment, procession, an inchoate offspring is generated that is a likeness of its progenitor but in a state of potentiality. The procession from the One, for example, results in the generation of the Pre-Intellect, which is still only potentially the Intellect. The second moment, reversion, is what accounts for this potentiality being led to a state of activity: by reversion the Pre-Intellect becomes the genuine

[8] E.g. Porphyry *AG* 14.3 (54.12–13 Kalbfleisch); Proclus *In Parm.* 792.7–8 Steel; cf. Plotinus III.7.11.23–4. And see the references in the following note.

[9] See Porphyry *AG* 10.5–6 (47.16–48.5 Kalbfleisch) with 14.3 (54.12–15 Kalbfleisch); Proclus *In Parm.* 792.7–18 Steel; Asclepius *In meta.* 404.9–31 Hayduck; Simplicius *In phys.* 313.5–27 Diels and *In cat.* 244.1–4 Kalbfleisch; Elias *In Porph. isag.* 85.3–7 Busse; Philoponus *In de an.* 306.2–8 Hayduck. Somewhat more ambiguous statements to this effect can be found in Syrianus *In meta.* 36.22–30 Kroll; Ammonius *In de int.* 250.6–7 and 26–30 Busse with *In Porph. isag.* 48.3–7 and 104.32–105.8 Busse. There is no comparable explicit statement of the doctrine in Plotinus, but his remarks on plants (discussed below) suggest that he was on the same track. All of these passages are analysed in greater detail elsewhere (see above note 4).

[10] Aristotle *De gen. animal.* 730b15–22. To be sure, Aristotle is prepared to call the menses 'seminal' and in *De gen. animal.* 4.3 outlines his account of how the female, too, may contribute to the form of the offspring, but this in no way affects the point here that the male seed is the efficient cause and sole initiator of embryological development.

[11] See Galen *De sem.* 86.20–5 De Lacy (4.536 Kühn), and see De Lacy's note *ad* 86.24–5 and Nickel 1989: 40–9, esp. 46–7.

Intellect. When this framework is applied to embryology, the theses (iii) and (iv) above follow. The male's emission of a seed is likened to procession, with the form-principles in the seed still being in a state of potentiality, and this potentiality is led to actuality by the mother at conception and throughout the process of gestation. Thus, the male and the female are on a par in so far as each corresponds to one of the two moments of the One's creative activity.

In the light of this new conception of the female's role in embryology we should expect a comparably new understanding of the earth's relation to vegetation, since the Neoplatonists on balance shared the traditional view of the embryo's relationship to the mother being analogous to a plant's relationship to the earth. (I say 'on balance', because, as we shall see below, Porphyry is somewhat anomalous in this respect.) This means that whereas the dominant theory prior to this Neoplatonic revolution held that the earth contributes only material nourishment to plants, just as the mother supplies nourishment to the embryo, the Neoplatonists ultimately arrive at the view that the earth not only supplies nourishment but more importantly makes a psychological contribution to plants only by virtue of which plants may be said to have vegetative souls. What is more, this contribution should account not only for the generation of plants but also for their vital activity throughout their lives since, unlike animals, plants maintain a lifelong relationship to their 'mother'.

With these expectations in mind, let us now turn for confirmation to a selection of passages in which Neoplatonists discuss the relationship that obtains between plants and the earth. This examination will proceed chronologically, beginning with Plotinus, who has relatively little to say about the specifics of *generation* of plants,[12] but who in *Ennead* IV.4.22–7 (treatise 28) seeks clarity on the broader issue of the relationship between the souls of plants and the soul of the earth. His investigation begins with a clearly formulated *aporia*:

> In plants are there two things – the echo, as it were, [of soul] in their bodies and that which supplied this echo, which is called the 'appetitive' part in us and the 'vegetative' in plants? Or is this supplier located in the earth, since there is soul in the earth, with the echo coming from it being located in the plants?[13]

[12] At least at one point he does describe the earth as the generator of plants. See *Enn.* III.8.1.4: τὴν ἐν φυτοῖς φύσιν καὶ τὴν ταῦτα γεννῶσαν γῆν.

[13] Plotinus *Enn.* IV.4.22.1–5 (treatise 28): Ἐπὶ δὲ τῶν φυτῶν ἄρα ἄλλο μὲν τὸ οἷον ἐναπηχηθὲν τοῖς σώμασιν αὐτῶν, ἄλλο δὲ τὸ χορηγῆσαν, ὃ δὴ ἐπιθυμητικὸν μὲν ἐν ἡμῖν, ἐν ἐκείνοις δὲ φυτικόν, ἢ ἐν μὲν τῇ γῇ τοῦτο ψυχῆς ἐν αὐτῇ οὔσης, ἐν δὲ τοῖς φυτοῖς τὸ ἀπὸ τούτου;

Here Plotinus is raising the question of the self-sufficiency of the psychic principle in plants, and he does so by distinguishing the vegetative soul proper from its 'echo' (τὸ οἷον ἐναπηχηθὲν), which is what he more commonly refers to as a 'trace' or 'footprint' (ἴχνος) of soul, though the term 'echo' is actually better suited to his purpose. For while both terms indicate causal derivativeness, 'echo' more clearly expresses the sense of derivativeness that Plotinus is aiming at, which is that the continued existence of the effect is dependent on the continued existence of the cause. He appears confident that individual plants possess at the very least this echo of the vegetative soul but expresses his uncertainty as to whether the source of this echo, the vegetative soul itself, should be attributed to them or rather to the earth. In the latter case, Plotinus' position would appear to be that the echo or trace of the earth's vegetative soul that is present in an individual plant can itself account for all of the plant's biological activities (growth, nutrition, reproduction), but also that this echo is itself not a self-sufficient psychological entity in so far as its causal efficacy as a psychological entity and indeed its very existence is derived from the earth's vegetative soul and thus depends upon its maintaining a connection to the earth.[14]

To be sure, one can find many passages in the *Enneads* in which Plotinus says or suggests that plants possess vegetative souls,[15] and even that the principle of their souls is in some sense located in their roots.[16] Yet the above *aporia* is no mere rhetorical device. When Plotinus ponders whether plants should be seen as having their own vegetative souls or merely traces of the earth's soul, he is not asking a question to which he already knows the answer, as is confirmed at the beginning of IV.4.27, where he returns to the *aporia*:

> If, then, [the earth] supplies the generative [soul] to plants – either [by giving] the generative [soul] itself, or else the generative [soul] is in the earth and it is a trace of this [soul] that is the [soul] in plants – plants would be like flesh that is already ensouled and receive, if [the earth] has [the generative soul], the generative [soul], too, in themselves. And when

[14] For an illuminating recent discussion of this trace of soul in Plotinus see Noble 2013. Noble takes issue with my identification of this trace with *physis* (271 n.36), because the trace is in fact at a lower ontological level than *physis* (or vegetative soul). This much is true, but Plotinus' view appears to be that this trace may be called a vegetative soul in plants in so far as it is the proximate agent responsible for the activities associated with the vegetative soul, which depends on its maintaining a connection to its source, the vegetative soul in the earth. See below.

[15] See, e.g., *Enn.* III.3.7.10–24, III.4.2.22–4, III.8.10.10–14, IV.3.4.26–7, IV.4.28.59–60, IV.7.8⁵.28–36 (on which see Tornau 2005), IV.7.14.6–8, V.2.1.22–7 and V.2.2.4–15.

[16] See, e.g., *Enn.* III.1.4.5–6, III.8.10.10–14, IV.4.1.29–31, IV.7.8⁵.28–36, V.2.2.4–15, VI.8.15.33–6.

> this [generative soul] is present in them, it gives to the body of the plant the very enhancement by which [the plant] differs from what has been cut off and no longer a plant but only wood.[17]

It is striking that Plotinus here continues to hedge on whether the immanent principle of soul in plants should be described as a vegetative soul proper or a trace, but the main point that he is interested in making is that regardless of terminology whatever psychic principle plants possess comes to them from the soul in the earth. This relation of the psychological dependence of plants on the earth is brought out well by his analogy of plants to ensouled flesh, which strongly suggests that he is thinking of plants as parts of a larger organic whole, the earth. Neither can remain ensouled if it is cut off from its psychological source. Moreover, he appears to believe that even if the psychic principle in plants is *sensu stricto* a trace, in some sense plants may still be said to have vegetative souls.[18] This terminological vacillation is easily explained by the difference a plant's connection to the earth can make. The idea would be that as long as the plant is organically connected to the earth, its trace of soul is effectively equivalent to a vegetative soul.[19]

Later Neoplatonists address this issue of the earth's psychological contribution to plants with more precision and with a greater emphasis on the earth's role in the generation of plants. This development is probably due in no small part to the influence of Porphyry's focused account of embryology in his *To Gaurus on how Embryos are Ensouled* (*Ad Gaurum*), in which the Neoplatonic embryological theory set out above receives its first systematic articulation. If so, it is interesting to note that Porphyry does not himself explicitly subscribe to the view that the earth makes a

[17] *Enn.* IV.4.27.1–7: Εἰ οὖν τοῖς φυτοῖς δίδωσι τὴν γεννητικήν – ἢ αὐτὴν τὴν γεννητικήν, ἢ ἐν αὐτῇ μὲν ἡ γεννητική, ταύτης δὲ ἴχνος ἡ ἐν τοῖς φυτοῖς – καὶ οὕτως ἂν εἴη ὡς ἡ σὰρξ ἔμψυχος ἤδη καὶ ἐκομίσατο, εἰ ἔχει, καὶ τὴν γεννητικὴν ἐν αὐτοῖς τὰ φυτά. ἐνοῦσα δὲ δίδωσι τῷ σώματι τοῦ φυτοῦ ὅπερ βέλτιον, ᾧ διαφέρει τοῦ κοπέντος καὶ οὐκέτι φυτοῦ, ἀλλὰ μόνον ξύλου. The exact sense of the apodosis is difficult to determine, in part due to the qualification εἰ ἔχει in line 4. Theiler emends εἰ to ἤ. I follow Brisson in taking the earth to be the subject of ἔχει in line 4. Armstrong sees plants as the implicit subject, but in the next line Plotinus makes clear that plants *do* have this soul (ἐνοῦσα). It is tempting to emend εἰ ἔχει to τῷ ἴχνει: 'and receive by means of this trace the generative soul, too, in themselves'.

[18] The implicit subject of ἐνοῦσα in line 5 must be ἡ γεννητική, and here I agree with Brisson (Brisson and Pradeau 2003–10, IV: 259 n.259 *ad* IV.4.27.1) that this term is being employed broadly to refer to what is responsible for the activities of nutrition, growth and reproduction. In the following chapter Plotinus also describes trees as possessing τὸ φυτικόν (IV.4.28.59–60), despite his unwillingness to decide between these two alternatives here.

[19] Explained in this manner, Plotinus' understanding of the psychological status of plants is very similar to Porphyry's understanding of that of embryos in the *Ad Gaurum*. See Wilberding 2008b: 415–16.

psychological contribution to the generation of plants. This is because Porphyry rejects the traditional analogy between the embryo's relation to the mother and a plant's relation to the earth. Instead, in *Ad Gaurum* 10.1–2 he compares the embryo's relation to the womb to that of a scion grafted onto a rootstalk. The generation of the embryo, he says, is

> as much as possible like that of plants, with the father emitting the seed and the mother collecting it in order to nourish it – not merely in the manner of earth for the supply of nourishment nor in the manner in which she furnishes milk alone to offspring after birth – but rather resembling in a way [buds] that are being grafted and added onto [the base of another plant]: <the> power in the womb forms a natural unity with the seed, and it is by this power that the base [*viz.* the womb], which has its own nature, and the part being added on [*viz.* the seed], which has its proper nature, make up a kind of mixture in order to form the single nature of the part that has been successfully added on [*viz.* the embryo].[20]

In this passage Porphyry is working towards establishing thesis (iv) that the female's nature or vegetative soul is the agent responsible for initiating and maintaining the process of the embryo's formation, and he revises the traditional comparison of the embryo to a plant rooted in the earth in favour of a grafted scion in order to illustrate this causal relationship better. As he envisions it, the soul principle of the rootstalk forms a psychological unity with the soul principle in the scion, and this relationship is what accounts for the vegetative activities of the grafted scion. Elsewhere in the *Ad Gaurum* he describes this relationship in terms of the former soul principle serving as a 'captain' (κυβερνήτης) over the latter, both setting it in motion and providing a contribution to its formal development.[21] This is precisely the type of relationship that later Neoplatonists see obtaining between plants and the earth, yet Porphyry carefully distinguishes between the rootstalk–scion relationship and the earth–plant relationship apparently because he sees this psychological union obtaining only in the former case. The latter case, as he describes it, involves merely the supply of nourishment. In other words, he is conceiving of the plant's relationship to the earth along the traditional

[20] Porphyry *AG* 10.1–2 (46.13–20 Kalbfleisch): εἰ μὴ πᾶσα ἡ τῶν ἐμβρύων γένεσις τῇ τῶν φυτῶν ὡς ἔνι μάλιστα ἔοικε, τοῦ μὲν πατρὸς καθιέντος τὸ σπέρμα, τῆς δὲ μητρὸς συμβαλλομένης εἰς τὴν ἔκθρεψιν οὐ μόνον γῆς τρόπον εἰς χορηγίαν τῆς τροφῆς οὐδ᾽ ὃν τρόπον κυηθεῖσι χορηγεῖ μόνον τὸ γάλα, παραπλησίως δέ πως τοῖς ἐγκεντριζομένοις καὶ ἐνοφθαλμιζομένοις, συμφυομένης τῷ σπέρματι <τῆς> [following Kalbfleisch's suggestion in the critical apparatus] ἐν τῇ μήτρᾳ δυνάμεως, ὑφ᾽ ἧς τό τε ὑποκείμενον ἔχον ἰδίαν φύσιν τό τε ἐνοφθαλμιζόμενον ἔχον τὴν οἰκείαν, κρᾶσίν τινα συνίστησιν εἰς μίαν τοῦ ἐνοφθαλμισθέντος φύσιν.

[21] See esp. Porphyry *AG* 10.4–6 (47.5–48.8 Kalbfleisch) and 14.3 (54.12–15 Kalbfleisch).

Aristotelian and Galenic lines. This, however, does not sit well with his own conception of seeds. For he clearly subscribes to (i)–(iv) and considers the seed to be a collection of form-principles in a state of potentiality requiring actualisation from an external agent possessing them in actuality, which raises the question of how the form-principles in plants' seeds are brought to actuality. Porphyry does not address this issue in the *Ad Gaurum*, perhaps because his focus here is on human embryology. If he had, he might have reconsidered his Aristotelian conception of the earth's relation to plants and recast the earth in a role comparable to that of the rootstalk. This is in any case exactly what subsequent Neoplatonists did.

Proclus, for example, launches into an examination of earth in his commentary on Plato's *Timaeus*,[22] stirred by Plato's labelling of the earth not only as 'our nurturer' (τροφὸν ἡμετέραν) but also as 'the first and oldest of the gods generated within the cosmos' (πρώτην καὶ πρεσβυτάτην θεῶν ὅσοι ἐντὸς οὐρανοῦ γεγόνασιν).[23] Proclus relates these accolades to the contributions that the earth makes to all living things on the earth, underlining that this is not limited to material contributions but includes a psychological component as well, and he singles out plants as being particularly dependent on the earth for their life force: 'For how can it be that plants live as long as they remain in the earth but die once they have been detached from it, unless this mass of earth is full of life? Rather, it must be universally accepted that wholes are ensouled prior to their parts.'[24] And in his commentary on Plato's *Parmenides* Proclus also integrates the earth into his account of the generation of plants. This occurs at a point in the commentary where Proclus is aiming to establish the necessity of there being Forms of all living things, which leads him to consider the role of higher causes in biological generation. He begins with a brief account of human reproduction which follows the Neoplatonic embryological theory outlined above.[25] Proclus places particular emphasis on theses (iii) and (iv), insisting that the female's nature must be responsible for leading the form-principles in the seed to a state of actuality, and that in order to do so her nature must possess these form-principles in actuality.[26]

[22] Proclus *In Tim.* 3.134.9–136.29 Diehl. [23] Plato *Tim.* 40b8–c3.

[24] Proclus *In Tim.* 3.135.25–28 Diehl: πόθεν γὰρ ἐν γῇ μὲν ἔτι μένοντα τὰ φυτὰ ζῇ, ἀποσπασθέντα δὲ νεκροῦται, μὴ καὶ τοῦδε τοῦ ὄγκου τῆς γῆς ζωῆς ὄντος πλήρους; δεῖ δὲ καὶ καθ' ὅλου λαμβάνειν, ὅτι τὰ ὅλα πρὸ τῶν μερῶν ἐψύχωται.

[25] Proclus *In Parm.* 791.21–795.6 Steel. For a more detailed examination of the implications of this passage on human reproduction, see above note 4.

[26] Proclus *In Parm.* 792.7–18 Steel.

He then extends this thesis, first to other species of animal,[27] and then to plants and the earth:

> But does the nature in animals possess form-principles, but not [the nature] in plants, too? No, in these too there is ordered generation, and the lives of plants reveal how they are brought to perfection in accordance with ordered causes. It is clear, then, that by the same argument their natures, too, already contain the form-principles that are made visible.[28] But let us now ascend to the single nature of the earth, which in like manner generates all things that 'breathe or creep upon the earth'. Does not the earth's nature, then, possess in a much prior manner the form-principles of things that grow? What other source [of form-principles] could there be, especially (καὶ) in those cases in which we observe generation being brought to completion from something of a different kind, e.g., in cases of generation from putrefaction? What is the source of generation in these cases? And how is it that at different times different species of plants grow in the same place without human intervention? Clearly it is because the entire nature [of the earth][29] possesses the form-principles and creative powers of all these things in itself.[30]

Here Proclus states in no uncertain terms that the earth's nature is responsible for generating plants. The two phenomena that Proclus selects to demonstrate the necessity of the earth's involvement, namely so-called 'spontaneous' generation and the local fluctuation of vegetative species, might give the impression that the earth is actively involved only in certain anomalous cases of plant generation, namely those in which no seeds are involved, but this impression is misleading, since Proclus makes clear that the earth is involved in the generation of '*all* things that breathe or creep upon the earth'.[31] These two phenomena are singled out because the

[27] Proclus *In Parm.* 792.19–26 Steel.

[28] τοὺς φαινομένους λόγους: cf. τὸ φαινόμενον εἶδος at 792.13 Steel.

[29] I agree with d'Hoine 2006: 54 and Zekl 2010: 228 in taking this to refer to the nature of the earth. Cf. Luna and Segonds 2007–11, III.2: 130: 'La mention de la nature universelle est un peu étonnante, car on s'attendrait plutôt à lire "la nature de la terre."' As they rightly note, Proclus is working his way up to universal nature, which he arrives at only in 793.16 Steel.

[30] Proclus *In Parm.* 793.26–793.11 Steel: Ἀλλ' ἆρα ἐν μὲν ζώοις ἡ φύσις ἔχει τοὺς λόγους, οὐχὶ δὲ καὶ ἐν φυτοῖς; ἢ καὶ ἐν ἐκείνοις ἡ τάξις τῆς γενέσεως καὶ οἱ βίοι τῶν φυτῶν δηλοῦσιν ὅπως κατὰ τεταγμένας αἰτίας ἐπιτελοῦνται· δῆλον οὖν ὡς κατὰ τὸν αὐτὸν λόγον καὶ ἐκείνων αἱ φύσεις προσειλήφασι τοὺς φαινομένους λόγους. Ἀλλ' ἐπὶ τὴν μίαν ἀναδράμωμεν λοιπὸν φύσιν τὴν τῆς γῆς, τῶν πάντων γεννητικὴν ὁμοίως, ὅσα γαῖαν ἐπιπνέει τε καὶ ἕρπει. Ταύτην τοίνυν οὐ πολλῷ πρότερον ἔχειν τοὺς τῶν φυομένων λόγους; ἢ πόθεν καὶ ἐφ' ὧν μὴ ἐκ τοῦ ὁμοίου τὴν γένεσιν ἐπιτελουμένην ὁρῶμεν, οἷον τῶν ἐκ σήψεως; πόθεν δ' οὖν ἐπὶ τούτων ἡ γένεσις; πῶς δὲ ἐν ταὐτῷ τόπῳ φυτῶν ἄλλοτε ἄλλα γένη φύεται χωρὶς ἀνθρωπινῆς ἐπιμελείας; Ἢ δῆλον ὡς τῆς ὅλης φύσεως λόγους ἐχούσης καὶ ποιητικὰς τούτων ἁπάντων ἐν ἑαυτῇ δυνάμεις.

[31] This is a very selective quotation from Homer. In fact, what Homer says is 'Man is the most wretched of all that breathes and creeps upon the earth'. Cf. [Plato] *Axiochus* 367d8–e1; Plutarch *Mor.* 496b and 500b.

earth's agency would seem to be nearly indisputable in non-seminal cases of generation.[32] Unfortunately, Proclus' view of how the earth is involved in normal seminal plant reproduction is left obscure in so far as he makes no explicit mention of the *seeds* of plants and their development. Nevertheless, given his description of human generation above, we should expect him to say that the seeds of plants also contain their form-principles only in potentiality and therefore require the earth as an agent of actualisation, and this would certainly go a long way towards explaining why the earth needs to have the form-principles of plants 'in a much prior manner'.[33]

It is in Simplicius' commentary on Aristotle's *Physics* that we find the clearest expression of this doctrine. In his discussion of Aristotle's remarks on formal causes in *Physics* 2.3 Simplicius begins to address the manner in which nature functions as a cause in biological generation.[34] Here we find him advancing the four core theses of Neoplatonic embryology,[35] and he places particular emphasis on the potentiality inherent in the seed and the necessity of an external agent to actualise the seed. As he says, 'everything that comes to be in a state of actuality from a state of potentiality must be led to the state of actuality by something in a state of actuality'.[36] This leads him to conclude that the earth's nature works together with the natures of plants in the same way in which the female's nature works together with the male's nature in human generation:

> The true and proximate creative/efficient cause in the case of animals is the maternal nature and the paternal nature, while in the case of plants it is the nature of the wheat and of the earth, since the form pre-exists in actuality in the father and the mother and in the form-principles established in actuality within the earth, by which what is in a state of potentiality is led to actuality.[37]

[32] Hence my translation of καὶ in 793.7 Steel as 'especially'. We must bear in mind here that while the Neoplatonists accepted the phenomena associated with what Aristotle terms 'spontaneous' generation, they denied that such cases of generation were genuinely spontaneous. For Neoplatonists no form of life, no matter how far down the *scala naturae*, can emerge from mere material causes without the guiding influence of form-principles. See Wilberding 2012.

[33] A further problem is that Proclus obscures the special role that the earth must play in plant reproduction by insisting that the earth is active in the generation of animals, too, as was also the case in *In Tim.* 3.134.9–136.29 Diehl. I explore some of these problems elsewhere (see above note 4).

[34] Simplicius *In phys.* 310.20–314.24 Diels; cf. Aristotle *Phys.* 194b26–29.

[35] There are some difficulties in piecing together Simplicius' remarks on human generation that do not need to be examined here, since his views on vegetative generation are unambiguous. I discuss Simplicius' views on human generation elsewhere in more detail: see Wilberding 2008b and above, note 4.

[36] Simplicius *In phys.* 312.34–6 Diels. Cf. Simplicius *In de cael.* 168.2–15 Heiberg.

[37] Simplicius *In phys.* 313.9–13 Diels: τὸ δὲ ποιητικὸν κυρίως καὶ προσεχῶς αἴτιον ἐπὶ μὲν τῶν ζῴων ἡ μητρικὴ φύσις ἐστὶ καὶ ἡ πατρική, ἐπὶ δὲ τῶν φυτῶν ἡ τοῦ πυροῦ καὶ τῆς γῆς, τοῦ εἴδους ἐνεργείᾳ

For Simplicius the earth serves as the mother of plants not merely in the sense of providing them with nourishment; more importantly, the nature of earth, containing the form-principles of all vegetation, serves as the agent of actualisation in seminal generation. Elsewhere he makes clear that he sees the earth's psychological contribution to vegetative life persisting throughout their lives. We are justified in attributing life and soul to the earth, even though it appears to be inert, on account of its vital contribution to plants.[38] Like Plotinus and Proclus, he views plants as parts of the earth that cannot survive the separation from their whole, since their life derives from the whole,[39] yet this dependence relation does not inhibit Simplicius from classifying plants as being themselves ensouled.[40]

In Damascius' *Problems and Solutions Concerning First Principles* (*De principiis*) we find a particularly striking account of this understanding of the plant–earth relationship. He introduces it in the course of setting out a series of *aporiai* surrounding one of the central tenets of later Neoplatonic metaphysics, namely that whatever proceeds from a higher cause must proceed while remaining in that cause.[41] Damascius here is searching for a way to elucidate this paradoxical tenet without trivialising or falsifying it, and it is in this vein that he examines the earth–plant relationship as a possible analogy:

> Perhaps what is proceeding or has already proceeded has a cause in its producer (ἐν τῷ παράγοντι), and this is what it means to remain in that: the proceeding thing has its root in it [*viz.* the producer][42] like a root from which the proceeding thing grows. For even the plant shoots up from the earth while remaining in the earth *via* its root. But this suggestion, when subjected to careful scrutiny, appears unreasonable. For the cause is not itself the thing being produced by the cause. Rather the cause is primary and the thing produced is derivative.[43] For the cause of what is derivative is seated in what is primary. For even the cause of what is produced is a kind

προϋπάρχοντος ἔν τε τῷ πατρὶ καὶ τῇ μητρὶ καὶ ἐν τοῖς τῆς γῆς κατ' ἐνέργειαν ἑστῶσι λόγοις, καθ' οὓς τὰ δυνάμει εἰς ἐνέργειαν ἄγεται.

[38] See Simplicius *In de cael.* 489.22–5 Heiberg: εἰ δέ, ὅτι κατὰ τόπον ἀκίνητος ἡ γῆ, διὰ τοῦτο ἄζως αὐτῷ δοκεῖ καὶ ἄψυχος, πρῶτον μὲν αἰσχύνεσθαι δεῖ, εἰ τὰ μὲν φυτὰ τὰ ὑπὸ τῆς γῆς ζωούμενα ζῆν λέγομεν καὶ ἔμψυχα εἶναι, αὐτὴν δὲ τὴν γῆν ἄζων καὶ ἄψυχον.

[39] See Simplicius *In phys.* 628.7–11 Diels.

[40] E.g. Simplicius *In de cael.* 3.2–4 Heiberg, 16.12–17 Heiberg; Simplicius *In phys.* 262.13–17 Diels, 1061.4 Diels, 1257.30–1258.3 Diels.

[41] See Damascius *De princ.* 2.117.4–123.12 Westerink. Cf. Proclus *El. Th.* § 30.

[42] At 2.118.21 Westerink prints ἐν αὑτῷ for the MSS's (and Ruelle's) ἐν αὐτῷ, incorrectly in my opinion. As the ἐν τῷ παράγοντι (2.118.19–20) and the ἐν αὐτῇ (viz. the earth) (2.118.22) make clear, the cause is 'rooted' in the producer that remains and not in that which proceeds.

[43] Not, as Combès would have it, 'mais ce qui produit est deuxième, tandis que ce qui est produit est premier'. See Kühner and Gerth 1904: 264 Anmerkung 1.

> of substance of the producer, and is not a part of the produced but of the producer, nor is it analogous to the root, but to the form-principle of the entire plant, the very form-principle that as a whole is substantiated in the nature of the earth (ἐνουσιωμένῳ τῇ φύσει τῆς γῆς).[44]

The suggestion under examination is that the rootedness of plants can provide a coherent illustration of this paradoxical causal tenet: What proceeds nevertheless remains in its producer because it has its cause rooted in its producer, just as a plant grows and separates itself from its producer, the earth, because it has its cause, namely its roots, firmly planted within the earth. This suggestion should come as no surprise, as the root's status as a 'principle' of the plant can be traced back to Plato, who also uses it as an analogy to explain higher psychological connections. In contrast to Aristotle, who designated the roots as the 'upper' part of the plant, not because he viewed them as the seat of the plant's soul, but only because they are analogous to the heads of animals in terms of their being the part with which nourishment is drawn in,[45] when Plato describes human beings in the *Timaeus* as 'celestial plants' he establishes the connection between the root and the principle of soul that would seem to prepare the way for seeing the root as a metaphysical mediator to a still higher principle.[46] This metaphysical conception of a root as a mediating principle that links an inferior being to the superior being that generated it develops into a standard metaphor in Neoplatonic metaphysics, where it becomes a commonplace to describe one ontological level as being 'rooted' in its parent cause, upon which its ontological preservation depends.[47]

[44] Damascius *De princ.* 2.118.19–119.6 Westerink: μήποτε οὖν τὸ προϊὸν ἢ προεληλυθὸς αἰτίαν ἔχει ἐν τῷ παράγοντι· καὶ τοῦτό ἐστι τὸ μένειν ἐν ἐκείνῳ, τὸ ῥίζαν ἔχειν ἐν αὐτῷ [with MSS (see above note 42). Westerink emends to αὑτῷ] τὴν αἰτίαν ἀφ᾽ ἧς ἀποφύεται· καὶ γὰρ τὸ φυτὸν ἀπὸ γῆς ἀναβλαστάνει μένον ἐν αὐτῇ κατὰ τὴν ῥίζαν. ἢ καὶ τοῦτο φανεῖται ἄλογον, εἴ τις ἀκριβῶς αὐτὸ βασανίσειεν· ἡ γὰρ αἰτία οὐκ ἔστιν αὐτὸ ἀπὸ τῆς αἰτίας παραγόμενον, ἀλλὰ τὸ μὲν δεύτερον, τὸ δὲ πρῶτον. ἐν γὰρ τῷ πρώτῳ ἵδρυται τοῦ δευτέρου ἡ αἰτία· καὶ γὰρ οὐσία τις τοῦ παράγοντος ἡ τοῦ παραγομένου αἰτία καὶ οὐ τούτου μέρος, ἀλλ᾽ ἐκείνου, οὐδὲ ἀναλογεῖ τῇ ῥίζῃ, ἀλλὰ τῷ λόγῳ τοῦ ὅλου φυτοῦ ὅλῳ γε ὄντι καὶ ἐνουσιωμένῳ τῇ φύσει τῆς γῆς.

[45] Cf. Aristotle *De an.* 412b3–4 and *IA* 705a29–b8.

[46] 'For it is from heaven, the place from which our souls originally grew (ὅθεν ἡ πρώτη τῆς ψυχῆς γένεσις ἔφυ), that the divine part suspends (ἀνακρεμαννὺν) our head, i.e., our root, and so keeps our whole body erect' (*Tim.* 90a7–b1). This metaphor of human beings as plants rooted in a different element should be compared to the Hippocratic testimony recorded in the Anonymous Londinensis papyrus (5.35–6.44, translated and discussed in Nutton 2004: 58–9), which describes human beings as plants rooted in the air by their nostrils.

[47] Some examples: Damascius *De princ.* 1.76.11–14 Westerink, 1.99.10–11 Westerink, 1.100.11–14 Westerink, 1.109.5–7 Westerink; Damascius *In Parm.* 2.98.13–14 Westerink; [Iamblichus] *Theol. arith.* 32.13 De Falco and Klein; Proclus *In Parm.* 1116.12–13 Steel, 1193.1–5 Steel; Proclus *El. Th.* § 144 (126.28–9 Dodds); Proclus *In Tim.* 1.209.21–4 Diehl, 1.210.8–9 Diehl, 1.247.10–13 Diehl; Simplicius *In phys.* 628.12–14 Diels. Proclus appears to even see the root as an instance of the

So the rootedness analogy is a worthy candidate for Damascius' consideration, yet he urges us to conclude that, as an illustration of the paradoxical thesis, it is not completely satisfactory. His dissatisfaction lies in the conception of the root as a *cause* of the plant. This won't do because the root is a part of the plant; that is, it is a part of the product itself and so cannot be the cause of that product. Rather, the cause of the product must be a part of the producer, or as he says, it should be 'a kind of substance' (οὐσία τις) of the producer. In this connection he explains to us what the genuine cause of a plant is, namely the form-principle (λόγος) of the entire plant, and he corrects our understanding of plants' relationship to the earth in terms of this form-principle. For he says this form-principle is 'substantiated' (ἐνουσιωμένω) in the earth's nature.[48] His meaning appears to be that the plant's form-principle, which is responsible for creating the plant and maintaining its existence, is strictly speaking a part of the earth and not of the plant; and this, I would suggest, is best understood along the same lines that we saw above. Because the earth possesses the form-principles of all vegetation in actuality, it is the agent that both leads the potential form-principles in seeds to actuality and keeps them in this state of actuality.

The lesson to be learned from all of these passages would appear to be that according to the Neoplatonists separating whole or parts of plants should render them lifeless and unable to grow or nourish themselves, and so by way of a conclusion, we should consider the glaring empirical counter-example, namely, that plants do continue to live when separated from the earth. After all, we routinely remove plants from the earth and then after some time replant them in another location or even in pots of soil, and these relocated plants appear to be no less alive. Moreover, new plants can even be generated by taking slips and shoots from existing plants and planting them in the soil. Such phenomena urge us to reject any claim to the effect that the life status of individual plants is wholly dependent on

so-called 'law of mean terms'. It mediates between the earth and the plant as a part of the plant that is itself 'earthy', which for Proclus follows the pattern of his own henology, according to which each divinity is linked to the One via a henad (*In Parm.* 1050.6–12 Steel). This new conception of roots also leaves noticeable traces in the discussions of Aristotle's remarks on plants in the commentary tradition. When Philoponus, for example, comes to Aristotle's claims about the roots of plants being analogous to the head of animals (*De an.* 416a4), he goes well beyond the Aristotelian text in his comments, explaining that the root houses the plant's principle of motion (*In de an.* 276.29–31 Hayduck). Likewise, in [Simplicius] *In de an.* 112.27–9 Hayduck, the roots are counted among the ἡγεμονικοῖς μορίοις.

48 Ahbel-Rappe's translation – 'since it is *pervaded* by the nature of the earth' – obscures this metaphysical relation. Cf. Combès' 'substantialisée'. Cf. Damascius' use of συνουσίωσθαι elsewhere, e.g. *De princ.* 1.51.15–23 Westerink, 1.105.20–5 Westerink, 2.16.7–11 Westerink.

their maintaining a connection to the earth. Aristotle was well positioned to account for these phenomena. Since on his view individual plants possess their own vegetative souls which are wholly present throughout their bodies and require the earth only for nourishment, detached (parts of) plants should be expected to survive as long as they can make do without nourishment.[49] In fact, the Neoplatonists also acknowledge these phenomena,[50] and they have the resources to explain them. Plotinus, to take one example that we have seen, describes plants as possessing a 'trace' or 'echo' of soul, but one of the features of his theory of soul traces is that, although not entirely self-sufficient, they do persist for a limited time in the body after the connection to their source has been severed. This can be witnessed in some of his reflections on the consequences of the soul's separation from the human and other animal bodies at death:

> Why, then, supposing that the body is like something warmed, but not like something illuminated, does it not have any trace of life when the other soul has gone out of it? It does have it for a short time, but it fades quickly, just as the things which are warmed when they go away from the fire. There is evidence for this in the growth of hair on corpses, and the growth of their nails, and the living creatures which move for a long time after they have been cut in two; for this is probably the trace of life still present in them.[51]

Given that the vegetative soul trace in animals, even when severed from the soul that sets it in motion, persists for a short time, during which it continues to perform its characteristic activities of growth, nourishment and reproduction, we should expect the same to hold of the vegetative soul traces in plants separated from the earth and parts of plants cut off from

[49] See *De an.* 411b19–30 and 413b16–21 on the conclusions to be drawn from plants' and insects' survival of division. In *PN* 467a18–30 Aristotle adds that plant cuttings can regenerate lost organs, which is not the case with insects.

[50] See, e.g., Plotinus IV.3.8.44–60 (cf. III.3.7.11–28) and V.2.2.10–18; Philoponus *In de an.* 167.9–17 Hayduck, 200.10–201.32 Hayduck, 238.1–239.6 Hayduck; [Simplicius] *In de an.* 63.3–27 Hayduck, 79.13–27 Hayduck, 101.25–7 Hayduck, 101.35–102.2 Hayduck. In some of these passages the detached part is simply said to possess soul, which would seem to be in serious tension with the claims above regarding the necessity of maintaining a connection to the earth. But this talk of soul is best understood as a reference to *traces* of soul. Themistius, for example, in *In de an.* 31.18–19 Heinze describes the cut-off parts of plants and insects as δοκεῖ τὴν αὐτὴν ἔχειν ψυχὴν τῷ εἴδει, but later he prefers to describe them as having only a trace (ἴχνος) of soul (*In de an.* 45.34–6 Heinze).

[51] Plotinus *Enn.* IV.4.29.1–7 (Armstrong translation): Πῶς οὖν, εἴπερ τῷ θερμανθέντι τὸ σῶμα ἔοικεν ἀλλ' οὐ τῷ φωτισθέντι, ἐξελθούσης τῆς ἄλλης ψυχῆς οὐδέν τι ζωτικὸν ἔχει; Ἢ ἔχει ἐπ' ὀλίγον, ἀπομαραίνεται δὲ θᾶττον, ὥσπερ καὶ ἐπὶ τῶν θερμανθέντων ἀποστάντων τοῦ πυρός. Μαρτυροῦσι δὲ καὶ τρίχες φυόμεναι ἐπὶ τῶν νεκρῶν σωμάτων καὶ ὄνυχες αὐξόμενοι καὶ ζῷα διαιρούμενα ἐπὶ πολὺ κινούμενα· τοῦτο γὰρ τὸ ἔτι ἐγκείμενον ἴσως.

their wholes.[52] In the commentary on Aristotle's *On the Soul* that has come down to us under Simplicius' name,[53] we find an example of a Neoplatonic author negotiating a comparable account of these phenomena without specifically invoking the terminology of traces. Although at certain points in the commentary this author freely adopts Aristotle's language of the plant cutting possessing soul,[54] at one point he tweaks Aristotle just enough to re-establish the earth–plant dependency relationship, namely when Aristotle states that the soul in each plant is actually one but potentially many.[55] The commentator takes this opportunity to explain that this potential plurality of vegetative souls becomes an actual plurality only once the connection to the earth has been restored. He acknowledges that the soul of the whole plant is present in the cutting 'to some extent' (μέχρι τινὸς), but at the same time he significantly reduces the psychological contribution it makes to the cutting. Rather than directly providing the cutting with vegetative soul, it has been demoted to a mere precondition of the cutting's receiving vegetative soul. As he puts it, it 'makes the cutting suitable for the reception of its own soul when it has been planted'.[56]

[52] Cf. also Plotinus *Enn.* III.4.6.40–5, on which see Wilberding 2012: 210–11. We can find Philoponus putting forth a similar view in *In de an.* 17.9–19 Hayduck. This same account could presumably be extended to account for the prolonged lives of potted plants. The earth in the pot retains enough soul to serve as the plant's source of life for some limited time, after which new soil would be required. There is, however, no evidence that the Neoplatonists concerned themselves with such matters.

[53] On the authorship of this commentary see the introductions in Blumenthal 2000; Steel 1997; Steel and Ritups 2013; and Urmson and Lautner 1995.

[54] Cf. [Simplicius] *In de an.* 63.3–27 and 79.13–27 Hayduck.

[55] *De an.* 413b18–19.

[56] [Simplicius] *In de an.* 101.35–102.2 Hayduck: Οὐ τῆς μιᾶς εἰς πλείους ὁμοίας μεριζομένης (μεριστὴ γὰρ ἂν εἴη ὡς ὁμοιομερές τι σῶμα), ἀλλὰ τῆς ἐν τῷ φυτῷ καὶ τῷ ἀποτμηθέντι παρούσης κλάδῳ μέχρι τινὸς καὶ ἐπιτήδειον αὐτὸν ποιούσης πρὸς ἰδίας ψυχῆς ὑποδοχήν, ἐπειδὰν φυτευθῇ. δυνάμει οὖν πλείους αἱ ἐν ἑνὶ φυτῷ ὡς διαιρεθέντι κατὰ κλάδους καὶ φυτευθέντι πλείους ἐνυπάρξουσαι.

CHAPTER 10

Astrology and the will in Porphyry of Tyre

Aaron P. Johnson

As the author of an *Introduction to Ptolemy's Tetrabiblos* and a work on oracles evincing a substantial interest in the relationship between horoscope casting and divination, the third-century Platonist Porphyry of Tyre has frequently been deemed a philosophical promoter and legitimiser of astrology for some if not all of his intellectual career.[1] This assessment does not, however, sit well with a position he developed in the less well-known work *On Free Will* (literally, *On What is Up to Us*),[2] which survives in fragments preserved by Stobaeus, as well as discussions and scattered indications in other works (especially the *Letter to Anebo* and the *Against Nemertius*). This position prizes the freedom of human agency, thus placing a significant moral responsibility upon the soul, and offers a much more critical approach to astrology.

The argument can be (and has been) made that Porphyry possessed a broadly coherent vision of the relationship of astrology and human agency throughout his corpus and that his *Introduction to Ptolemy's Tetrabiblos* and *Philosophy from Oracles* do not straightforwardly endorse a belief in the causative effect of the stars and need not, therefore, be relegated to an early 'superstitious' phase of Porphyry's intellectual development.[3] Assuming this coherency thesis to be correct, a more precise delineation of Porphyry's articulation of agency, especially in the *On Free Will*, is necessary in order

[1] On Porphyry's conception of astrology in these works see Johnson 2013a: 113–21 (on the *Phil. orac.*), 159–164 (on the *Intro. Ptolem. Tetrab.*). Astral phenomena would find a place in other works as well: e.g. *De antro nymph.* 22–8; *Comm. Tim.* frs. 12 and 79 Sodano.

[2] It may be objected that a concept of 'free will' is not necessarily equivalent to a concept of 'what is up to us' in antiquity. While the objection is well grounded (especially before the late Stoics, such as Epictetus), the two concepts are arguably identifiable in Porphyry's works, this one in particular. I have thus opted for *On Free Will* as an alternative to the more unwieldy *On What Is Up to Us*. The literature on the development of a concept of the will in antiquity is large; I have found particularly illuminating the following: Kahn 1988; Bobzien 1998; Sorabji 2004a; see also Dihle 1982; Frede 2011.

[3] *Pace* Amand 1973: 163–9. For general formulations of dividing more or less sharp pre-/post-Plotinian stages in Porphyry's thought see Bidez 1913; Hadot 1960; Romano 1979. For emphasis on broader coherence see Smith 1987; Johnson 2013a.

to appreciate his understanding of the nature of the soul in its embodied state below the stars and to identify his contribution to the development of a notion of free will in late antiquity. Though this work now survives in only a few fragments (all preserved in the *Anthologia* of Stobaeus),[4] there are significant hints that Porphyry pushed exploration of what we might call the freedom of the will[5] in ways distinct from those of Plotinus and the Middle Platonists.

Alcinous, Pseudo-Plutarch's *On Fate* and Maximus of Tyre had begun to develop Platonic formulations of a position that sought to protect the freedom of the soul against notions of causal determinism or astral fatalism, while Plotinus had made an incisive and innovative advance in bringing together the cluster of concepts of freedom, the voluntary, self-determination and will in his treatise on the problem of whether the One had free will (*Enn.* VI.8).[6] In the sparse fragments of his *Against Nemertius*, Porphyry, too, had united these concepts as part of his defence of a doctrine of providence.[7] That work may have been written under the influence of Plotinus' treatise.[8] The *On Free Will* marks a greater independence, however – and this in spite of the fact that it was clearly written during or some time after his time in Rome under Plotinus.[9] It may even have been composed after Plotinus' death in 270, and thus after one of Plotinus' last treatises, *On Whether the Stars are Causes* (*Enn.* II.3 (treatise 52)), had been composed.

[4] Unless otherwise noted, all fragment numbers follow those of Smith 1993. Wilberding 2011a: 123–4 may be correct in his argument that the fragments attributed to a commentary on the *Republic* derive instead from the *On Free Will* (after all, the content of those fragments is entirely limited to the Myth of Er). Since Proclus, our source for those fragments (which are only paraphrases and not verbatim fragments proper), does not name a particular work of Porphyry, it may also be the case that he draws on more than one Porphyrian treatise: frs. 184, 186 and 187 can easily be identified with the material of *On Free Will* (especially since fr. 187 has close parallels to *On Free Will* fr. 271, though probably only a few lines of the former should be assigned to Porphyry himself, rather than to Proclus' own elaborations); but fr. 182 could come from the *On the Styx* (and with it fr. 183, which seems to belong to a Porphyrian passage contiguous with the material in fr. 182, since both directly critique the Epicurean Colotes); frs. 185 and 185a could come from his *Comm. Tim.* (where material from the *Rep.* could be invoked to supplement his discussion of the *Timaeus*); and even fr. 186 could derive from the *Comm. Tim.* (which also discusses the relationship between the moon and the intellect), rather than the *On Free Will*. For a critique of the attempt to elide the fragments of the *Comm. Rep.* with those of the *On Free Will* see Romano 1979: 170–1.

[5] For further conceptual distinctions see Bobzien 1998: 133–4.

[6] See, variously, Dihle 1982; Kahn 1988; Sorabji 2004a; Frede 2011.

[7] See *C. Nemert.* frs. 276–7.

[8] There are no clear internal indications of date in the *C. Nemert.*; for a brief introduction to the fragments see Johnson 2013a: 38–9.

[9] The only secure indication of date is the fact that the *On Free Will*'s addressee, Chrysaorius, was a Roman senator and was also the dedicatee of Porphyry's *Introduction*. Bidez 1913: 59–60 supposed that the *On Free Will* had been written during Porphyry's time in Sicily; Romano 1979: 113–14 located it during his time in Plotinus' school.

The following analysis will first examine Porphyry's sometimes complicated exposition of Plato's Myth of Er in the *On Free Will* in order to determine his notion of the range and limits of the agency of the soul in both its disembodied and embodied states and its ramifications for human responsibility. It will then move to consideration of Porphyry's account of a force that would seem to be in tension with freedom of the will, namely the stars.

Porphyry's concept of free will: the argument of the fragments

The problem as stated in the opening lines of the first fragment (fr. 268.1–33), which clearly comes from the preface of the original work,[10] arose from and remains (at least in the four extant fragments) centred upon the words of Plato in the Myth of Er (*Rep.* 10.616b–621b). There, Plato had provided a narrative in which souls, already formed by experiences and character traits of their former lives, came to choose their next lives. Their choices were made in order according to lots, and the Fates were involved at every stage: Lachesis gave their allotments, Clotho spun their destiny, and Atropos made it binding. Then, a 'certain inescapable (*anapodrastos*)[11] daemon' was assigned to each to hold them to their choice and they were led under the throne of Necessity, drank from the streams of Lethe ('Forgetfulness'), and descended through the heavenly circuit as they fell into bodies.

There were two key phases in this narrative where the self-determination (*autexousios*) of the soul was threatened. First, the souls came to their choice of future lives already morally formed and predisposed in particular ways so that it might seem that deterministic forces were in place that hindered a full and open freedom of choice. Second, if Necessity and the Fates were so closely involved before, during and after their choice of lives, then it seemed to allow little room for freedom in their embodied lives. 'Since these things have been spun out, necessitated and confirmed by the Fates, Lethe and Necessity in this way, and since the daemon attends [us] and guards [our] Fate (*heimarmenē*), of what are we the masters (*kurioi*)?' (fr. 268.28–31).[12] Not only did Plato's narrative inclusion

[10] Rather than being the commencement of a section of a larger work, as supposed by Deuse 1983: 149; Wilberding 2011a: 124.

[11] The term does not occur in Plato's text, but may have been prompted by Porphyry's (possible) reading of Pseudo-Plutarch *De fato* 1.568d; Alexander of Aphrodisias *De fato* 2.166.2; Alexander of Aphrodisias *Mant.* 180.1; Plotinus *Enn.* IV.3.13.1 (treatise 27). On the 'natal daemon' see Porphyry *Ep. Aneb.* frs. 83–92 Saffrey-Segonds; for general discussion see Alt 2005.

[12] All translations of Porphyry are mine (that of Wilberding 2011a may be consulted with profit).

of these forces threaten self-determination and mastery over one's life, but it seemed to contradict Plato's own assertions that sought to protect human moral responsibility: 'Virtue is free, which each will have more or less by honouring or dishonouring it' (*Rep.* 10.617e3–4, quoted with slight modifications at fr. 268.32–3).

Porphyry claims that the problems arising for a notion of freedom in the first phase (before the souls make their choices of future lives) could be left aside since a plausible defence could be made in answer to them. It was the second phase that deserved careful attention. The remainder of this and the other fragments seek to answer these difficulties through sustained attention to details in the Platonic passage. Indeed, the fragments of *On Free Will* remain ever close to the text of Plato and refuse to provide a systematic account of freedom and human agency, or any detailed criticism on philosophical grounds of causal determinism (for instance, his explicit criticism of the Stoics limits itself to showing that they do not use the term *bios* with the same range of significations as Plato does). In spite of the lack of a robust systematic account, these fragments nonetheless indicate significant contributions by Porphyry to the ongoing discourse on freedom and determinism in antiquity. In particular, they seem to offer the lengthiest treatment of the Myth of Er within the context of the problem of determinism (at least before the fifth century). The Myth of Er had been invoked in earlier discussions, but not at the length or with the thoroughness of Porphyry's account.

The exegetical argument Porphyry raises in the first fragment runs as follows. One must recognise that Plato uses the term 'life' (*bios*) in two ways, or rather, as designating two levels of specificity.[13] As just noted, in contrast to the Stoics who posited only a single sense of the word *bios* as a rational life-form, *logikē zōē*, Plato had used *bios* to refer to irrational animals as well, such as swans, lions and nightingales. On the one hand, then, were the generic lives of humans (male and female), dogs and so on; on the other were the lives of hunting-dogs, tracking-dogs, table dogs and guard dogs, or alternatively the lives of a king or soldier.

This second level of 'life' was necessarily contingent upon the first, and the choices in a soul's determining a first-level and second-level life were made according to the two phases mentioned above. When the souls were gathered at the time of the pre-embodied allotments the choice of the first-level life was given: at that point a soul could choose to become a dog,

[13] For discussion see Deuse 1983: 148–59; Wilberding 2011a: 124–32.

a swan, a man or a woman. It was after the entrance into bodies[14] that souls would be faced with the choice of the particular kind of life they would live. Free choice of a second-level life was limited to humans (as rational animals). For non-rational animals, such as a dog, their second-level life depended upon nature or their owner. No allowance for the possibility of choice in animals is discernible in this or the other fragments. In earlier thinkers the faculty of choice was limited to rational animals, that is, humans, and thus it was natural to assume an absence of self-determination in all other animals.[15] For Porphyry, however, animals did possess reason (*logos*) since they communicated in their own sort of language (*logoi*). At least this was the argument of the third book of his *On Abstinence from Eating Meat*.[16] One would like to know whether Porphyry had sought to make allowance for some limited capacity in animals for choice in the original *On Free Will*, but the fragments give us little indication that he did (though see below).

In a manner similar to animals, the bodily beauty or parents of a human (corresponding to the nature or owners of a dog in his example) could not be chosen. But the acquirement of skills or expertise in various areas of knowledge, or the pursuit of political office, 'depend on what is in our power, even if some of them, like political offices, tyrannies, and demagogies, are difficult to obtain because they are also bound by external hindrance, for which reason it is difficult to obtain them and not easy to get rid of them' (fr. 268.69–73). Even if some were attended with difficulties, they depended on a 'deliberate choice' (*prohairesis*) to pursue or practise them or not, though the outcome was not 'entirely up to us' (*eph' hēmin*: fr. 268.75). It has, however, 'entirely been granted' to use these second-level lives in either a good or bad manner (fr. 268.78–9).[17] Porphyry's distinction of first and second levels of particularity in the signification of 'life' and the freedom of choice (though not uninhibited in its results) of the second-level life of humans became the means of ameliorating Plato's talk of Necessity and the Fates from any strong sense of fatalism. Necessity and the Fates restricted the embodied soul from changing from the first-level life it had chosen (though that choice was now forgotten): for instance, even if a man acted or dressed like a woman, he would remain a man (fr. 268.90–4). Thus, while one's familial, genetic

14 Or possibly even during the descent; see below.

15 Cf. Origen *De princ.* 3.1.2–3; Origen *De orat.* 6.1.

16 See Sorabji 1993: 80–5.

17 This claim is resonant of the so-called Cylinder Argument of the Stoics; see *SVF* 2.974 (*apud* Cicero *De fato* 39), with Dihle 1982: 103; Bobzien 2001: 258–71. On the notion of 'using sense impressions' see Epictetus *Discourses* 1.1.7, 1.20.5, with Kahn 1988: 252–4.

and bodily traits were not up to the soul (whether in its pre-embodied state or not), the fact that one was a human man or a human woman had been up to the soul (at least initially).

This fragment represents a sort of dual compatibilism (if we can avoid anachronism in the use of such a label).[18] First, elements of embodied life, such as being born a human woman, which seemed entirely out of the control of the person, had been the result of a choice that was, as it turned out, entirely within the control of the person before their embodiment.[19] It was similar to a person who chose to jump off a bridge but was unable to reverse the choice once the action that arose from that decision had been commenced, and indeed, had forgotten that the choice had once been theirs at all.[20] Second, the forces of Necessity that ratified the pre-embodied choice were not involved in or responsible for the pursuits and moral quality of the second-level lives, at least of rational humans. Some elements of embodied life (such as one's parents) were the result of external causes or chance, while others (such as pursuing the philosophic life) were up to us for whom there is a 'free will' (*aneimenon {eleutheron} to ethelousion*: fr. 268.98),[21] which can arrive at 'self-determined deeds' (*autexousia erga*) even in the embodied state (fr. 268.101–4).

Consideration of these last claims, which occur in the final two sentences of the fragment, hint at Porphyry's broader psychology explored in the (probably later) *Sentences*. The work was something of a series of paraphrases of key passages in Plotinus' *Enneads*, which nonetheless exhibits a certain independence of thought from the Neoplatonic master.[22] In the *Sentences* Porphyry presents a doctrine of two levels of activity (or of power) of the soul: one persists in its own indivisible existence, while a second is made to 'subsist' in bodies through the soul's 'inclination' (*rhopē*, on which see below).[23] Elsewhere these are named 'life-giving activities'

[18] It may be objected that our notion of compatibilism is foreign to antiquity. I am, however, unsure what 'our' notion of compatibilism might be (as well as 'our' notion of free will), and have adopted the term here to signify, not an equivalence to any modern notion, but the complicated situation that Porphyry envisions.

[19] At least, provided that Porphyry could marshal a compelling defence of free will at this phase (as he had claimed he could at fr. 268.9–11) in spite of Plato's reference to the souls' coming to their choice already shaped by previous lives.

[20] See Alexander of Aphrodisias *De fato* 29.199.28–9.

[21] Meineke bracketed *eleutheron*, while Heeren conjectured *eleutheron te*; see app. crit. at Smith 1993 *ad loc*. Both the MSS and Heeren's conjecture certainly make the claim for free will stronger, but are not necessary for my current reading of the fragment.

[22] See Schwyzer 1974.

[23] Porphyry *Sent.* 4 Lamberz; see also *Sent.* 3, with Smith 1974: 1–3. Cf. Porphyry *Pros Boethon* frs. 245.18 and 247.10 Smith.

(*zōtikai energeiai*) and are arguably identifiable as double 'lives' (*zōai*) of the soul (one 'in itself', the other 'in relation' to bodies).[24] Even though the *On Free Will* fragment distinguishes two phases for the choice of two levels of lives that 'map' onto the soul's relative distance from the embodied state, the dual-*bioi* interpretation cannot be identified with the dual-*zōai* approach. Instead, the double-*zōai* doctrine designates the hypostatic condition of the soul and seeks to protect the ontological unity of the soul even in the midst of the individuation of particular embodied souls. In other words, Porphyry adopts the Plotinian doctrine of the 'undescended' soul.[25]

It would seem that it is precisely this notion that is assumed in his claim in the *On Free Will* that, in the case of the human soul, *qua* soul, there is a 'free will' (*aneimenon {eleutheron} to ethelousion pros men ta tēs psuchēs*)[26] as it is in itself and not bound by body. At the same time, *qua* living thing and soul–body composite, the will is 'loosened'[27] since in this respect it possesses impulse (that is, that which executes an action of or through the body) – but it is an impulse 'appropriate to self-determined acts'. In other words, Porphyry envisions grades of unfetteredness of the will as it is focused upon things of the soul or things of the composite. It seems likely that this passage presumes the broader Plotinian psychology of the complex negotiation of the soul as undescended and as embodied, as Porphyry had expressed it in his *Sentences* and *On the Powers of the Soul.* This statement is, at any rate, the clearest claim for freedom of the will in the fragments of the *On Free Will.* Importantly, it seeks to contextualise the freedom with respect to the soul in itself or in relation to bodies. Its distinguishing of the soul's freedom in relation to bodies would continue to be formulated in the following fragments.

The second rather brief fragment emphasises that the 'self-determination' (*autexousion*) appropriate to a human is in the human's own power. Even if impressions or appearances (*phantasiai*) bear a high level of persuasiveness upon the soul, it remains in the latter's power to choose 'to assent' to these impressions or not, since the impression does

[24] Porphyry *On the Powers of the Soul*, fr. 253.110–22, with Smith 1974: 3.

[25] Plotinus would emphasise the doctrine of the 'undescended' soul in his own treatment of astrology: *Enn.* II.3.9 (treatise 52).

[26] Why did Porphyry adopt *to ethelousion*, a term rarely (if ever) used in philosophical discussions of the will? It may be that he sought to avoid *thelēma/thelēsis*, a favoured term in Plotinus *Enn.* VI.8 (treatise 52) (esp. sections 13, 16 and 18), because of its Christian overtones; cf. Sorabji 2004a: 6. Cf. Porphyry *Pros Boethon* fr. 247.9–10.

[27] I agree with Wilberding in following the manuscript witness *lelumenon*, rather than the emendation *lelumenou* (which Smith adopts): Wilberding 2011a: 149 n.19.

not drag it along like a 'puppet on a string' (*neurospastei*, fr. 269.7–11).[28] It is difficult to avoid the suspicion that Porphyry may have read Origen's discussion on free will at *On First Principles* 3.1 and is obliquely reflecting the latter's language here. Origen had written: 'But if anyone should say that the impression from without is of such a sort that it is impossible to resist it whatever it may be, let him turn his attention to his own feelings and movements and see whether there is not an approval, assent and inclination of the governing faculty of the soul towards a particular action on account of some persuasiveness' (*De princ.* 3.1.4; trans. Butterworth modified).[29] Like Origen, Porphyry adopted the Stoic (Chrysippean) concept of freedom in terms of 'assenting' to sensible 'impressions' that possessed a sort of persuasiveness, which the soul was nonetheless free to resist or accept.[30] Significantly, Origen had applied the term 'inclination' (*rhopē*) to the soul as a means of emphasising that the soul was not passively drawn towards the impressions, but that it had a quality that predisposed it to follow after those impressions. We shall see below Porphyry's inclusion of *rhopē* to account for a soul's choice of a second-level life. Although he had attacked Origen in his *Against the Christians*,[31] it seems that the Christian philosopher may nonetheless have had some influence in the basic formulation of the relation of souls and impressions in Porphyry's otherwise unique two-choice account of lives.

The third fragment confirms the overall picture of different levels of lives and different phases of choices. The disembodied souls are allotted the choice of first-level lives; once they have chosen they are bound to that choice, and subsequent freedom is allowed only at the secondary more particularised level of lives. The range of this subsequent freedom depends upon the constitution of the life chosen. An irrational animal has a singularity of character and only little movement of its self-determination. A human, on the other hand, has 'much intellect and much movement' with respect to self-determination (fr. 270.12–15). That which moves a human's faculty of self-determination is its own self, although it may be led 'according to the desires arising from its constitution (*kataskeuē*)' (fr. 270.15–18).[32] Again a sort of compatibilism is expressed in these fragments: the body and bodily impulses coexist with a capacity for free will, even as they can mutually impede or overcome the other. The cause of human

[28] For discussion of the notion of *phantasia* in other works of Porphyry see Sheppard 2007. On the rather rare *neurospastei* ('being a puppet drawn on a string') in other literature see Johnson 2013b.
[29] For general discussion see Scott 1991: 113–49, esp. 145–7; Jackson 1966.
[30] On the Stoic notion of assent see Bobzien 2001: 240–1, *passim*.
[31] Porphyry *C. Christ.* fr. 39 Harnack; see Johnson 2012.
[32] Cf. Origen *De princ.* 3.1.5.

agency is the human's free will. Even in those instances where the performance or completion of an action is thwarted by external factors, the will or determination to choose between opposing options remain in the soul's power.

As in the first fragment, this third fragment takes for granted that animals are irrational and lacking in a full freedom of choice. The reasons for this are here, however, spelled out more clearly. The souls that have descended now bear a faculty of self-determination (*to autexousion*) that is 'appropriate to the constitution' of the living being they have chosen. At this point the distinction between humans and irrational animals seems to become somewhat less rigid. He states: 'Having become souls of living things ... they bear the faculty of self-determination (*to autexousion*) related to the condition of the living thing, and in some there is much mental ability (*polunoun*) and much movement (*polukinēton*), as in the case of a human; but, in others there is little movement and singleness of character (*monotropon*), as in the case of nearly all the other living things' (fr. 270.11–15).[33] Porphyry's language of relative speed of movement and thought (which may recall, though without verbal allusion, the myth of the creation of living things in Plato's *Protagoras*) allows for some animals to be more like humans in their quickness and intelligence. Thus, Porphyry's compatibilism is formulated as a graduated compatibilism, such as one would expect in the framework of a Platonic ontological hierarchy. Among living things there is an interplay of freedom of decision and action within the restrictions, relative to a range of bodily constitutions and their impulses and desires. It seems that here would be the conceptual mechanism for allowing differing levels of rationality in animals such as was posited in the *On Abstinence*, though again we can only suggest the possibility since the fragment ends there.

Significantly, in this and the earlier fragment that remarked on the exclusion of free will from animals, Plato's description of the transmigration of souls into animals is taken literally. It may be that the problem for later Platonists of transmigration into animals, in that a rational soul could not easily be conceived as inhabiting an irrational life, was not yet formulated as a problem at all in Porphyry, who possessed a more flexible interpretative approach.[34] The fragments of the *On Free Will* offer a picture of rational souls freely choosing human or animal lives, which could then be lived out with greater or lesser degrees of acuity,

[33] Cf. Porphyry *De abst.* 3.7.2, 3.8.7–9; Origen *De princ.* 3.1.3.

[34] See Smith 1984. Cf. Deuse 1983: 148–59.

depending on their particular 'constitutions' and capable of resisting or succumbing to the desires and impulses of embodied life. In any case, Porphyry's adoption of the doctrine of the undescended soul (noted above) would have allowed for such inclusiveness, since that concept allowed for a soul to illumine an animal life without being 'in' the animal body.[35]

While the bulk of the final fragment addresses the issue of astrological determinism, and thus will be discussed later, it offers details relevant for further delineating Porphyry's conception of the levels of freedom allowed for the levels of lives, the choices of which belonged to the different phases of a soul's existence. First, an additional limitation seems to be placed upon the free will of embodied souls because it is tainted by an 'inclination with respect to one of the lives here when the soul was still outside [the body]' (fr. 271.16–19).[36] It seems, in fact, that Porphyry has forgotten his claim in the prologue that a proper account could be made in defence of free will in the face of Plato's assertion that the souls came to their choice of lives already shaped by their previous lives (fr. 268.9–11). Here, Porphyry seems to grant the assertion as it stands as a valid consideration limiting, or 'tainting', the souls' freedom of choice. Indeed, Porphyry identifies the 'inclination' (*rhopē*) of the soul with Plato's term 'choice' (*hairesis*). The inclination here seems to designate a sort of pre-embodied desire or some other motivational factor determining, at least partially, the decision that the soul is about to make.[37] It may be significant that Porphyry opts for the vague *rhopē* and avoids the term *prothumia*, which elsewhere in the fragments of this work is limited to desires or impulses that occur in the embodied life. Could his description of the 'tainting' of the soul's free choice here hint at a larger negotiation with the text of Plato to render it more amenable to a third-century Platonist doctrine of free will?[38] We can only suggest the possibility, any examination of which is caught up short by the limits of our fragmentary remains.

Second, Porphyry adopts the so-called hypothetical or conditional conception of Fate, which had been articulated in Middle Platonist accounts, particularly those of Alcinous and Pseudo-Plutarch.[39] Evincing no verbal dependence on these thinkers, Porphyry nonetheless conveys

[35] See Sorabji 1993: 191–3. [36] Cf. Plotinus *Enn.* IV.3.12.37–8 (treatise 27).

[37] Cf. Porphyry *Sent.* 3.4.27 Lamberz; Porphyry *De abst.* 1.30.2, with Smith 1974: 1 n.2; Clark 2000: 134 n.103.

[38] Such a doctrine of free will would not necessarily be positive; cf. the negative notion of free will in Plotinus' earlier works, e.g. *Enn.* V.1 (treatise 10); II.9 (treatise 33); see Atkinson 1983: 4–10. In this context *rhopē* occurs at *Enn.* IV.8.5.26 (treatise 6).

[39] Alcinous *Didask.* 26; Pseudo-Plutarch *De fato* 570C–E; see Boys-Stones 2007.

close conceptual ties to the parallel they drew between the laws of Fate and the laws governing a state. Just as the laws of a state did not causally predetermine a person's actions but only prescribed proper action and threatened punishment for infractions, so too Fate only prescribed a set of conditionals: if a person does *x*, then *y* necessarily will follow. Porphyry provides the example of a soldier. If one chooses the (second-level) life of a soldier, then he will face the things attendant upon that life (presumably long marches, battles and possible death). It was not necessitated that he choose the life of the soldier; the battles and so on only followed necessarily upon the free choice of the person who decided on such a life. The soul could have chosen otherwise.[40] This hypothetical conception of Fate is parallel to the notion of Fate's ratification of first-level lives noted above: the disembodied soul chose a life to which Fate and Necessity held it. Now, after choosing a second-level life, the embodied soul was seen as being held to its own previously uncompelled choice.

Throughout the fragments Porphyry has developed an argument for a position that we may designate a compatibilist free will approach. The argument posits a freedom of action, which is nonetheless restricted by one's nature or by chance; a freedom of decision, which is nonetheless tainted by one's inclination or impulse; a freedom of will, which is nonetheless active only in a manner appropriate to one's constitution. Furthermore, greater freedom is seen further from the physical world, which was a place characterised by greater impediments to freedom and the threat of becoming enslaved to one's impulses. We cannot be certain that Stobaeus, when referring to Porphyry's work as *On What is Up to Us*, was reporting the title assigned to the work by Porphyry or was providing his own designation of the content of the work (or at least the portion of the work that mattered to him). If *On What is Up to Us* was Porphyry's own title, it is clear that he used it as a springboard for discussing more wide-ranging issues of the embodied soul's freedom and responsibility. Importantly, he clearly expresses a notion of the will as a distinct faculty (the unique *to ethelousion*, or *to autexousion*, or even, possibly, *prohairesis* as a faculty that governs choices, *haireseis*), with greater or lesser grades of limitations placed upon it in accordance with the particular phases of a soul's existence (outside bodies, or within various types of embodied lives).

[40] See also Porphyry *Comm. Tim.* fr. 28 Sodano.

Freedom under the stars: 'soft astrology' and human agency

In addition to Necessity, the Fates, and allotments, Plato's Myth of Er also made mention of the 'circuit' of stars, the 'revolution' of the heavenly vault, and the eight embedded whorls, which was clearly a model of the planetary orbits. These become a subject of concern for Porphyry only in the fourth fragment. Since he had already allowed for some degree of causation outside the person, which could impede the accomplishment of a person's decision (such as to attain high political office if one had been born into an obscure family), it would have theoretically been possible for him to argue for some causative role of the stars in a human life. After all, even Plotinus had (remarkably) made allowance for such astral causation in a discussion that explicitly recalled the Myth of Er.[41] After criticising notions of 'hard astrology' (the modern label for notions of astral determinism) in favour of 'soft astrology' (the modern label for notions of astral signification of future events without any causal relation to those events),[42] Plotinus had remarked on the stars' connectedness to the universal whole in such a way as to allow a sort of astral causation of human events in so far as the stars were intermediate in the ontological hierarchy between the world soul and individuated embodied souls.[43] He had already granted stars and planets power over certain physical changes (for instance, the effects of heat and cold);[44] now a greater range of events seemed to be placed within a causal relationship to the stars through *sumpatheia*.[45]

If Porphyry had read Plotinus' essay on the stars by the time he composed his *On Free Will* (a possibility not excluded by chronological considerations, though it was one of Plotinus' last treatises), there is no indication that he followed his master in allowing this sort of astral causation. Instead, he limited himself to the strict adoption of soft astrology in ways that make one wonder if he was not intentionally seeking to improve upon Plotinus' approach. After expounding his concept of hypothetical Fate, Porphyry asks where Plato might have derived his ideas in the Myth of Er. Then follows a sentence troubled by textual difficulties, as may be seen from the emendations recorded in Smith's *apparatus criticus* for this passage, but whose basic sense

41 *Enn.* II.3.9 (treatise 52).
42 On the distinction of 'hard' and 'soft' notions of astrology see Long 1982.
43 *Enn.* II.3.13–14 (treatise 52). 44 *Enn.* II.3.2 (treatise 52).
45 For discussion of this and the other passages where Plotinus treats astrology see Adamson 2008. For general treatment of Plotinus' cosmology and its background see Wilberding 2006.

seems clear.[46] Plato, in Porphyry's estimation, stood in opposition to the Egyptian belief in astrological determinism. Astral configurations, he averred, did not

> compel (*anankazein*) the lives to be of such a kind for the souls going into birth through the horoscopic degree (*moira*), in the way the constellations are at that time; but rather, when the souls are carried to the horoscopes according to their internal dispositions and when they see the lives, <which> the constellations indicate (*sēmainousin*),[47] inscribed in the heavenly region as if in a painting, those [souls] having made a choice are also capable of not living this way because of freewill (*to autexousion*). (fr. 271.46–55)

The 'disposition' (*diathesis*) of the astral–planetary configuration is not determinative of lives; on the contrary, it was the 'dispositions' of the souls themselves (in other words their desires and will apart from astral causation) that was determinative of the souls' lives.[48] In this Porphyry directly counters the admission of Plotinus that the stars were in charge of both their own dispositions and the dispositions of souls.[49]

Such a 'soft' approach to astrology was not new among Platonists, and even the image of the astrological signs being spread across the heavens like an inscribed painting or book had been adopted by thinkers important to Porphyry.[50] What seems unique in this account is his attempt to explain how the souls' choices and the stars' significations come to correspond to each other. In response to the standard question of sceptics (beginning at least with Carneades)[51] as to how exactly simultaneous births could occur for so many different lives (men, women, dogs and so on), Porphyry described the souls as descending through the heavenly spheres at different rates of speed and veering off to particular degrees of the zodiac in accordance with their choice of lives.[52] But here his account takes an odd turn, for he claims that these differences in the souls' descents were a result of their being moved 'in accordance with their desires for certain of

[46] For a translation that more closely follows the emended text printed by Smith see Wilberding 2011a: 145; for a different translation based on a more limited adoption of the various conjectures see Johnson 2013a: 343, with n.72.

[47] Cf. Porphyry *Comm. Tim.* fr. 12 Sodano.

[48] Cf. Porphyry *Sent.* 29 Lamberz; Plotinus *Enn.* IV.3.12.37–9 (treatise 27).

[49] *Enn.* IV.4.31.25–30 (treatise 28), II.3.11.4 (treatise 52); Adamson 2008: 275, 285.

[50] For the comparison to a book or painting see Origen *Comm.Genes. apud* Eusebius *Prep. Evang.* 6.11.63; Plotinus *Enn.* II.3.7.5–7 (treatise 52); III.1.6.20–2 (treatise 3).

[51] See Amand 1973: 165.

[52] Porphyry had elsewhere connected the soul's descent into bodies with astral phenomena (see *De antro nymph.* 22–8, with Anghelina 2010), and had even remarked on the difference of speed in their descent through the spheres (see *Comm. Tim.* fr. 79 Sodano); cf. the account of Nigidius Figulus at Augustine *Civ Dei* 5.3.

the second [level] of lives' (fr. 271.70–1). The examples he provides are only of first-level lives, and the discussion in the earlier fragments would have led one to suppose that determination of second-level lives would not commence until after the soul had joined its body. Because he then goes on to describe how Justice bears the soul to the degree of the zodiac appropriate to the soul's particular character formation (fr. 271.75–6),[53] it would seem that Porphyry has in mind here only those elements of the second-level life that a person would otherwise attribute to 'chance', such as the sort of family into which one was born. Porphyry explains: 'Justice is called Chance (*Tychē*), since she is a cause undisclosed to human reasoning' (fr. 271.78–9).[54] He then remarks on the Egyptian doctrine of beneficent and maleficent degrees of the zodiac; but the comment on Justice's role in the souls' descents is sufficient to show that even those factors seemingly most out of our control, such as family, are also accounted for in terms of choice. Such choices are not free in a random or completely open sort of way, but are freely made within the context of the individual soul's character. Porphyry can conclude:

> Plato, therefore, posited that the kind of appearances (*schēmata*) [of the stars] indicate the lives (*bious*); but they do not further necessitate them, rather the [souls] who have made their choice live and have the sequence of things inscribed, just as it necessarily indicates. A cause, therefore, of the movement (*phora*) into the horoscope belongs to a woman's, a man's or some other animal's choosing; the choice of the second [type of] life is a cause also of the [soul] being held down to this horoscope; the arrangement of the stars, which is ordered in accordance with the horoscope, shows the [second-level life] inscribed. (fr. 271.87–95)

Necessity is a force not external or contrary to one's faculty of choice, but that which holds the soul to its prior choice.[55] But how much does Porphyry suppose to be inscribed in the astral configurations? Surely, he does not seem to have in mind the minutiae of daily life. But what of life's larger questions: marriage, children, travel, business, political career? No word on the exact scope of astral significations is offered in the fragments of this work. Several fragments of his *Philosophy from Oracles* that directly address astrological divination, however, provide some clues. In a discussion exhibiting the claim that the gods (or daemons) who delivered oracles depended upon astrology in making their predictions, a small number of oracles are quoted as evidence. They predict the sex of a baby about to be born (*Phil. orac.* fr. 333.5–9); a plague (*Phil. orac.* fr. 333.13–14); either a war

[53] Cf. Plotinus *Enn.* IV.3.13.1–6 (treatise 27).
[54] Cf. Alexander of Aphrodisias *De fato* 8.174.1–28.
[55] Cf. Plotinus *Enn.* IV.3.13.1–6 (treatise 27).

or plague ('ashen evils of black strife': *Phil. orac.* fr. 334); a plague that would be fatal to the consulter of the oracle (*Phil. orac.* fr. 335); the military inclinations of one person, and death in battle of another (*Phil. orac.* fr. 336).

This does not allow for much elaboration on the range of phenomena that Porphyry took to be appropriate for astral signification, and in any case it is not entirely safe to interpret fragments of one lost work by fragments from another, especially when their composition may have been separate from each other by many years.[56] But there is nothing in the astrological fragments of the *Philosophy from Oracles* that appears to be at odds with his specific claims about 'soft' astrology in the *On Free Will* or the general philosophical framework of the latter work. The sex of the human baby fell within the first level of lives. War fell within the second-level lives, as did the military inclinations of the subject of one of the oracles, inclinations that would have prompted the soul to veer off to the corresponding degree of the zodiac during its descent into a body according to the *On Free Will*. Plagues and other causes of death were not treated in the latter work, but they could have found a place in material that is now lost.[57]

The gods of the oracles were not the causes of the things predicted, but only the readers of astrological signs. In fact, they were themselves said to be 'under the fates (*moirai*)' (*Phil. orac.* fr. 337.4–5) or even 'under Fate (*heimarmenē*)' (*Phil. orac.* fr. 342.6–7) when they descended below the stars.[58] In a final passage at the end of the fourth fragment of the *On Free Will* Porphyry provides a different set of considerations that would have removed divine culpability from what may have seemed to be the 'evils' of human life. The passage is included not so much as a defence of the divine, however, but as an extension of Porphyry's two-level approach to the choices of lives in the text of Plato to that of Homer. The fate that no man, good or bad, could escape once the 'first things' should happen (declared by Hector at *Iliad* 6.488–9) applied to the first-level choice of life. For, claimed Porphyry, once chosen the life was inescapable and could not be changed, though it was up to the soul to live 'with virtue or wickedness' (fr. 271.107–14).

[56] For problems of determining the date or even relative chronology of the *Phil. orac.* with respect to Porphyry's other works see Johnson 2013a: 16–17, 24–7.

[57] For an account of the time of death as determined by Providence see *C. Nemert.* frs. 279–80, 282.

[58] It seems that Porphyry would have accepted Pseudo-Plutarch's division of primary, secondary and tertiary levels of providence. Only the third level would be held by daemons, and these would have been deemed to be 'in accordance with Fate' (*De fato* 573a). That Porphyry understood the 'gods' of the *Phil. orac.* to be a designation for 'daemons', see Johnson 2013a: 96–100.

The second-level life was expressed in the lines of Homer's *Odyssey* where humans are said to blame the gods for evils, 'but they too/ have grief beyond their portion for their arrogance' (*Odyssey* 1.32–4). The words 'but they too' showed, for Porphyry, that gods contributed 'something' to humans, 'even if the majority are because of them [humans]' (fr. 271.119–21). Or, if the gods were to be blamed for providing the archetypes of lives in the first place, this could be resolved by 'their making the souls free-willed (*autexousious*) and entrusting to them the choice of lives, but they, because of hastiness [in making their choice][59] and ignorance, "have grief beyond their portion for their arrogance"' (fr. 271.123–6). The instilling of free choice in souls was thus part of a providential means of giving moral responsibility to humans.[60] This sentiment would recall that of Plato: 'The fault belongs to the one who chooses; God is blameless' (*Rep.* 10.617e4–5).[61]

Conclusion

Porphyry's motivation for grappling with the issue of free will was not merely the desire to show the compatibility of Plato's text with a later Platonist doctrinal position. The issue of moral responsibility seems to have lain at the heart of his exegesis. Virtue and moderation were raised in the first fragment (fr. 268.32–3, 77–9), and the avoidance of extremes while aiming at the mean is explicitly advised in the opening lines of the fourth (fr. 271.2–5). Moral responsibility was a common concern in Platonist articulations of the freedom of the soul. Porphyry is noteworthy in formulating a two-life approach that could maintain an exegetical closeness to Plato's Myth of Er and offer fresh insights into how a doctrine of free will could be espoused in the face of the 'scientific' claims of contemporary astrology while avoiding Plotinus' slippage into a notion of astral causation. Restricted though we are by the fragmentary state of Porphyry's discussion, we are able to glimpse the sort of distinctive exegetical manoeuvres that he was known for among his peers and later readers as both a philosopher and philologist.

59 Instead of taking *epidromē* as an external assault upon the soul (as I had previously understood it: Johnson 2013a: 346), it seems best to see Porphyry as making an allusion here to Plato *Rep.* 10.619d5, where those souls who suffered punishments are said to have not made their choice of life 'from hastiness' (*ex epidromēs*).

60 See *C. Nemert.* frs. 276–7.

61 Porphyry leaves this sentence out of his quotation (at fr. 268.32–3) of the material directly before it in the text of Plato.

CHAPTER 11

Proclus on the ethics of self-constitution

Michael Griffin

Introduction: individuality and consciousness

Individual organisms, such as Socrates, seem to be real, unified beings if anything is:[1] this struck most ancient philosophers as common sense, despite some metaphysical quarrels.[2] Already for Aristotle, ὁ τις ἄνθρωπος was the primary exhibit of an οὐσία that exists by nature, displays unity in number, form and function, and is obvious to everybody.[3] When Chrysippus[4] and Plotinus[5] elaborated hierarchies of compound beings according to a criterion of unity, ranging from armies and artefacts to plants and animals, the coherence of a living, thinking organism crested

[1] Sorabji 2006 and Gill 2006 offer, from different vantage points, wonderfully learned surveys of ancient Greek notions of selfhood; Sorabji's survey also ranges into Indian and modern analytic philosophy in detail. I think it is reasonable to say, in Parfit's (1984: 210–11) terms, that Greek philosophy as a whole tended to be non-reductionist about the self. Modern metaphysicians, especially those sympathetic to Aristotle (such as Fine 1994 and Koslicki 2010), also treat the unity of organisms as a central target for any promising explanation of composition. Even mereologists who endorse a form of unrestricted composition also still feel obliged to explain how animal organisms are prime, or common-sense, instances of individual beings, and those who doubt that there are such 'ordinary things' at all still feel obliged to give an account or paraphrase of what it is to be an organism.

[2] At least since Democritus, metaphysicians were prepared to argue that Socrates and Bucephalus 'exist' merely by convention, but in reality only their simplest elements or particles are real (fr. 68 B9). Meanwhile, paradoxes about diachronic identity – such as the Ship of Theseus (Plutarch *Life of Theseus* 22–3) and the Growing Argument (Epicharmus fr. 23B2 Diels–Kranz) – intrigued thinkers from a wide variety of traditions. On one interpretation, even Plato may have entertained such doubts about composite beings (Mann 2000), though it would perhaps be going too far to call even this version of Plato a compositional nihilist. For modern versions of compositional nihilism, consider for example Unger 1979 and van Inwagen 1995.

[3] Thus ὁ τις ἄνθρωπος ἢ ὁ τις ἵππος can be used as examples of individual οὐσίαι (*Cat.* 3, 1b4–5), as individuals that are 'one in respect of number' as well as form (*Metaph.* 5, 1016b31–3), and exist naturally (*Phys.* 2.1, 192b9–11). See for example Charlton 1994.

[4] Climbing from lifeless rocks united only by natural tenor (ἕξις), to plants bound by both tenor and nature (φύσις), and our form of life (ψυχή). See Long and Sedley 1987: §47.

[5] *Enn.* VI.9.1: first weak arrangements (choirs, flocks and armies), then artefacts that display continuity (houses and ships), and finally plants and living organisms that exhibit genuine unity and really participate in existence (as ἓν ὄντα).

the ladder. To be a living being or ζῷον appeared to entail a robust, structured unity: one could not be alive without being, in some strong sense, a synchronic and diachronic *one*.

Ancient philosophers also found a greater degree of unity in specifically *human* existence and persistence – in 'personhood'[6] – than the coherence that might be attributed to a stone, plant, or even another animal. To be a unified person intuitively seemed to demand the capacity for reason, especially the possession of second-order desires and beliefs,[7] by means of which we might craft a coherent character out of our choices, actions and opinions. Aristotle regarded such self-guided moral action (πρᾶξις) as unique to human beings (*Eth. Nic.* 6.2, 1139a17–20), who can craft our choices to serve an end with reason or λόγος (*Pol.* 1332b3–5). The Stoics, elaborating this intuition, treat human reason as a higher form of unity than the natural coherence of a plant or the bare 'tenor' of a stone (*SVF* 2.458; Long and Sedley 1987 47PQ).

The Neoplatonic tradition, on its face, endorses both of these views: human persons are organic unities, and this unity involves our capacity for self-reflective reason. Proclus, for instance, stresses that every human psyche is a unity that is self-animated, self-constituted and self-reflexive (*El. Th.* §§ 43, 186, 189). Plotinus remarks that 'we', the genuine ἄνθρωπος, are an authentic being possessing unity and presiding over acts of reasoning, belief and attention (*Enn.* I.1.7–8), and he also argues that individual persons such as Socrates are Forms in Intellect (*Enn.* V.7; cf. Wilberding 2006: 47–8). But in this chapter I would like to argue that this natural interpretation of the Neoplatonist view is only provisionally right. Socrates certainly *ought* to be a conscious individual (and perhaps Socrates himself, as the paradigmatic sage, was exactly that). But he had to *work* at it first: he was obliged to make a conscious unity of himself through a process that might be described as self-constitution, or self-causation (cf. Proclus *El. Th.* §§ 44, 83, 189). We don't get our unity for free; unification is a matter of effort, a virtuous achievement. Here, I would like to suggest that Proclus regards this project of *self-constitution*, or self-unification, as an important source of normativity; thus Proclan 'metaethics' might be profitably compared to the recent work of Christine Korsgaard (2009), who, like Proclus, draws on conceptual resources from Plato and Aristotle. But Proclus' project is also interestingly different.

[6] Important recent treatments in English of the concept in antiquity include Gerson 2003; Gill 2006; and Sorabji 2006.

[7] Discussed further below. For the notion of human personhood as defined by the possession of second-order desires (desires about our desires) and beliefs see Frankfurt 1971.

Roadmap

We begin with a brief overview of a recent argument for self-constitution as a moral goal, leading to what Korsgaard has called the 'paradox of self-constitution' (2009: 35): how can I create myself unless I am already there in the first place? I suggest in the next part of this chapter that Plotinus' two-storey model of the human person, his division of 'us' (ἥμεις) into an undescended soul and a life-form living in the world, furnishes the basic components of a solution (a solution, at least, that could tempt someone who already endorsed the backbone of Neoplatonic metaphysics). The subsequent sections explore a later Neoplatonist development and adaptation of that solution: given the basic framework of the *scala virtutum* developed by the Neoplatonists after Porphyry, human progress towards the Good can be understood as an ascent on a ladder of unity, where, moreover, the unified self is *already there* in eternity before its temporal action of self-constitution. Thus Proclus might be seen as an ancient contributor of a constitutivist model of metaethics, one whom we could better understand by comparing the similar views of Korsgaard and others, but one whose picture of selfhood and solution to the 'paradox' of self-constitution is deeply intertwined with Neoplatonic metaphysics.

Setting the stage: unity and normativity

Korsgaard argues that 'to be a living thing is just to be self-constitutive ... a living thing is a thing that is constantly making itself into itself' (Korsgaard 2009: 41). More broadly:

> The task of self-constitution, which is simply the task of living a human life, places us in a relationship with ourselves – it means that we interact with ourselves. ... The only way in which you can constitute yourself well is by governing yourself in accordance with universal principles which you can will as laws for every rational being. (Korsgaard 2009: xii)

This kind of agency for self-constitution also helps to explain our persistence through time, our psychological unity or connectedness (cf. Korsgaard 1996: 369–77). Moreover, it attempts to offer an account of normativity in human life: just as 'a house with enough cracks will crumble, and cease to be a house altogether' (2009: 28), in the same way it follows that 'bad actions, defective actions, are ones that fail to constitute their agents as the unified authors of their actions' (2009: 32), which cause us not to be ourselves, to fail to be unified, reasoning human agents.

For Proclus too, to find our unity is to find our good (*El. Th.* §§ 9, 13; *In Alc.* 1.3–3.2), and to escape from the rending force of multiplicity is to escape our danger (*In Alc.* 6.12–17.8, 43.7–44.11). But there are also differences from Korsgaard's view, which I will try to illustrate briefly here.

Like Korsgaard (2009: 20), the Neoplatonist ethics of self-constitution faces a puzzle, which Korsgaard calls 'the paradox of self-constitution': how can you constitute yourself *unless you are already there*? Korsgaard herself finds this to be a false dilemma:

> [The dilemma] assumes that the endorsement of our identities, our self-constitution, is a state rather than an activity. If self-constitution were a state then we would be stuck on the horns of this dilemma. Either we must already have constituted ourselves – in which case the self would be full and determinate. Or we must not have done so yet – in which case the self would be empty. But we don't have to choose between these two options, because self-constitution is not a state that we achieve and from which action then issues. (Korsgaard 2009: 44)

On this view, the dilemma only appears when we fail to recognise that self-constitution, and indeed human *being*, is a process or activity, rather than a state. 'Being a giraffe is doing something: a giraffe is, quite essentially, an entity that is always making herself into a giraffe' (2009: 36). That approach rests on Korsgaard's interpretation of Aristotle. Broadly, Korsgaard works within an Aristotelian teleological model. According to this model, 'to be an object, to be unified, and to be teleologically organized, are one and the same thing' (2009: 28).

The Neoplatonists accept the foundations of that model, which they also locate in Aristotle, but they attempt to go beyond it. As Proclus puts it (*In Alc.* 2.1–3), 'Aristotle asserts' that 'both being and the good [or unity, τὸ ἕν] spring from the same hearth and primal spring'. Thus whichever feature explains a thing's existence is also the feature that explains its individuation, unity and good. And thus far, Proclus might agree that Korsgaard has rightly understood Aristotle. But for Proclus, Aristotle only has part of the story. The individuation and *good* of a being actually spring from a source above and beyond its raw existence (see *El. Th.* §§ 1–13). Thus Proclus and his contemporaries augment the traditional Aristotelian list of four causes of a teleological process (formal, material, efficient, final) with others, including the *Paradigmatic* – a cause answering to the function of Platonic forms, the paradigm of the *Timaeus* (see Simplicius *In phys.* 3.13, 6.31).

On this new metaphysical model, the truly unified self doing the constituting can be viewed as an eternal paradigm who was there 'all along', while the process of constitution and its result, the constituted self, can be viewed as a temporal outcome of that being's action. This framework allows the Neoplatonists to step around the dilemma that Korsgaard calls the 'paradox' of self-constitution. For Proclus and the later Neoplatonists we should highlight a distinction between our *essence* and *activity*, between character and action. The soul's character – its self – is eternal, but its activity – its persistent work of self-constitution or self-expression – is temporal (*El. Th.* § 191.3). Eternally or essentially, we have already constituted ourselves: but in temporal activity we are always at work on this project. Our achievement can be measured by the degree of unity, or individuation, that we have achieved, stretched on a spectrum between the divisibility of bodies, on the one hand, and the sheer unity of τὸ ἕν, on the other; the more unified we are, the more successful we have been at the virtuous project of self-constitution, as τὸ ἕν is the goal and agent of our completion (*El. Th.* §§ 12–13). Thus τὸ ἕν is the ground of our being or essential character, and also the goal of our temporal action: for the Neoplatonist it is the ultimate ground of normativity.

Plotinus' two persons

We might proceed by comparing Plotinus' view of personhood, foundational for later Neoplatonist accounts.[8] Plotinus employed the plural pronoun 'we' (ἥμεις) to indicate the rational soul. And 'come what may', Plotinus argued, 'the soul will be at peace, turned to itself and resting in itself' (*Enn.* 1.1.9). But ordinary sensations – pleasure and sadness, longing and fear (1.1.1), simply needing food and drink (1.2.5; cf. Aristotle *Eth. Nic.* 10.8; Sorabji 2006: 35) – would surely cause us some disturbance. So how would Plotinus explain our ordinary, day-to-day clamours, sensations and wishes: are they really part of us? Should we, like an Epictetan sage, cast many of these affections aside and resist the urge to identify with them (*Disc.* 3.3.14–19; cf. Sorabji 2006: 182)? More seriously, we might wonder how *we* – the real us – might ever *be held to account for a mistake* (1.1.12), or credited for an achievement.

Plotinus feels these concerns. In response, he picks apart two different descriptions of what 'we' are, suggesting that the true human being or

[8] This will be an extremely compressed account of a very complex question, but for more on Plotinus' view of the self and soul see O'Daly 1973; Gerson 1994: III.2, VII.2; Sorabji 2006: 36–7; Remes 2008; and on the 'undescended soul' in its wider contemporary religious context in the third century, Narbonne 2008.

ἄνθρωπος remains separate, dwelling safely in the intelligible world as the human animal, the ζῷον, goes about its daily affairs.

> 'We' is used in two senses, either including the beast or referring to that which even in our present life transcends it. The beast is the body which has been given life. But the true human is different, clear of these affections; he has the virtues which belong to the sphere of intellect and have their seat actually in the separate soul, separate and separable even while it is still here below. (*Enn.* I.1.10.7–8, trans. Armstrong 1966–88, I: 115)

Thus there are always two of us: one reasoning being subsisting above in eternity, and a persistent shadow that trudges along in sublunar time, like the shade of Homer's Heracles (*Odyssey* 11.601–2).[9] The former is *genuinely* who we are, the fringe of our intellect (*Enn.* IV.3.32) like the edge of sunlight, the natural self-expression of real, unadulterated unity that is the god in us, the god we truly are (I.2.6). The two souls are like neighbours, with the higher soul exerting a positive moral influence on the lower soul by its proximity and company (I.2.5). It turns out that our experience of 'ordinary' or conventional day-to-day affections is a by-product of this higher, rational soul's 'casting a shadow' as it gazes downward, or outward, at ordinary life (I.1.12), projecting its values into matter and valuing what is exterior over the interior source of its own projection (V.8.1, V.1.1). As soon as we turn our attention fully upward or inward, and recognise that the source of our values is within us or 'above', that shadow simply vanishes. But it takes a kind of *work* to get 'up' to that separate state. In particular, it requires an abandonment of multiplicity and turn to unity.

> For the higher soul also flies from multiplicity, and gathers multiplicity into one and abandons the indefinite; because in this way it will not be [clogged] with multiplicity but light and alone by itself; for even here below, when it wants to be in that higher world, while it is still here below it abandons everything that is different. (IV.3.32.19–23, trans. Armstrong trans. Armstrong 1966–88, IV: 135)

This 'turn' to unity is no small achievement. Plotinus alludes to *Republic* 10, 611d–612a – where Socrates describes 'knocking off the encrustrations' of the sea-god Glaucus to see his essence – to illustrate the process of finding the true or authentic soul within. From the day-to-day point of view it is a struggle, although from the upper soul's vantage point it is simply a matter of shifting attention to the intelligible (*Enn.* I.1.12.29–32). Philosophers are able to ascend to the intelligible realm and identify with

[9] See Plotinus *Enn.* I.1.12; Plutarch *De fac.* 944E–945B; Sorabji 2006: 101.

this highest part, through its 'ultimate fringe ... tied to the world below' (III.4.3.21–7) by a kind of 'ascent' of consciousness (see Chlup 2012: 25). But this movement towards self-unification is framed as a cognitive act, almost instantaneous, and it would seem misleading to frame it as a change of being – rather, it involves a change of mind, a kind of realisation that amounts to *noticing what one really is.*

Turning upward: later Neoplatonism's ladder of virtues

Later Neoplatonists, following Iamblichus (see *De an.* 1.7 = Stobaeus *Ekl.* 1.49.32.78–84), did not subscribe to Plotinus' doctrine of an undescended soul. They held that the human soul descends *completely* into becoming, though once here we continue to possess innate images of the higher world within (Proclus *In Parm.* 948.12–30; cf. *In Tim.* 3.334.10–14 and *In Alc.* 227.19–21) – a version of the Platonic recollection thesis, explicated by the Neoplatonic rule that 'all are in all, in a manner appropriate to each' (Proclus *El. Th.* § 103). Plotinus had it right that the soul stands on the cusp between multiplicity and unity, between becoming and being, and possesses the capacity to turn its attention to either extreme: we are eternal in essence and temporal in activity (*El. Th.* § 191.3), perpetually existent yet subject to becoming (§ 192.1–2). Moreover, it remains characteristic of the later Neoplatonist outlook to describe this momentum as one from multiplicity to unity.

Since the Neoplatonists came to regard the soul as a higher being fully descended into the lower cosmos, they naturally emphasised the importance of the 'turn upward' or 'inward', if anything even more sharply than Plotinus: the soul in the world carries a 'great cosmic responsibility' to mediate between the higher and the lower (Chlup 2012: 29). The emphasis fell less on distinguishing two facets or persons within the soul, and recognising which is really *us* – our real being, namely the 'ultimate fringe' or illumination linking us to the world below (Plotinus *Enn.* III.4.3.21–7) – and more on bringing the soul's temporal *activity* into alignment with its essential *being.*

The soul's temporal activity might be virtuous or non-virtuous, and the 'turn' of the soul coincided with the formal beginning of the philosophical curriculum, and a shift from pre-philosophical modes of virtue to authentically philosophical excellence.[10] Following Porphyry's systematic adaptation of Plotinus' treatise *On Virtue* (*Enn.* 1.2),

[10] See Dillon 1996a; Hadot 1996; Edwards 2000.

Iamblichus and his successors developed a complex 'ladder' of virtues that talented human beings might climb.

Developing virtue

First, they envisage 'pre-philosophical' virtues that belong to us either (1) by our natural constitution (φυσικὴ ἀρετή, over which we have little control, as a lion is *bound* to be courageous and an ox temperate) or (2) habituation (ἠθικὴ ἀρετή), which might be fostered by absorbing myths and stories and rhetoric (such as the Pythagorean *Golden Verses* or the *Handbook* of Epictetus, although we might also envisage the use of moral myths in rhetorical schools as serving this function). Then, when we embark on philosophy, or become a 'philosopher' in training, we come to develop (3) civic virtue (πολιτικὴ ἀρετή), which works on the right organisation of our own soul and the souls of our fellow citizens, placing reason in charge over spirited emotion and appetite, but still looking primarily to the outer world and our actions in it (cf. Olympiodorus *In Phaed.* 20.4; here we read works such as Plato's *Alcibiades* and *Gorgias*), and then (4) get to work on 'purifying' or separating the soul from the body so far as possible (reading works such as Plato's *Phaedo* to achieve purificatory virtue or καθαρτικὴ ἀρετή), culminating in (5) the contemplative philosopher who surveys the intelligible realm with θεωρητικὴ ἀρετή (through works such as the *Philebus* or *Parmenides*), and (6) the 'theurgist' who is identified with the divine (Olympiodorus *In Phaed.* 8.2.1–20).

The hierarchy of the virtues in Proclus' Athenian school can be summarised as follows:

Below Philosophy (innate or trained)
  Natural (*phusikos*)
  Habituated (*ēthikos*)
Philosophical
  Constitutional/Civic (*politikos*)
  Purificatory (*kathartikos*)
  Contemplative (*theoretikos*)
Beyond Philosophy (divine, inspired)
  Paradigmatic
  Hieratic[11]

[11] This reconstruction is based on that of Westerink 2010: XXXIX–XL.

The second series of virtues – constitutional, purifying and contemplative – fall within the rubric of 'philosophy'. Virtues that accrue to us *earlier* in life are pre-philosophical, arising from innate talent or habit, while virtues that accrue to us *after* the basic curriculum of philosophical study is complete are treated as lying 'beyond' philosophy, stemming from a divinely inspired source. But the philosophical virtues themselves must be trained with the aid of an able teacher. The Neoplatonists argue that a teacher is essential in making the first step towards the 'reversion' of the soul towards the intelligible, as one (possibly Iamblichean) passage in Simplicius suggests:[12]

> Neither are significant expressions wholly separate from the nature of beings, nor are beings detached from the names which are naturally suited to signify them. Nor, finally, are intellectual concepts extraneous to the nature of the other two; for these three things were previously one, and became differentiated later. For Intellect (νοῦς), being identical with realities and with intellection (νόησις), possesses as one both beings and the intellectual concepts of them, by virtue of its undifferentiated unity (ἀδιάκριτος ἕνωσις), and there [sc. in the intelligible world] there is no need for language. . . .
>
> When, however, the soul has fallen into the realm of becoming, it is filled with forgetfulness, and requires sight and hearing in order to be able to recollect. For [the soul] needs someone who has already beheld the truth, who by means of verbal expression (φωνή) uttered forth from the ἔννοια also moves the concept within [the soul of the student] which had until then grown cold. This, then, is how the need for φωνή came about . . . νοήσεις join the learner's concepts to those of the teacher. . . . When νοήσεις are set in motion in the appropriate way, they adjust themselves to realities, and thus there comes about the knowledge of beings, and the soul's spontaneous ἔρως is fulfilled (*In cat.* 12.13–13.4, trans. adapted from Chase 2003: 27–28).

This passage, drawn from Simplicius' *Categories* commentary, focuses on the effects on the student of beginning the Aristotelian curriculum with the *Organon*, which trains the soul, now fallen fully into the world, to use the reason innate in language to recover its birthright. After this kind of basic Aristotelian training, which continues with the study of natural forms and ethics, the philosophical virtues should be trained through the study of the Platonic dialogues with an able teacher, in a reading curriculum laid down by Iamblichus:

[12] On this interesting passage and its attribution see Hoffmann 1987b. Griffin 2012 also attempts to build a case for attribution to Iamblichus.

First Cycle of Ten Dialogues

Civic (*politikos*)

1. Introduction: *Alcibiades*
2. *Gorgias*

Purifying (*kathartikos*)

3. *Phaedo*

Contemplative (*theoretikos*)
. . . of names (*onomata*)

4. *Cratylus*

. . . of thoughts (*noēmata*)

5. *Theaetetus*

. . . of realities (*pragmata*)
. . . which are natural (*phusikos*)

6. *Sophist*
7. *Statesman*

. . . which are divine (*theios*)

8. *Phaedrus*
9. *Symposium*

Culmination

10. *Philebus*

Complete Dialogues On Nature

11. *Timaeus*

On Theology

12. *Parmenides*[13]

Within this scheme, what are called the 'political' (or civic) virtues constitute the first level of philosophical excellence. As Proclus and commentators who follow him argue, what is under consideration here is the correct organisation of the *parts* of the soul under the guidance of reason, or λόγος (see, e.g., Olympiodorus *In Alc.* 4.15–5.1): inspired by *Republic* 4, this organisation of the parts of the soul into a structured whole is the most elementary kind of philosophical unification available to the Neoplatonist

[13] This reconstruction is based on that of Westerink 2010: XXXIX–XL.

student. (It is also, in broad terms, one inspiration for Korsgaard's 'constitutivist' model of normativity; it is mainly the Plato of *Republic* 4 and the tripartite soul whose conceptual resources are put to use in her model of normativity and virtue). But for Proclus this kind of structured whole can only be the beginning of the ladder: higher forms of unity follow with the remainder of the curriculum, as the soul purifies itself, turns 'upward' to contemplate more fully unified and eternal Forms, and finally achieves genuine unification or ἕνωσις. This distinction is important, since it is these higher, eternal forms of unification that allow Proclus' self to 'pre-exist' the process of self-constitution.

Self-knowledge and the journey from multiplicity to unity: Proclus on the *Greater Alcibiades*

As the First Cycle of Ten Dialogues above illustrated, at the beginning of the Platonic task of philosophical self-development Iamblichus and Proclus followed earlier Platonists who favoured the *Greater Alcibiades* as an introduction to philosophy,[14] crediting that dialogue's emphasis on self-knowledge.[15] Proclus, following Iamblichus, adapts this basic story of self-transformation to his metaphysical scheme, beginning with the question of *what kind of being* we are. He explains the importance of studying the *Alcibiades* as follows:

> The most valid and surest starting-point . . . is . . . the discerning of our own being. If this is correctly posited, we shall in every way, I think, be able more accurately to understand both the good that is appropriate to us and

[14] Albinus (*Prologue* 15–16), for instance, had already argued that philosophy should begin with its portrayal of Alcibiades, the paradigm of a naturally adept beginner. The dialogue helps the young student to 'change direction, turn inwards, and recognise what he should be caring for', namely his soul (5.16–17).

[15] The narrative arc of the *Alcibiades* treats the intellectual and moral journey of its titular protagonist as Socrates 'turns' him to philosophy on the verge of a public life of political affairs (τὰ πολιτικά). Alcibiades is first made to recognise that his innate advantages and talents (φύσις) do not make him a statesman (πολιτικός), but that his education has been purely 'natural' or φυσικός in character (*Alc.* 106e, cf. 119c). Having recognised this, he turns next to investigate, with Socrates, what is 'better' in matters of public life (109c), and seeks a kind of excellence (ἀρετή) which is πολιτικός, that is, 'constitutional' or civic, whether for a state or for an individual soul (ψυχή, 126d). This ἀρετή proves to require a τέχνη which 'cares for' oneself in order to better oneself (127e), and that requires self-knowledge, the recognition that the 'human being is the soul alone' (129c–130a) – and especially the ἀρετή of the soul (133b), which, on the analogy of the eye's own excellence (132b), is σοφία, and is also godlike. Should we succeed in securing this ἀρετή, and turn from our trust in φύσις (cf. 122c) to the condition of the true statesman (πολιτικός) and beyond, to 'wisdom' in the soul itself, we will have obtained a certain 'likeness to god' (cf. *Theaet.* 176b), we will become pleasing to god, and we will benefit the polis as a whole – with divine support (*Alc.* 135d).

> the bad that fights against it. Of the things that are, as it is natural for each one to differ in being, so also their completion varies in different cases, according to their descent in the scale of being. Either both being and the good proceed from the same heart and final spring, as Aristotle asserts (*EN* 1096a23) . . . or the good has come to things from one source, higher and more holy. (Proclus *In Alc.* 1.3–2.3, trans. O'Neill 1965: 1)

It is essential to discover what kind of entity the soul *is* in order to know what is best for it – how to foster its virtue and bring it to completion. This determination needs to be made against the complex canvas of Proclan metaphysics:

> It is surely necessary . . . to ascertain their being before their completion; for perfection is not of itself, but of the being by which it is participated. This, then, must first be considered, whether it is one of the undivided beings or of those divided in association with bodies or of those in middle rank; whether it is of those that are eternal or those that subsist for all time or all those that come to be in some portion of time; and whether it is of those that are simple and established prior to all composition, or of those that are composite indeed, but forever composed 'in indissoluble bonds', or of those that can be resolved again into the elements from which they were composed. (Proclus *In Alc.* 4.3–11, trans. O'Neill 1965: 2–3)

Every soul, in fact, 'is intermediate between the indivisible principles and those which are divided in association with bodies' (Proclus *El. Th.* § 190). This is what we are, and Proclus also argues that Alcibiades can be portrayed as symbolic of ourselves, as the soul experiencing coming-to-be:

> Each one of us and of mankind in general is more or less clearly subject to the very same misfortunes as the son of Kleinias. Held bound by the forgetfulness incident to generation and side-tracked by the disorder of the irrational forms of life, we do not know ourselves, and we think we know many things of which we are unaware, by reason of the innate notions present in us according to our being. (*In Alc.* 7.1–6, trans. O'Neill 1965: 5; cf. *El. Th.* §§ 194–5)

This entity, symbolised by Alcibiades, naturally the companion of Socrates but 'dragged down' by multiplicity and irrationality like Plotinus' lower soul tempered by its wiser neighbour (*Enn.* 1.2.5), is the rational soul. What is its situation, or 'misfortune'?

According to the analogy of the extremes we must relate Alcibiades to the rational soul, to which are still attached the emotions and the irrational powers, as it were plotting against the life of reason and like the Titans attempting to rend it; but the intellect, like Athene, is set above, keeping it from sinking in the scale and tending to implication in matter. For it is the function of Athene to preserve life undivided, 'for which reason Pallas Athene is called "saviour"', but of the Titans to divide it and to entice it to the process of coming-to-be' (Proclus *In Alc.* 43.4–44.4).

This soul can be rescued, and the imagery of this rescue can be portrayed mythologically in a number of idioms. For example, the character of Socrates in the dialogue, as he 'turns' Alcibiades to philosophy, can be seen as the soul's saviour:[16]

> Think of Alcibiades as twofold, both as a soul simply and as a soul using a body. For these two are not the same: likewise neither the steersman and the man in himself nor the driver and the subject; and in general the unparticipated is other than the participated, what is by itself is other than what is observed along with another, and the transcendent differs from what has received its rank in conjunction with some other. Now ... Alcibiades is understood in two ways ... insofar as he is a soul, Socrates preserves the analogy of the intellect toward him, but insofar as he is a soul using a body, that of the good spirit; and the same person is a spirit as regards man, and intellect as regards soul. (Proclus *In Alc.* 45.13–22, trans. O'Neill 1965: 28–29)

Thus the 'upper' soul is unparticipated by multiplicity, and the 'lower' soul is participated by multiplicity. The core process of change is one of movement from multiplicity to unity; but now we might see this as less a *cognitive* process or change of mind (or being), and more a *practical* process or change of *activity*, once the mythological language is correctly translated:

> As, too, the intellect (*nous*) is always active in our regard and ever bestows the light of intelligence, both before we incline to irrationality and when we live with the emotions and after these have been stilled by us, but we are not always conscious of it except when, freed from the many waves of temporal process [cf. Porphyry *Vit. Plot.* 22.25.31–5; Plotinus *Enn.* V.1.2, 15], we anchor our lives amid some calm (for then intellect is revealed to us and as it were speaks to us, then what was formerly silent and quietly present gives us a share of its utterance), so also the divine lover is both presented to the beloved before the many lovers and with them and after them, but in silence and quiet and forethought alone; but when they have left off he gives a share of conversation to the beloved, offers him mutual intercourse and reveals his identity and that his love is provident, of the form of the good and elevating, not like that of those many lovers, divided, deficient, implicated in matter and concerned with mere images. (Proclus *In Alc.* 44.11–45.6, trans. O'Neill 1965: 28)

[16] I have explored the allegorical use of Socrates in this dialogue and others in Griffin 2014.

We find ourselves as the rational soul suspended between unity, represented by the mind (νοῦς), and multiplicity, represented by the diversity and divisibility of bodies. In the *Alcibiades*, as the starting-point of the 'ladder' of philosophical virtues discussed above, we are invited to pursue civic or πολιτικὴ ἀρετή, to work towards the harmony and coherence of the parts of our soul under the guidance of reason. Alcibiades signifies our own condition; Socrates signifies the mind; Alcibiades' 'vulgar' or common lovers represent the divisibility 'below' us. Now it is up to us, first, to get control of the diversity of the parts of the soul and bring them under the common banner of λόγος; this is what civic virtue is, what the *Alcibiades* especially trains students to do,[17] what Plato's *Republic* 4 appears to enjoin, and roughly speaking the inspiration for the kind of constitutive metaethics worked out by Korsgaard. But for Proclus and his contemporaries, as I have suggested above, it is important that this is only the first step in the process of ascent on the ladder of virtues. It is also in our power to shift our gaze away from temporality altogether, 'upward' towards greater unity and eternity, and ultimately towards our own causes – towards the inner architecture of νοῦς itself, toward οὐσία. As Damascius puts it, 'the soul can choose three different ways: of ruling the lower, of finding within itself the principles of its actions, or of looking up towards causes higher than the soul' (*In Phaed.* 1.74).

Proclus on unity

So far we have worked with a relatively generic concept of unity, and a loosely described process of 'unification' or ἕνωσις that the Neoplatonists prescribe for the soul. I would like to dwell, briefly, on the kind of unity that Proclus has in mind, drawing primarily on the *Elements of Theology*. The opening propositions of the *Elements* discuss 'The One' (τὸ ἕν), the fount of Neoplatonic metaphysics.[18] This is something that every multiplicity participates (in order to avoid an infinite regress where nothing has unity), but it is not identical with any of its posterior participants, neither groups that it unifies nor more mysterious entities called 'units' (*henads*) (*El. Th.* §§ 1–6; cf. Siorvanes 1996: 167–79; Chlup 2012: 112–18). It is a challenge to get clear about what Proclus' One, or the Neoplatonic One in

[17] Proclus and Iamblichus are both at pains to stress that this is not *all* it does: it contains all philosophy 'as if in a seed' (Iamblichus fr. 1 Dillon). Damascius, at least according to Olympiodorus, stresses its civic σκόπος or focus more heavily (Olympiodorus *In Alc.* 5.13–15).

[18] See for example Dodds 1963: §§ 1–6; Chlup 2012, 48–111; Proclus' 'higher' metaphysics or theology are surveyed in detail in the introduction to Westerink and Saffrey 1968.

general, might be: it is beyond being, identical with the Good for all beings, and unnameable. A broad array of vocabulary is used about it, and a great deal of later systematic theological thought in other systems of thought transfer that language to explain the nature of God.

The opening propositions of the *Elements* suggest that Proclus and his colleagues are not exactly treating 'The One' as a *thing* – there are plenty of really existing 'ones' in the universe, the ἑνάδες (§ 2), but they all depend upon τὸ ἕν, the subject of § 1, which might be most cleanly described as a *principle of individuation* that is not itself any particular individual being. τὸ ἕν is offered as the basic criterion for explaining identity or individuation. As such, it brings Proclus down on one side of the debate whether there is any such thing as a primitive 'thisness' of things, apart from their qualitative constitution or description. Proclus thinks that things do have a 'thisness', and it ought to be called τὸ ἕν: in a modern doxography he would fall on the former side of the question, with Scotus, Kant and Peirce and against Leibniz, Russell and Ayer (see, e.g., Adams 1979: 5).

According to the first proposition of the *Elements*, what makes a plurality into a unified whole *just is* τὸ ἕν – 'participation' in τὸ ἕν as a principle of individuation is a necessary condition for unity, and τὸ ἕν is the first thing that there is (cf. Proclus *In Parm.* 696.32 ff., with Dodds 1963: 188): 'For suppose a manifold in no way participating τὸ ἕν. Neither this manifold as a whole nor any of its several parts will be one; each part will itself be a manifold of parts, and so to infinity; and of this infinity of parts each, once more, will be infinitely manifold' (*El. Th.* § 1). By the same token, for Proclus, to be a real, unified individual is not just a matter of possessing a unique bundle of qualities, but of a further feature: bearing a participation relationship to τὸ ἕν. Everything that bears such a relationship can in some sense be 'picked out' as a unity; without that relationship, nothing can really be 'picked out'. Proclus' primary argument that this *must* be so relies on the intuition that *without* a principle like τὸ ἕν, reality would be structured around an indefinite infinity of infinities. From this point of view, the intuitive function of τὸ ἕν is to explain how some subject of discourse is *definite* – how it is something delimited, not ἄπειρον, and something rather than nothing at all (§ 1.10–14). Based on that opening proposition, it might be most promising to treat τὸ ἕν as a principle of individuation: it is what makes this *this* and not something else, what makes me *me* and not someone else.

Everything that participates in τὸ ἕν, individuation, is both one and not-one: it cannot be individuation itself, since by definition the participation relationship requires a differentiation of participant from participated (§ 2).

But we can trace a hierarchy of degrees of individuality, from the real primary individuals or ἐνάδες (which in Proclus' system answer to the role of the gods), to a range of unified groups (ἑνώμενα). It might be reasonable to treat ἕναδες as 'individuals', which are distinguished by their uniqueness or ἰδιότης, which they also communicate to 'lower' groups.[19]

Entities in Proclus' system, as we have already noticed from the *Alcibiades*, are ranked according to their degree of participation in individuation. But the role of τὸ ἕν is not just descriptive, but also prescriptive: things *are* unities, and it's *good* for them to become more unified until they are fully individuated. The *good* of human beings is to be completed in this way, and the target of their completion is ἕνωσις, to become authentic unities or fully individuated entities (§ 9): thus τὸ ἕν can also be called 'the Good' for all things (§ 13). Now some entities, such as bodies, are entirely dependent on external forces for their unity and individuation (§§ 15, 16); others, such as intelligences (§ 169) and henads (§ 114), are already unities; but souls, as intermediates, are *self-sufficient* or *self-unifying* agents (§§ 9, 44, 83, 189, 190). The being or existence of a soul is derived from itself as it reverts upon itself (§ 189), and it constitutes itself in the process, aspiring to itself as its own source of existence (§ 43); thus souls are in a sense *self-individuating*, although they depend on higher principles of individuation beyond themselves. This 'individuation' is the soul's endeavour to make itself 'complete', to realise its own unity or Good. In this whole process it is again important to stress that the real unity of the soul, its henadic and intelligible essence and individuality, are always there in eternity, acting paradigmatically and in a sense 'behind the scenes' as the soul engages in its work of self-constitution.

Conclusion: self-unification as a moral goal

Korsgaard argues, interpreting Plato, that 'the kind of unity that is necessary for action cannot be achieved without a commitment to morality' (2009: xii): the source of normativity for human beings just *is* our attempt to bring ourselves into unison, like the soul of the *Republic*, brought into harmony under the aegis of reason. 'The principles of practical reason serve to unify and constitute us as agents, and that is why they are normative ... normative principles are in general principles of the unification of manifolds, multiplicities, or, in Aristotle's wonderful phrase, mere heaps, into objects of particular kinds (*Metaphysics* 8.6, 1045a10)' (Korsgaard 2009:

[19] Butler 2005; Chlup 2012: 115.

27). As I suggested above, Proclus would agree with the heart of this claim: normative principles are principles of unification. Hence the One, as the goal of each being's completion (*El. Th.* § 9), is the Good (§ 13). We can be held to what Korsgaard calls 'constitutive standards' (2009: 28), and 'bad actions, defective actions, are ones that fail to constitute their agents as unified authors of their actions' (2009: 32). Thus Korsgaard argues that

> Being a giraffe is doing something: a giraffe is, quite essentially, an entity that is always making herself into a giraffe. . . . To be a giraffe is simply to engage in the activity of constantly making yourself into a giraffe . . . leading the life of an unhealthy giraffe is not a different activity from leading the life of a healthy giraffe. It is the same activity, badly done. (2009: 36–7)

Again, I think that Proclus would broadly agree, but only to a point. Being a giraffe is not *only* doing something: that is the *activity* or ἐνεργεία, but the being has to be distinguished as separate from, and prior to, the activity (see again *El. Th.* § 191.3). For Proclus the individuality of a thing, its participation in τὸ ἕν, is not identical with its being, but comes *before* its being. And again, the being of the thing lies *before* its teleological process of completion – say, the being of the soul and its activity of self-constitution and self-animation. Thus Proclus' system naturally digests the question 'how can I constitute myself before I exist?' that Korsgaard introduces (2009: 20, 35), employing the explanation that my being and henadic individuality exist in eternity, before any temporal action or ἐνεργεία occur, while that latter, temporal action is the natural expression or emanation of my real being.

In conclusion, we have found that Proclus regards unity as the ground of the human good – a unity that requires self-reflection, reason and self-constitution. In seeking the ground of normativity in unity (and the explanation of human defects and flaws in the lack of unity), Proclan ethics might, I think, be profitably compared to the constructivist project of philosophers such as Korsgaard, whom I have brought into contrast with Proclus here. Proclus invites human beings, in a nutshell, to 'create ourselves' in a sense rather like Korsgaard's. But the difference is also crucial: unlike Korsgaard (and Korsgaard's Aristotle, and for that matter Korsgaard's Plato), Proclus stresses that we are already a unity in one sense, in our eternal being or essential *character*, and we are only a multiplicity in respect of our temporal *action*. The soul exists in eternity and acts in time. Moreover, from the later Neoplatonic point of view, its unity and individuation are prior even to its existence.

In this way Proclus' system comes with a built-in metaphysical solution to the 'paradox of self-constitution': we are already 'there' before the activity of self-constitution begins. This 'solution' comes with the implication that souls on earth – each one an embodied expression of a unique and eternal unity – are obliged to do work, the kind of work that Alcibiades is invited to do as he begins to climb the ladder of virtues, in order to become real wholes *in action* and not only in character. That is what is good for us: that is what brings us to completion. As the passages from the *Alcibiades* commentary illustrate, Proclus regards the discovery of this fact as the goal of philosophical education, as we may choose to focus the soul's attention on uniting its inner diversity (civic virtue), or withdrawing from diversity altogether (purificatory virtue), or committing its gaze to the unity that pre-exists its project of self-constitution: all three of these activities are, for Proclus, essential for the project of 'creating oneself' as a fully fledged person.

CHAPTER 12

Deficient causes: Augustine on creation and angels

Gillian Clark

> Do not look for the efficient cause of a bad will (*voluntas*): it is not efficient but deficient, because the will also is not effective but defective. To defect from that which supremely exists, to that which exists less, is to begin to have a bad will. But wanting to find the causes of these defections, when the causes are (as I said) not efficient but deficient, is like someone wanting to see darkness or hear silence. Both are familiar to us, darkness only through the eyes and silence only through the ears, but by lack of form (*speciei privatione*), not by form. So let no one seek to know from me what I know that I do not know, unless it be to learn not to know that which must be known as impossible to know. . . . Our mind sees intelligible forms by understanding, but where they are defective, it learns them by not knowing. For *who understands failings?* [*delicta quis intelligit?* Ps. 19:13] (8) But this I know: the nature of God can never, anywhere or in any part, be deficient; and those things which are made from nothing can be deficient. In so far as they have being and do good things (for then they are doing something), they have efficient causes; in so far as they are deficient and therefore do bad things (for what are they then doing but futility?), they have deficient causes.
>
> (*Civ. Dei* 12.7–8)[1]

Unde malum, where does badness come from? Who created it, what causes it? In *Confessions* (7.3.5) Augustine asked 'Who put and sowed in me this root of bitterness [Heb. 12:15], when I was wholly made by my most sweet God? If the devil is responsible, where does the devil come from? If he by his perverse will was made from a good angel to a devil, where did that bad will in him come from by which he was made a devil, when the angel was

Warm thanks to the editors and readers, to the participants in the Creation and Causation seminar, to Karla Pollmann for comments on a draft, and to Stephen Clark for advice on Plotinus. Mark Edwards (this volume) and James Wetzel (2012b) offer different, and inspiring, approaches.

[1] All translations are my own. For *City of God* I have also consulted Combès 1959 (books 6–10), Bettenson 1972, Combès 1994 (books 11–18), Dyson 1998 and Babcock 2012 (books 1–10). Titles of Augustine's works follow Fitzgerald 1999.

made entirely by the good creator?' His attempts to answer such questions began in *De libero arbitrio* (2.20.54) at least thirty years before he embarked on *City of God*, where he returns to them in characteristic style. There are word-plays, more obvious in the Latin, on efficient and deficient; knowing and not knowing; doing something and doing, literally, nothing. We should not dismiss word-play as the result of rhetorical training: it seizes attention, and Augustine uses it to make people think. What exactly do we mean by 'nothing'? If an 'efficient cause' is a cause that gets something done, what counts as 'something'? Exegesis, similarly, makes people think why the text uses a particular word or number (Augustine was evidently one of those for whom numbers are a language), what exactly the words mean, and what more they signify. The passage quoted above has an example. Augustine cites a psalm-verse, *Delicta quis intellegit?*, which modern translations interpret as 'Who understands his own faults?' He uses it to demonstrate that faults or failings are not intelligible.

'Deficient cause' is a word-play that makes us think, but what does it contribute to debate on creation and causation? *City of God* can be a frustrating text, because Augustine's aim is precisely what he says (11.1). First, it is to refute (books 1–10) the enemies of the City of God, who think that many gods should be worshipped for blessings in this life or the next. Next, it is to discuss (books 11–22) the origins, course and destined ends of the two cities: the city of God, known to us from scripture, which worships the true God who is its founder (*conditor*, which also means 'creator'), and the earthly city which prefers its own gods. Sometimes this aim required Augustine to challenge rival views of how the universe and its inhabitants came to be: that the universe was not created but is coeternal with God; that the universe is cyclical; that God created the universe, not because it is good, but as a way of dealing with evil; that God created the universe, but out of pre-existent matter; that souls are not created out of nothing, but are actually or potentially consubstantial with God. But Augustine did not recognise an obligation to engage in careful understanding and balanced discussion of the other views he encountered.

Augustine's reading of philosophy and theology was limited, partly because he had little time in his earlier career as a teacher of literature and rhetoric and even less when he became a priest and then a bishop, partly because his Greek was not fluent. He learned about philosophy chiefly from Cicero. He read Aristotle's *Categories* as a rhetoric student (*Conf.* 4.16.28), but does not show familiarity with the debate on whether it is logic or metaphysics. Ideas came from his years as a Manichaean 'hearer', and from the sermons of Ambrose of Milan, who included unacknowledged material from Basil of Caesarea.[2] At Milan Augustine was also given translated Platonist works (*Conf.* 8.2.3). *Confessions* shows their impact, but debate continues on what exactly they were (O'Donnell 1992, II: 413–18) and how far Augustine understood them, given that he was not familiar with the works of Plato and Aristotle which Plotinus and his students knew as well as Augustine came to know his favourite books of scripture (O'Brien 2012; Yudin 2013). He cited (*Civ. Dei* 13.14) Cicero's partial translation of the *Timaeus* (27d–47b), but not the partial translation (to 53c) and commentary Calcidius wrote around 320. In 415, when Orosius asked him for a response to Origenist theology, he read *De principiis* in the Latin translation of Rufinus or of Jerome, and Origen appears briefly in *City of God* when his ideas require correction (O'Daly 1999: 261–2). Augustine also responded to questions. But when he refers to 'other interpretations' of scripture (e.g. *Gn. litt.* 2.4.7–5.9), he rarely attributes them or engages with them in detail. Perhaps he read more than he cited; but when Cicero and Manichaeans and representative or unnamed Platonists feature in *City of God* it is as targets, not as dialogue partners (Clark 2007).

Scripture, not philosophy, was Augustine's starting point, because he held that human reason is not enough and that scripture has divine authority:

> That God made the world, there is no one we can believe more securely than God himself. Where did we hear him? Nowhere, for now, better than in the holy scriptures, where his prophet says 'In the beginning God made heaven and earth'. Was this prophet there when God made heaven and earth? No, but the wisdom of God was there, through which all things were made, which 'also passes into holy souls and makes them friends of God and prophets' [Wisdom 7:27], and inwardly and soundlessly tells them His works. (*Civ. Dei* 11.4)

Augustine takes it as fundamental truth that canonical scripture (the books accepted by the consensus of churches: *Doc. Chr.* 2.24–8) is divinely inspired and is therefore true, consistent and authoritative. Scripture tells us that God made the world, and that the wisdom through which the world was made also inspired scripture. So Augustine's conclusions had to agree with all of scripture, and his preferred working method was to engage with scripture as he had once engaged with Virgil, connecting words and associations across his favourite books. Just as commentary on Plato was

[2] Perhaps Ambrose discussed Basil's argument (*Hex.* 3.8) that there really are waters above the heavens? *Gn. litt.* 2.4.7 does not say who held this view; the Latin translation by Eustathius Afer was *c.* 440.

for Platonists a way of thinking, so for Augustine reflection on scripture was a way of thinking, not an evasion of thinking.

Augustine knew that we cannot just say 'It says in the Bible . . . '. We have to be clear what the Bible says, and Augustine used his experience as a teacher of literature in his work as a preacher and exegete. He knew that the first task is to establish the correct text, and that it is not always obvious what the text means. He held that scripture is accessible to everyone, but also profound enough for anyone, and that scripture trains our minds and directs us to hidden truths by making us think about puzzling statements or apparent contradictions or unanswered questions (*Civ. Dei* 11.19). He warned that there are constraints on thinking and expression. Non-Christian philosophers can speculate as they please (*Civ. Dei* 10.23), but careless talk by Christians costs lives, because it may have disastrous consequences for belief. Christian philosophical reflection on creation and causation must be compatible with scripture. If we have to choose between Moses and Plato, Moses wins – not just because he came first and Plato may have borrowed from him (*Civ. Dei* 11.21), but because he is the divinely inspired author of the book of Genesis. This is not to say that if science (natural philosophy) seems not to agree with the Bible, so much the worse for science. In *De Genesi ad litteram*, his fullest commentary on the creation narrative, Augustine says that exegesis must make sense in relation to rationally established knowledge (*Gn. litt.* 2.9.21). Otherwise Christians will just look stupid (1.19.39), and non-Christians will not trust scripture on questions where it is impossible to establish a scientific theory, such as resurrection and eternal life.

Scripture does of course exclude some theories of power and structure. If God created out of nothing all that there is, and any power that acts in the created universe does so at his command or with his permission, then it cannot be the case that matter is coeternal with God; or that the universe is not governed by divine providence; or that the upper levels are so governed but the lower levels, the damp and earthy bits, are left to chance (*Gn. litt.* 5.21.42); or that a rival power is responsible for evil (*Civ. Dei* 11.21). But that still leaves plenty to think about. Some people did not accept that God made the world and that God's wisdom inspired the scripture which tells us so. Some accepted both these claims, but still asked when and how and why God made the world, and where it all went wrong, so that now we see evil and suffering. Augustine therefore returned many times to the creation narrative of Genesis 1–3 (Pollmann 2007).

While he was working on books 1–10 of *City of God* (begun 412/13, finished by 417), Augustine completed his twelve-book *De Genesi ad litteram* (*c.* 401–16), which is deliberately not an allegorical interpretation, but is also not a historical account of 'what actually happened'. Augustine thought that events, as well as words, in scripture often have a hidden significance, indicating realities we cannot see, and that the creation narrative of Genesis is uniquely significant (Fiedrowicz 2002; Pollmann 2007). *De Genesi ad litteram* is the text most often cited in discussions of Augustine on creation and causation (e.g. Williams 1999). This chapter follows another route: *City of God* on the creation of angels. In *Confessions* Augustine's concern was 'how do I go wrong?', not 'where did the devil come from?' In *City of God*, seeking the origins of the heavenly and earthly cities, he discussed the creation of the angels who are their citizens together with human beings. Scripture testifies to holy angels who worship God and carry out his will, and to bad angels, led by the devil, who resist God's will and try to lead humans astray. But where in the Genesis narrative is the creation of angels, and what caused some of them to turn away from God?

The second part of *City of God* begins by affirming the authority of the single, divinely inspired *scriptura* over the multiple *litterae* produced by people (Vessey 2012a): 'We speak of the city of God, to which that scripture bears witness, which, not by the random movements of minds but by the disposition of supreme providence, surpasses all the writings of all the peoples, and by divine authority has subjected to itself all kinds of human intellects' (*Civ. Dei* 11.1). Scripture inspires love of the city of God and of its founder, but the citizens of the earthly city prefer the false gods: 'the impious and arrogant gods who are deprived (*privati*) of the immutable light of God which is common to all, and so, reduced to a kind of needy power (*egenam potestatem*), pursue their own private privileges (*privatas potentias*)'. Augustine returns later (11.13) to the loss of light and to the word-play on *privatus*; here, his point is that false gods want worship for themselves, whereas the good and holy gods worship God. He takes for granted his earlier explanation (9.23 and briefly in 10.1) that scripture, as well as Platonist philosophers, may call angels 'gods'; he also takes for granted that the false gods of pagan literature and cult are the same as the rebel angels of scripture.[3] He briefly explains that books 1–10 answer the enemies of the holy city, and that he will now discuss the origins, course and destined ends of the two cities which are intermixed in this world.

[3] It is surprisingly difficult to find a clear statement, but in *Civ. Dei* 18.18 'avenging evil angels' punish Diomedes for his crimes at Troy.

This introduction ends 'and first I shall say how the origins of these two cities began with a difference among angels'.

In fact he does not reach the angels until 11.7. (Book divisions are Augustine's; chapter divisions were added by medieval editors.) Book 11 opens, as is usual in *City of God*, with a replay of themes from the previous book, where Augustine had also returned to the question of creation. Book 11 is often read without reference to book 10, because its opening is a second preface distinguishing 1–10 from 11–22. These two parts are often published in separate volumes and studied by different groups of readers. Books 1–10 interest classicists in search of Varro on Roman religion, Porphyry on Platonist philosophy, and late antique reception of classical texts; books 11–22 offer more theology and exegesis. But there was no break in Augustine's work schedule (Orosius *Hist.* 1 prol. 11), and there is continuity from book 10 to book 11.

Book 10 begins by reminding readers that Augustine had (in book 8) chosen the Platonists as the best philosophers with whom to discuss the achievement of happiness after death, because Platonists understand that the happiness of the immortal soul requires participation in the light of the immutable God, who created the soul and the world. Augustine's sentence structure here allows unwary readers to think, for a while, that Platonists too say that God created the soul and the world.[4] The mistake of the Platonists is to allow the worship of many gods, whereas the holy angels (10.7) want us to worship God. Throughout book 10 Augustine contrasts the activity of angels, who carry out the will of God, with the activity of demons, who cannot (as some Platonists claim) purify the soul or mediate between human and divine. The only true mediator is Christ (10.24), who is prophesied in scripture. But Platonists, who rely too much on human reason, do not accept the incarnation (10.29), and do not accept what 'divinity', i.e. scripture, says: that 'the soul is not coeternal with God, but was created and had not previously existed', [*divinitas*] *animam quoque ipsam non Deo coaeternam, sed creatam dicit esse, quae non erat* (10.31). Augustine does not spell out the contrast between a soul which God created from nothing, and a soul which is a fragment of the divine or which is capable (10.29), once purified, of becoming consubstantial with God.

[4] *Sapere potuerunt licet immortalem ac rationalem vel intellectualem hominis animam nisi participatio lumine illius Dei, a quo et ipsa et mundus factus est, beatam esse non posse*: 10.1; also 10.2 with reference to Plotinus, but see *Enn.* v.2.

Platonists do not accept that the soul was created, because, they say, anything that has not always existed cannot exist for ever: *nisi quod semper ante fuisset, sempiternum deinceps esse non posset* (10.31) (That is, it cannot exist for ever unless it cannot *not* be.) But Plato himself says that God created the universe and the gods within it: they come into being and have a beginning, but, by the will of the creator, they will not have an end (*Tim.* 41b). The Platonists counter that this is a beginning not of time, but of *substitutio*, which here must mean logical, not chronological, dependence. Augustine presents their example in the form he uses for a quotation, but he does not specify 'in their own words', and it is often difficult to distinguish a quotation from Augustine's version of 'this is what they say'. If it is a quotation, the author has not been identified (see Sorabji, Chapter 4 in this volume).[5] 'It is like this, they say: if a foot had always for eternity been in the dust, a footprint would always be under it. No one would doubt that the footprint was made by the pressure of the foot; but neither would be prior to the other, although one was made by the other' (10.31). Augustine does not spend time on this example. Instead he asks 'so must we say that if the soul always was, its wretchedness also always was?', and once again uses the argument that this is not what the Platonists themselves say. Porphyry acknowledges that the blessedness of the soul will be more secure after the experience of evils, so blessedness has a beginning in time, yet will exist for ever. The conclusion, for Augustine, is 'let human weakness yield to divine authority'.

Book 10 ends (10.32) with affirmations that preview the second part: scriptural prophecies of the coming of Christ, the growth of belief in the true God, the final judgement, and the reign of the city of God. Book 11 begins with the supremacy of scripture, and continues (11.2) with the remarkable fact that people can, in thought, go beyond the mutable world to the immutable God, and learn from God that he alone created all there is. God speaks through the impact of truth on the rational mind, which is in the image of God. But the mind is weakened by dark and long-standing faults, *vitiis quibusdam tenebrosis et veteribus invalida est*, and cannot endure the light. So we need the help of Christ as mediator between God and human; and we need divinely inspired scripture as witness to what we cannot see (11.3). Augustine moves swiftly to the visible world and to belief, on the evidence of scripture, in God its creator (11.4). The first verse of Genesis, 'In the beginning God created

[5] Courcelle 1943: 174 n.3 suggested Porphyry, but the evidence is weak.

heaven and earth', raises a question about beginning: why did God create heaven and earth then and not before?

Augustine rejects any suggestion (by persons unnamed) that the world is eternal without beginning, and therefore was not created. This is ruled out by scripture, and by the order and beauty of the world. Others (also unnamed) recognise that God created the world, but say that it does not have a beginning in time, so that 'in some scarcely intelligible way' it was always made. This avoids the charge that the immutable God changed because he had a new idea, but leaves the problem of explaining how the soul changed: if the soul is not created but is coeternal with God, how could it change from blessedness to its present wretchedness? Augustine does not refer back to his discussion in 10.31; here, as there, his aim is to show that scripture is right, not to debate with Platonists. The unnamed philosophers of 11.5 agree that God created the world, but ask what Christians say about the time of the world. Augustine asks in return what they say about its place: 'why then and not before?' is like 'why there and not elsewhere?' There is neither time nor space beyond God's creation. Augustine moves rapidly through his argument (best known from *Conf.* 11) that the world was made not in time, but together with time. Time begins 'in the beginning' with creation, because creation provides the change and movement that makes time possible, as in the days of creation in Genesis which have morning and evening (11.6). But how can we understand these days? Equally rapidly, Augustine moves (11.7) through the interpretation of the 'days' which he set out more fully in *De Genesi ad litteram*. Here, at last, we come to the angels.

The six days of creation are not days as we understand them, measured by the sun, for the sun was created on the fourth day. So when God makes light and separates it from darkness, and there is evening and morning (Genesis 1:3–5), either this was corporeal light in some region far beyond our sight, or 'light' here means 'the holy city in the holy angels and blessed spirits', as when Paul says (1 Thess. 5:5) 'You are all children of light, children of day: you are not of night and darkness' (see further *Gn. litt.* 1.3.7–1.5.11). What then are the evening and the morning? Evening knowledge (twilight) is knowledge of the creation, morning knowledge (full light) is praise and love of the creator (as in *Gn. litt.* 4.24.41). Scripture says that there was evening, not that there was night (no light), because knowledge is not wholly lost provided the created being loves the creator. So there is one 'day' when the created light comes to know itself (evening) and praises the creator (morning), and so on through the days. God's rest on the seventh day is to be understood (11.8) as the rest of those who find their rest in God, as in the figure of speech 'the theatre applauds'; this will be discussed later (11.31, 22.30).

De Genesi ad litteram offers a different interpretation of God's rest, to explain how it is true both that God rested on the seventh day and that, as Jesus said (John. 5:17), 'my Father is working until now': briefly, God will not create any more *naturae*, but continues to sustain and guide his creation unfolding in time and space (Blowers 2012: 153–9). But in this part of *City of God* Augustine is concerned with angels, and provides (11.9) one of his many signals that he has not forgotten where he is in the argument. He is discussing the origins of the holy city, and the angels form the greater part of that city. Angels are not explicitly mentioned in the creation narrative, but other passages of scripture show that God created them and allow us to deduce when he did so in the unfolding of creation. Perhaps they are the 'heaven' God created in the beginning; or, more probably, God's *fiat lux* creates the angels who immediately partake of the eternal light which is the wisdom of God. But some of them turn away. Bad angels are 'no longer light in the Lord, but darkness in themselves, deprived of participation in eternal light; for there is no nature of badness, but the losing of good has received the name of badness' (*nec iam lux in Domino, sed in se ipsis tenebrae, privati participatione lucis aeternae. Mali enim nulla natura est; sed amissio boni mali nomen accepit*: 11.9). Darkness is absence of light, but Augustine chooses the word *amissio*: badness is the losing of good, not the absence of good.[6]

There is one sole good (11.10), which is single (*simplex*) and immutable: this is God. This good created all good things, but they are created: they are made, not begotten, so they are not single and not immutable. Augustine briefly explains that the Trinity is single in that it is what it has, so it cannot lose what it has. Everything else can lose what it has. The soul will always be wise by participation in changeless wisdom, but wisdom is other than the soul, just as air which is always infused by light is other than light. (This is not to say, Augustine adds, that the soul is air, as some have thought; darkness of soul, like darkness of air, comes by loss of light.) The wisdom of God is single: when scripture says that wisdom is multiple, that is because in this one wisdom are the invisible and immutable *rationes* of all the visible and mutable things made through wisdom. Augustine does not use here the much fuller discussion

[6] Williams 2000: 105 translates 'evil is not some kind of object, but we give the name "evil" to that process in which good is lost'. To keep the connection with 'bad', I translate *malum* as 'badness' rather than 'evil'.

of wisdom and the *rationes* (more familiar in Greek as *logos* and the *logoi*) he offered in *De Genesi ad litteram*.[7]

God made nothing in ignorance: what he made, he knew. Here a medieval chapter division is unhelpful. Augustine wrote:

> This world could not be known to us if it did not exist, and it could not exist if it was not known to God. (11.11) These things being the case [*quae cum ita sint*], no way were those spirits, whom we call angels, darkness at first for any interval of time: as soon as they were made, they were made light.

The chapter-division makes readers ask '*which* things being the case?' from the range of discussion in 11.10, but Augustine means what he has just said: the angels did not start as darkness which was gradually illuminated, because this darkness is absence of the light of wisdom; it is not a *ratio* known to God. (Augustine returns in 11.20 to natural darkness, which God created.) The angels were illuminated so that they could live in wisdom and blessedness; but some turned away. These have a rational life without wisdom, and they were never as blessed as those who did not turn away, for blessedness must be secure and everlasting. The good angels (11.13) are now assured of this blessedness. Someone may say that the bad angels were never blessed, because Jesus said that 'the devil was a murderer from the beginning and did not stand fast in the truth' (John 8:44); moreover, 'the devil sins from the beginning' (1 John 3:8). But Augustine offers a possible interpretation: the devil was a murderer from the beginning of the human race, that is, from the time when there was someone to kill; also from the beginning, that is, from the time of his creation, the devil did not stand fast in the truth. He refused to be subject to his creator, and through pride, *superbia*, thought that he had his own power, *privata potestas* (11.13).[8] Augustine returns here to an important word-play. *Privata potestas* contrasts with the public power conferred on an official: it is 'private' as in 'private property'. It is also deprived: the false gods, impious and proud, are deprived of God's unchanging light and reduced to a kind of needy power, *egena potestas*, in pursuit of *privatas potentias* (11.1; Markus 1990). *Potentia* is the kind of power that is not official;

[7] Solignac (Agaësse and Solignac 1972, 1: 662) notes that Augustine uses *rationes seminales* only twice, about growth from seed, and most often uses *rationes causales* or *primordiales*. On Augustine and evolutionary theory see Blowers 2012: 156–9.

[8] Some thought the devil's first sin was envy: *Gn. litt.* 11.14.18, Lunn-Rockliffe 2013. Augustine opted for pride, because pride makes people envious but envy does not make them proud, and because scripture says that pride is the beginning of all sin (Ecclus. 10.15).

properly conferred *potestas* is not needy, because it is power to take actions which have an effect in law (Clark 2009).

Augustine insists that this interpretation of 'the devil was a sinner from the beginning' does not say, with the Manichaeans, that the devil 'has his own nature of badness, *natura mali*, from some opposing principle'. He adds a grammatical comment (11.14): 'he did not stand fast in the truth, because there was no truth in him' does not mean that the absence of truth was the cause of his not standing fast, but that his failure to stand fast showed that there was no truth in him. If the devil's *nature* is bad (11.15), there is no question of sin (Augustine does not immediately spell out that sin is a defect in nature, and in a nature of badness it would not be a defect); but scripture shows that the devil, like all natures, was made by God, whose creation is good. Logically, because *vitium* is a defect in nature, *vitium* is possible only where there is a nature not *vitiata*. So a bad choice, *voluntas mala*, which is contrary to nature and harmful, in fact testifies to a good nature (11.17). God puts even bad choices to good use; contraries enhance the beauty of the world as antithesis enhances a poem. Perhaps (11.19) the verse in which God divides light from darkness (Gen. 1:4) expresses God's foreknowledge that some angels would turn away; Augustine notes (11.20) that 'and God called the darkness Night' is not followed by 'and God saw that it was good', whereas the creation of lights in heaven, which separate natural light from natural darkness (Gen. 1:14–18) is so followed.

'God saw that it was good' does not mean that God saw what he had made and realised that it was good, any more than Plato meant (*Tim.* 37c) that God's happiness was increased by the created world. In 'God saw that it was good', God is teaching, not learning (11.21). God's knowledge is perfect and unchanging. 'There are three things we especially need to be told about a created being: who made it; how; and why' (11.21). The answers are God; by saying *fiat*; and because it was good. Plato too said that the world was created so that good works should be done by a good God (*Tim.* 30); perhaps Plato knew of this passage in Genesis, perhaps he understood from the visible world or learned from those who had.

'This cause, that is, the goodness of God in creating good things, this just and fitting cause, when considered with attention and devotion, puts an end to all the questions of those who seek the origin of the world' (11.22). Apparent evils have their own place, *cum omnino natura nulla sit malum nomenque hoc non sit nisi privationis boni*. The first phrase means 'there is no nature which is badness', or 'badness is not a nature', that is, not an *ousia* or *essentia*; a 'nature' is what it is to be something (12.2).

So 'this name [badness] is only of the lack of good'. There are people who think that badness is some kind of nature, so do not accept that the good God created it. Manichaeans do not believe that the universe was made because it is good: they think that rebellious evil forced God to make it, and that part of his good nature cannot be rescued from the struggle and will become a prison for his enemy. They should realise that God is immutable and cannot be harmed, and that the soul, which can change for the worse by *voluntas*, is not a part of God but was created by God. Origen (11.23) recognises that God created everything, including souls, and that souls sinned by withdrawing from God to a greater or lesser extent; but he mistakenly thinks that the world was created not because it was good, but in order to restrain evil by imprisoning sinful souls in bodies according to their sin.

Now Augustine returns to the three questions 'who, how, why' and the answers 'God, through his word, because it is good'. Is this a revelation of the Trinity? 'This requires long discussion, and we should not be pressed to explore everything in one *volumen*'; but Augustine does in fact discuss the Trinity (11.24–8) before a recall to his immediate concern: the City of God which is not away from its home (*peregrinata*: Clark 2004) in this mortal life, but is immortal in heaven; that is, back to the angels who do not desert God. Augustine has already shown how God divided them from the angels who became darkness. The holy angels know the Trinity and the immutable *rationes* of all created things (11.29); we know in part, and must work to understand. There are other interpretations of scripture on their creation (11.32), but rather than digress, Augustine moves to the two communities of humans.

Just as book 11 did not in fact begin with a discussion of angels, so book 12 does not in fact begin with a discussion of humans, but with a replay of themes from the previous book. Augustine wants to show that angels and humans are part of the same two cities, that is, the two *societates* (12.1), one of the good and one of the bad. God created all the angels, and they do not have different natures; the difference comes from *voluntates* and desires. Augustine does not discuss *voluntas* here, but two comments are needed. First, *voluntas* is often translated 'will' rather than 'wish' or 'choice', but Augustine does not use *voluntas* to mean 'will' as a faculty (Harrison 2006; Wetzel 2012b). He uses it to mean a wish to which we consent, so that it is something we will, not a vague 'that would be nice'. Bad wishes can of course build up a habit, so that someone has a generally bad will. Second, Augustine had already argued that all efficient causes are *voluntates*, which come from God or are permitted by God. He briefly surveyed causation when rejecting (book 5) the claim that fate or fortune gave the Romans their empire. 'Chance' causes (*fortuitae*) are hidden causes, which Christians ascribe to the *voluntas* of God or of spirits. Natural causes cannot be separated from the will of the creator of nature, that is, God. Voluntary causes are from God, or from angels or humans, both good and bad, or even from animals, if it is appropriate to call non-rational impulses *voluntates*. Bodies are subject to *voluntates*. God sets in order all *voluntates* and gives power to some; all power comes from God, but not all *voluntates*. Good *voluntates* come from God; bad *voluntates* do not, because they are against nature, which does come from God.

Book 12 reaffirms that good angels love God, the supreme and only unchanging good; the others 'took delight in their own power, as if they were their own good' (12.1). 'The things God made are indeed good, because made by him; but they are changeable, because they were made not of him, but of nothing, *de nihilo*.' Augustine does not discuss 'nothing', but, again, some comment is needed. One of his favourite Bible quotations is Wisdom 11:21, 'thou hast ordered everything in measure and number and weight'. Measure sets limits: it defines. Weight, like specific gravity, settles something into its proper place. Number gives *species*, which means 'how something looks'; it is the shape by which we recognise something, and it is beauty, so Cicero uses it, together with *forma*, to translate Platonic *idea*, and Augustine connects it with *modus*, 'way of being' (O'Donnell 1992, II: 46–51). Where there is no *species*, there is nothing. Augustine asks (*Conf.* 2.6.12) what delighted him in the theft of pears, *et ecce species nulla est*: the pears had no beauty, nor had the act, 'not even a kind of beauty which is defective and shadowy, when *vitia* deceive us' (*non saltem ut est quaedam defectiva species et umbratica vitiis fallentibus*).[9] There was not even a defective *species*, so there was nothing. There is good in the beginning of form, *inchoatio formae*, where you cannot yet see the measure and number and order of perfected form, but only the material on which the craftsman will work. God made the capacity for being formed, and that too is a good (*De vera rel.* 18.36). But take away even that good, 'take away all good, and what will remain is not something (*nonnihil*), but absolutely nothing' (*Lib.arb.* 2.20.54).[10] So in creation there could not be pre-existent formless matter, stuff that God

[9] Or 'when *vitia* fall short'?

[10] Augustine may previously have thought of 'nothing' as a sort-of-something which is no one thing: Rist 1994: 106 n. 46. But even that would be 'not solid stuff or indeed any kind of subject, but a beckoning, formless nullity, foreign to goodness' (Wetzel 2012c: 167 n. 1); see O'Brien 2012.

had to do something with: 'in the order of causes, not of time, the first thing to be made was formless and formable material, both spiritual and corporeal'.[11]

Angels were created from nothing and can change, but a rational nature can achieve blessedness by cleaving (*adhaerere*, 12.1) to God. Not to do this is obviously *vitium*, a failure or defect. *Vitium* harms nature, so is against nature, and the *vitium* shows what the nature is; blindness is a *vitium* which shows that eyes are for seeing. So (12.2) the rebel angels did not have another nature from another source. God gave existence, to a greater or lesser degree, to all the creatures he made out of nothing. To the nature which made everything, 'there is no contrary nature except that which is not' (*contraria natura non est, nisi quae non est*). Scripture speaks of 'enemies of God' (12.3), but they are enemies by *voluntas*, not by nature, and their *vitium* harms themselves, not God. Evils cannot exist by themselves. Corruption harms, and harm can be done only to good; so where there is no good, there cannot be anything to corrupt. Anything that is deprived of all good does not exist; so in so far as it exists, it is good (see *Conf.* 7.10.16–12.18).[12]

Why, then (12.6), did some angels make the bad choice to turn away from God, who supremely is, to themselves, who exist in a lesser degree? Augustine's reply is quoted at the start of this chapter. A bad will (*voluntas*) is the efficient cause of a bad action. He does not say that there is no efficient cause of a bad will, but says that the efficient cause is nothing, *nihil*: *mala voluntas efficiens est operis mali, malae autem voluntatis efficiens nihil est*. A *mala voluntas* is a defection from a greater to a lesser good, not to something that is bad in itself. There is nothing wrong with gold or with power as such: the problem is disordered love of gold or power (12.8). (Indeed, there was nothing wrong with the forbidden fruit: it was just forbidden, 13.20.) That which is nothing cannot be known, but Augustine knows (12.8) that the nature of God cannot in any way be deficient, whereas natures that were made from nothing can be deficient. The *mala voluntas* does not come from God; it cannot have a natural or essential cause, because the start of a bad *voluntas* is defection from God, and the cause of that defection is itself a deficient cause, *cuius defectionis etiam causa utique deficit* (12.9). But we must not say that there is no efficient cause of a good will, for that might suggest that the *bona voluntas* of the good angels

[11] *Gn. litt.* 5.5.13, trans. in Hill 2002: 282; Williams 1999: 252.

[12] This is moral corruption, not the 'generation and corruption' of the physical world: the decay of irrational and inanimate natures has its own beauty in the order of the universe: 12.4.

was not made by God, but is coeternal with God. Their will was made by God, as they were.

So the answer to *unde malum*, 'where does badness come from?', is that badness is not a separate principle, but is the name given to the losing of good. Losing of good happens by turning from God, the supreme good, to lesser goods. This defection from God is possible because God created natures from nothing, not from himself, so they are mutable, and cannot attain happiness unless they cling to God (12.1). But that does not explain why they defect. Augustine replies that we do not know: we cannot know, because that which is nothing cannot be known (Rist 1994: 106–7). Thirty years earlier he had recognised that this answer is not satisfying:

> You may be about to ask, since the will (*voluntas*) moves when it turns away from immutable good to mutable good, where that movement comes from. It is clearly bad, even though free will must be counted among goods because without it we cannot live rightly. Now if that movement, that is the turning away of the will from the Lord God, is undoubtedly a sin, we surely cannot say that God is the author of sin, so that movement will not come from God. Where then does it come from? If I reply that I don't know, you may be aggrieved, but it would be true. That which is nothing cannot be known (*sciri enim non potest quod nihil est*). (*Lib.arb.* 2.20.54)

But there is no better answer to give. Turning to lesser goods makes no sense, for angels or for humans; failings are not intelligible. The devil and the rebel angels took pride in themselves, who are less than God, and thought that the power they had was their own. When Augustine reaches the fall of human beings (book 14), he observes that in this there is no difference between angels, who are spiritual beings, and humans, who are soul and body, subject to passions. 'It is not by having flesh, which the devil does not have, but by living according to himself, that man became like the devil, for the devil chose [*voluit*] to live according to himself'(14.3). For humans, as for angels, *voluntas* is key (14.6). Humans experience emotions (*motus*), but these are *voluntates* for or against; those who live according to God, *secundum Deum*, love the good and have the right fears and desires, pain and gladness (14.9). Humans, as God created them, had a good will. The first bad will, which preceded all bad action, was a defection (*defectus*) from the work of God to its own works (14.11). It began in pride, and pride begins with pleasing yourself, both in the sense of being pleased with yourself and in the sense of doing what you please, rather than what is pleasing to God:

> What is pride but seeking (*appetitus*) for a perverse elevation? It is perverse elevation to desert the principle to which the soul should cleave, and to

> become and to be as it were one's own principle. This happens when someone is too pleased with himself; and he is too pleased with himself when he defects from that immutable good with which he ought to have been more pleased than with himself. That defection is voluntary (*spontaneus*), because, if the will (*voluntas*) had remained secure in love for the greater immutable good by which it was illuminated so that it could see and kindled so that it could live, it would not have turned away to pleasing itself, and thereby have been so darkened and chilled that the woman believed the serpent had spoken truth, and the man put his wife's instruction before the will of God and thought his transgression of the command was venial if he did not desert his life's companion even in companionship in sin. So that bad deed, that is the transgression of eating forbidden fruit, was done only by those who were already bad. For bad fruit would not have come except from a bad tree. Now it happened against nature that the tree was bad, because it would not have happened except by a fault (*vitium*) of the will, and that is against nature. But a nature could not be distorted by a fault unless it was made out of nothing. Consequently, that it is a nature comes from its being made by God; but that it defects from what it is comes from its being made from nothing. Yet man did not defect in such a way that he was absolutely nothing, but in such a way that, having turned to himself, he was less than when he adhered to the one who supremely is. So to be in oneself, that is to please oneself, having left God, is not to be nothing, but to draw near to nothing. (*Civ. Dei* 14.13)

What was true of angels is also true of humans: the angelic nature that turned away from God did not become nothing, but became less (12.6). Augustine says that only a nature made out of nothing could be distorted by a fault, and does not pause to explain why: is that because it can *not* be? The important point for him is that beings made out of nothing do not have to revert to nothing.

The origin of the two cities is in God's creation of angels out of nothing. The earthly city began when some angels turned to themselves and to lesser goods; this defect makes no sense, but is possible because angels were created out of nothing. God does not delegate creation to angels, for they are created beings and cannot themselves create (12.25).[13] God often acts through angels who carry out his will (10.7–8), and could have used angels to tell us all we need to know (*Doc. Chr. praefatio* 6.12–13). But God does not need angels to manage the universe and interact with its lower levels, to populate the aether like Porphyry's angels (10.8–9; *Timaeus* 24a), or to act as mediators like the *daemones* of Apuleius (8.16–22). Good angels subject their wills to God and direct us to God, not because they are purely

[13] So God creates Eve: *Gn. litt.* 9.15.26.

spiritual beings unburdened by a physical body, but because they have not turned away from God to lesser goods. Angels are not a fragment of the divine which will be reintegrated with God: they were created from nothing, yet they see God (11.29). So angels see and know in the way that we too shall see and know in the resurrection, when all *voluntates* are in accordance with God, and the soul and the resurrection body live in undisturbed union. Angels see the *rationes*, the potentialities of natures that make up the created universe; they see the work of the creator, both completed and continuing; they see how God's power informs natural causes and permits human and angelic wills to act as causes.

The important fact about creation is that God made it out of nothing. God did not have to make it, to do something with pre-existent stuff. Augustine's reflections on creation start from scripture, not from Plato, but Plato was right (11.21) that the reason for creation was so that good things should be made by a good God. The downside of creation out of nothing is that beings made out of nothing can slide towards nothing, turning towards the less good until there is no good left. The important fact about angels is that they too are created out of nothing, but some turn towards God. They see and understand and praise the works of God, and effortlessly do God's will. That is the right relation between creator and created; and that is what interests Augustine about creation and causation.

CHAPTER 13

Willed causes and causal willing in Augustine

Mark Edwards

Aetiology, or the study of causes, was not the sole preserve of philosophers in the classical era. Historians knew that a catalogue of unexplained events was not a narrative, and all physicians warned that the cure of symptoms would be short-lived if it did not address the origins of the sickness. As in philosophy, so in historical and medical writing it was common for the investigation of causes to be supported by a theory of causation: each of these disciplines had its Aristotle, its Thucydides or its Hippocrates, and each had its less rational antitype in the astrologer, the mythographer or the empiric. Christian theology, in referring every event in the present cosmos or beyond it to the omnipotent will of God, did not profess to have rendered any of the three disciplines obsolete: it did maintain, however, that there can be no sound philosophy without reverence for the Creator, no history without consciousness of the Fall and its divine remedy, and no health without a loosening of our bondage to corruption. In systematic theology, at its most abstruse, metaphysics is supplemented by *protology*, history is prefaced by *hamartiology* and physical healing is only one element in *soteriology*. Augustine was perhaps the first theologian to leave his mark on all three of these peculiar sciences, though he did not have a name for any of them; this chapter, accordingly, is divided into three parts. In the first I shall argue that, logically adroit though Augustine is in his analysis of volition, his account of our present bondage rests on principles that are distinctly theological. In the second I shall show that, while his understanding of history was pre-empted by the poets and historians of the Latin tradition, they offered no comparable explanation of the origin and ubiquity of sin. In the third I shall show that there was precedent in writing on the ensoulment of the embryo for some theory of the transmission of psychic traits from mother to infant, though only the traditions of the church, or Augustine's adaptation of them, could justify his belief that a child inherits the will to sin.

Augustine's *City of God* is a theodicy, in which the predestination of a few is the only compensation for the sin of all. Previous theodicies, pagan or Christian, had invariably presupposed the freedom of the will, and Augustine's later thoughts would not have been so seminal, or so odious to many of his contemporaries, had they not appeared to be so inimical to human liberty. The battle-lines between those who affirmed and those who were accused of denying the freedom of the will had already been drawn for centuries, and it is no surprise that the arguments proved to be infinitely repeatable. What seems stranger, at first sight, is that the butt-end of the same weapon is often turned against the original assailant, so that the theories of an astrologer such as Firmicus Maternus may be barely distinguishable from those of the Platonists whom he purports to be refuting (see Edwards 2012). Both agree, for example, that there are natural laws which stipulate that when one event occurs another will follow. Both would therefore insist that to act at random is not to act freely: in the common parlance of antiquity, chance and fate are both varieties of *to automaton*.[1] Both agree that human beings, in normal circumstances, are responsible for their actions, and that they cannot be held responsible when they are subject to coercion from without.

As our concern is only with Augustine, many questions raised by philosophical determinism, including many raised on his behalf by his admirers,[2] need not detain us. He was not a Hobbes, a B. F. Skinner or even a Jonathan Edwards: his arguments are always grounded in biblical or theological premises, and are always instrumental to the proof of a dogma that he thinks necessary to our salvation. His first attempt to demonstrate that all who are saved are saved by grace alone is his letter to Simplicianus of Milan, in which he argues, perhaps without precedent in early Christian literature, that Paul did not mean by 'vessels of wrath' and 'vessels of honour' those whom God *foresaw* to be worthy of wrath or honour but those whom he *elected* to make for one end or the other (*Ad Simp*. 1.2.3–7). His reasoning is grounded in the ineluctable logic of Paul's own dictum that it was not on the basis of works – and hence, presumably, not on the basis of works foreseen – that God loved Jacob and hated Esau (Rom. 9:13; cf. Mal. 1:6). Since (Augustine argues) we are saved by the merits of Christ, not by our own, it follows that we cannot owe our election to any choices that proceed from us without divine instigation. An ancillary argument, still presuming a faith in the God of the scriptures, is that his sovereignty

[1] On the role of chance in pagan historiography see Cochrane 1957: 474.

[2] He is credited with the invention of the will for example by Dihle 1982. For criticism see Irwin 1992.

would be abridged if one whom he wished to save were free to resist him. In his later writings against the Pelagians, Augustine finds it cogent to urge that Christ would have died in vain if we could be saved by our own exertions, and that unless we had inherited not merely Adam's sinfulness but his guilt, there would be no sin to purge by the baptism of infants (Gassi 1969). Against Julian of Eclanum, the last and most prolix of his Pelagian adversaries, Augustine also has a proof from experience, or rather from the repugnance of the Pelagian theodicy to experience: if salvation depended on the exercise of freedom, why are so many born in conditions where they cannot freely choose the good because it has never been shown to them (*C. Julian* 1.53, etc.)? The case for predestination, therefore, never rests on a philosophical proof of our subjection to the elements or superhuman powers, but on the assumption that God's justice and omnipotence will be manifest in the scriptures, the church and the order of the world.

Augustine therefore sets his face against any doctrine of predestination that does not presuppose a just and benevolent God. Even suppose, he urges in the second book of *On Christian Instruction*, that the prophecies of astrologers are reliable: why would any free man part with his wealth to become the slave of these mute signals, some of which did not even have a name before that of Julius of Augustus was bestowed upon them by senatorial fiat (*Doc. Chr.* 2.21.32)? Demons are all too ready to mislead us by the fulfilment of a prophecy, instilling a superstitious veneration for signs which in themselves have no more meaning than is imparted to them by custom (2.23.35). The astrologer is no wiser than an augur, a watcher of birds, for both assume that the ways of providence are made legible to us by less rational creatures (2.24.37). In the course of this demonstration that only an intelligence greater than ours could read the future, Augustine returns to the contrasting fates of Jacob and Esau, which (as he opines) could not have been ascertained from the positions of the stars at the time of birth (2.22.33–4). In the *City of God* he argues that the case of Jacob and Esau is one of many, and that even if an infinitesimal difference in the time of birth could account for such diversity of fortune in twins, it is clear that there must be cases in which humans are born simultaneously to different mothers and yet do not enjoy identical histories (*Civ. Dei* 5.1–7). He adds the caveat that, while no event is fated, all are foreknown by God (5.9). He denies that this divine prescience commutes our liberty of action (5.10), setting aside for the moment his own conviction that it was God's election rather than Jacob's merit that caused him to fare better than his brother. For all that, in the *City of God* as elsewhere, it seems that the most pernicious error of the astrologers is not their attenuation of human freedom but their failure to acknowledge the causal primacy of God.[3]

Zeal for the justice and sovereignty of God inform the writings of Augustine against the Manichaeans, who taught that all evil proceeds from matter, to which the divine itself has fallen captive, and that those who are born of matter alone without spirit are irrevocably destined to perdition. In his reply Augustine upheld our *liberum arbitrium*, or freedom of decision, with a vehemence that hardly seems consistent with his later doctrine of predestination. Nevertheless, his later thoughts are not simply the negation of his earlier thoughts, for even in *On Free Will* he draws the libertarian view into a logical conundrum, which is still perceived by many (cf. Moore 1912: 170–95) as an invincible rejoinder to those who demand not merely the freedom to act as we will but freedom in the act of willing:

> Willing is the cause of sin, and you ask for the cause of this willing: if I am able to discover this, will you not then ask me the cause of this very cause that has been discovered? And what will be the limit of the inquiry, what end will there be of interrogation and discussion, when there is no occasion for you to be going further back than the root in your inquiry? (*Lib. arb*.3.17.48)

Thus, if there is to be no infinite regress, it must be our nature that does the willing for us, and our liberty consists only in acting according to our natural propensity without external hindrance. This was already the teaching of the Stoics, for whom the word 'free' has the sense that pertains to it in the expression 'free fall', so that a cylinder rolling down a hill can be said to be free so long as its descent is unimpeded.[4] Augustine held that Adam was unencumbered by perverse appetite or by ignorance, and therefore had no natural propensity to sin. If he was none the less capable of sinning, the reason is that he was created out of nothing, and therefore prone to fall back into nothing when he forgot to seek the *auxilium*, or assistance, of his Creator.[5] He was all the more disposed to neglect it because he was free from weakness or deficiency. Not passion or his animal nature, therefore, was his undoing (as other Christians had supposed[6]), but the sin of pride, to which a perfect being will be more vulnerable than his inferiors. The

[3] On intentionality as a condition of signification see Markus 1996: 106–10.

[4] Cicero *De fato* 43.1. For a subtle elucidation of this, which does not deny assent a role in the actions of rational beings, see Bobzien 1999.

[5] On evil as non-being, and hence a negation of the unity of the Good, see Williams 2000. On the difference between the *auxilium* 'with which' Adam could not maintain his perfection and the *auxilium* 'by which' we are preserved from sin see *De corr. et gr*. 12.34.

[6] E.g. Gregory of Nyssa *De hom. op*. 18.4.

consequence of his Fall was that his reason became incapable of mastering his desires. The subjugation of reason by inordinate desire (*libido* or *concupiscentia*) is now the usual cause of sin in us, though seldom without the encouragement and complicity of pride (*Civ. Dei* 14.26 etc.).

It is often forgotten (and hence must be repeated here) that according to Augustine we inherit from Adam not only the *libido* or *concupiscentia* that engenders sin in us but the *reatus* or guilt of the sin that he committed.[7] That humans are indeed universally prone to sin would be obvious even without the explicit testimony of the saints from Paul to Adam; no precedent, on the other hand, has been found for Augustine's doctrine of hereditary guilt in earlier Christian literature except for a gnomic aside in a letter by Cyprian (see Beatrice 2013: 154), which is open to more than one interpretation (Letter 64.5). In the second part of this chapter I shall argue that a mixture of contrition and pride in Rome's view of its own history prepared Augustine and his Latin readers for this notion; in the third I shall argue that Augustine has predecessors who agree that the human tendency to sin is not only a common bequest from Adam but the inevitable concomitant of birth by sexual congress.

Sources of guilt

Adam's trespass, culpable only because it breached a divine prohibition, is represented in the biblical narrative as the cause of human mortality, of woman's pangs in childbirth and of man's unceasing obligation to labour for his own bread. It is not said to be the cause of the ensuing crescendo of sin, which begins with Cain's murder of his brother Abel and culminates in the ruin of Babel and the confusion of tongues. The killing of Abel is prompted by the rejection of Cain's sacrifice, which seems to imply that the pastoral farmer is dearer to God than the tiller of the soil (Gen. 4:5–8). Cain's progeny contribute in no small measure to the increase of sin, for if he (as we must suppose) was the first to employ a tool as a weapon, Tubal Cain was a smith 'in every work of brass and iron' (Gen. 4:22; cf. Ginzberg 1998: 118), and Lamech the second man to confess a murder (Gen. 4:23). If peace was to be symbolised in the prophets by the beating of swords into ploughshares, it was not forgotten that ploughshares can be beaten into swords (Isa. 2:4; Joel 3:10).

[7] See e.g. *C. du. ep. Pelag.* 1.13.27.

In Genesis 6 the giants appear on earth[8] at the time when the sons of God take wives from among the daughters of men. According to Philo[9] and others these sons of God were fallen angels, and in the apocryphal book of Enoch it is these apostates from heaven who sow iniquity on earth by creating the instruments of war and agriculture. The tower of Babel is another overweening work of manual enterprise, a parallel to the storming of heaven which Greek myth attributes to the giants.[10] Noah's ark was a nobler artefact, but this new father of the human race was also unlucky in his children, and the curse that he laid on his second son Ham resulted in the perpetual subjection of Ham's son Canaan to the descendants of his uncles Shem and Japheth (Gen. 9:22–5). While fratricide is not attested between the Flood and the Exodus, the supplanting of the elder by the younger is a common motif in the history of the patriarchs from whom Israel traced her right to the Promised Land. Ishmael is driven out by the mother of Isaac (Gen. 21:9–14); Jacob steals the birthright from his elder brother Esau (Gen. 27); Joseph's dream that his brothers will pay court to him comes true (Gen. 37:7 ff.), and his younger son Ephraim usurps the blessing intended for Manasseh (Gen. 48:13–14), thus ensuring that his name will be given to the northern kingdom of Israel when his descendant Jeroboam breaks away from the hegemony of Judah. Jeroboam, however, is a reprobate who sets up two golden calves in imitation of the one that had been erected, with Aaron's connivance, in defiance of Moses at Sinai (1 Kgs. 12:28; Exod. 32:4). It became a prophetic commonplace to exclaim that each generation, both in Israel and in Judah, had outdone the last in wickedness ever since that first secession in the wilderness. Not before New Testament times, however, do we hear the voice of an Israelite declaring that the patience of God has been abused too often and that his covenants with Abraham and Moses have been annulled.

It would not be impossible, therefore, to reduce the Old Testament to a series of antitypes, opposing Cain to Abel, Ishmael to Isaac, Esau to Jacob, Judah to the sons of Joseph, Ephraim to Manasseh. This binary scheme is applied to the entire history of the world in a Latin romance, the *Clementine Recognitions*, which assumed its present form at about the time of Augustine's writing of the *Confessions*. At 8.3 ten contrasting pairs are enumerated: Cain and Abel, the giants and Noah, Pharaoh and Abraham,

[8] On Cain as the forefather of the giants in the Anglo-Saxon epic *Beowulf* see Melinkoff 1979 and 1980.

[9] *De gigantibus* 2, where the text that he quotes reads 'angels' rather than 'sons of God'.

[10] Genesis 11:4–9; cf. Josephus *Antiquities* 1.3.1 and Genesis 10:8, which was often interpreted to mean that Nimrod was a giant.

the Philistines and Isaac, Esau and Jacob, Pharaoh's magicians and Moses, the tempter and the Son of Man, Simon Magus and Peter, the nations and the evangelist of the nations, Christ and the Antichrist. The duel of Peter and Simon in Rome, described in other hagiographic writings of the same or an earlier period, can be read as an epitome of the perennial conflict between the kingdom of God and that of this world, as Simon (the favourite of the emperor in this story) was reputedly worshipped as a local deity by the pagans of the capital,[11] while the Roman congregation believed that in the body of Peter it possessed the very rock on which the Saviour undertook to build his church.

Pagans had never heard of Cain, but according to Roman legend the blood of Remus sullied, or perhaps cemented, the walls of the rising city after he had accused his brother Romulus of cheating in a trial by ornithomancy. While Livy passes over this crime in haste in his *Histories* (*Ab urbe condita* 1.7), it is mitigated in Ovid by the introduction of a third party, Celer, who deals the fatal blow in the false belief that he is serving Romulus (*Fasti* 3.391, 4.51). The sixth of Horace's Roman odes begins with an apostrophe to the 'Roman', who, though *immeritus* (undeserving), is doomed to expiate the crimes of his fathers. What the crime is we are not told, and it quickly appears that the Roman is not quite guiltless: he can hope for a clean bill of innocence only when he rebuilds the temples which have been allowed to decay in the fratricidal age of civil wars. At the end of the ode we discover that the penalty of past sins is to go on sinning: 'our fathers' generation, worse than our grandsires, spawned in us a more evil brood, soon to produce more vicious offspring' (*Odes* 3.6.46–8). Historians in the imperial and late republican periods sometimes proposed a more recent date for the introduction of poison into the body politic. Perhaps the ancient habits of thrift and industry had been chased out of the city by the spoils that Manlius Vulso brought from Gaul in 187 BCE (Livy *Ab urbe condita* 39.6); or perhaps, as Sallust opines, the fall of Carthage in 146 had deprived Rome of a whetstone to martial virtue and moral vigilance (*Bellum Catilinae* 10).

Yet Virgil, in a text better known to Christians of antiquity than any yet cited here, alludes to a curse that is older than Rome and occasioned not by murder but by false dealing. In the new age of abundance, he prophesies, *pauca tamen suberunt priscae vestigia fraudis*, there will remain a few scars of the ancient perfidy (*Eclogue* 4.31): a new age will ply its oars on the deep and cleave the earth with its ploughshares, a new band of warriors will fill

[11] Justin Martyr *Apologia prima* 23, confusing Simon with the Sabine deity Semo Sancus.

the *Argo* and a second Achilles will be sent to Troy. Only then will the merchant and the mariner give up their trade and the earth bring forth without the travail of sowing and viticulture (32–41). It should be obvious that the *prisca fraus* is not the killing of Remus, or even the refusal of King Laomedon to honour his debt to Apollo and Poseidon after they built the walls of Troy: neither of these offences prompted the quest of the Golden Fleece, and it is hard to see why the erasure of their scars would inaugurate a new age of fecundity without tillage. It would seem that the primordial sin is the cause of human sweat and toil upon the reluctant earth, which is itself a further indignity rather than an expiation of the wrong. When Constantine cited Virgil as a herald of the Messiah and a witness to the Fall, this was far from the most perverse reading of the poem that has been offered in ancient or in modern times.

In Virgil's *Aeneid* a Trojan takes swift vengeance on an Italian who taunts him with effeminacy (9.621–37); in Latin historiography the positions are reversed, and it is the Roman, half a Trojan himself, who imposes civilisation with all its blessings and distempers upon the robust barbarian. The vanquished reply, unanswerably: 'they make a desolation and call it peace', exclaims the leader of the Britons in the *Agricola* of Tacitus (30), and in his *Annals* Caratacus proves more formidable in speech than he was in arms (12.37). Yet Tacitus will not forgive any emperor who fails to impose the Roman yoke on peoples whom he admits to be superior to the Romans in morality (*Annales* 1.11; *Agricola* 39–40). As Virgil proclaimed (*Aeneid* 6.851), Rome has a duty to rule, though that is no guarantee that she can rule without sin.

Augustine wrote the *City of God* to explain why God would allow the eternal city to be sacked by an army of professing Christians, abetted by a fifth column of Christian senators who had caused the removal of the Altar of Victory from the forum (see Croke and Harries 1982). It was not enough for Augustine to say, with his friend Orosius, that Christianity had in fact brought about a slight mitigation of the calamities to which the world was subject. Quoting the Sibyl although he has scoffed at Virgil (*Civ. Dei* 18.23; cf. *Conf.* 1.13.20), he undertakes to follow (and surpass) the poet's example by writing a teleological narrative which incorporates a history of decline. This history, for him as for the Roman censors of Rome, begins with a fratricide, though not the one that survives in their short memories. While venality, luxury and the spread of weapons play their part, his object is to show that these are not the roots of sin but the perennial flowerings of an endemic sinfulness, the origins of which are known to us only by revelation. To the author of this revelation civil strife is merely a symptom of the truceless warfare between two commonwealths, of which the one that is destined to prevail is not coterminous with any visible polity, or even with the church that has been held accountable for the sack of Rome.

In the *Confessions* Augustine sees the rivalry of twins, each fighting to own the breast that can feed them both, as proof of the congenital transmission of the sinful tendency (1.7.11). He has not yet formulated his infamous doctrine that we also inherit the guilt that Adam himself incurred by sinning, so that even before we have faults of our own to expiate, we are under sentence of eternal punishment for a fault that only the blood of Christ can purge (*Pecc. mer.* 1.18.23 etc.). He can, however, urge that children are frequently held responsible for the transgressions of their fathers in the Old Testament (Exod. 20:5 etc.). Some opine that this was also the regnant belief in Africa, where, according to John V. Taylor, a missionary would not find it so difficult as in Europe to persuade his audience of the son's liability to the father's punishment (Taylor 1963: esp. 97–8, 124, 188). But whether Augustine was culturally or ethnically an African[12] – or, for that matter, a Berber[13] – we could not hope to determine even if we could say what both these adverbs signify. By language, education and political allegiance he was a Roman, and we have seen that the Romans saw themselves as a people prone to vice, and all the more so because the gods required them to pay for crimes that they had not perpetrated. It could be argued indeed that Augustine's teaching on original sin is more consonant with the Roman than with the biblical view of history: although Ezekiel urges that the Israelites are all the more sinful because their fathers sinned before them (Ezek. 20), ancestral sin is more commonly invoked to explain why some other people is subject to, or at enmity with, the elect nation of God. The Roman, by contrast, holds that the malediction has fallen on the very people who have received their empire from the gods. Augustine likewise holds that the sin of Adam and the propensity to sin on one's own account are as much the patrimony of the saints as of those who are lost.

Romulus, the son of a Vestal Virgin who at the end of his life rose bodily into heaven after founding an asylum for ne'er-do-wells which its inhabitants extolled as an eternal city, was all too obviously the pagan counterpart of Christ. Yet Romans who know their own history, says Augustine, will

[12] Williams 1927 maintains that all ancient sponsors of this doctrine were either Africans or Copts, though there is little warrant for assigning Origen, Athanasius or Cyril of Alexandria to the second category.

[13] See Ferguson 1969. On the 'Punic' ancestry of Augustine, already urged to his discredit by Julian of Eclanum, see Elingsen 2005: 9–10.

perceive that Sallust combines a true diagnosis of Rome's distempers with a fallacious aetiology which presupposes a fall from innocence;[14] at the same time, even the poets have not divined that sin is a universal state, and that even fratricide – even the first, most heinous fratricide – can be only a symptom, not the cause, of moral or political imbecility. Two considerations show that Rome cannot owe her fortune to divine favour. On the one hand, the gods, if they had the power, would reward the virtuous, and would not permit a Regulus to perish in ignominy (*Civ. Dei* 1.24, 5.18); on the other hand, no god who deserved the name would permit his favourites to extend their wealth and power by the conquest of unoffending nations (5.17), and no polity that relies for its survival on coercion and torture can be of divine origin (19.6). Augustine has no quarrel with agriculture, but he says more clearly than anyone before him that Rome's claim to keep the peace of the world by *force majeure* is illusory so long as violence, camouflaged as justice, is not merely an occasional corollary but the necessary condition of civic peace.

Against his own arraignment of the Roman past, Augustine sets the biblical record in which the histories of saint and sinner are told concurrently. The Mosaic account begins with the creation, not only because that is the custom in barbarian histories, but because unless we grant that the world was created out of nothing, by God's grace and for his glory, we shall not understand the role of sin as a tolerated aberration within the divine economy. Ever since God divided light from darkness, separating his angels from the non-being which is all that we mean by evil, nothing but good has come from his hand, and nothing but evil from the disobedience of his creatures. Satan's sin was the first, but Adam's more deleterious to his human progeny. Its consequence is that no human being is free to do good without the assistance of God, and hence that no-one is saved who is not already chosen. Thus there are now two cities, or rather two types of city, one rising and falling on earth as a fugitive monument to pride and folly, the other growing more populous from age to age in the sight of God alone. Cain, the first rebel, is also the architect of the first city on earth (*Civ. Dei* 15.8), but the city of Abel is destined to outlive it, as we learn from the extinction of Cain's posterity in the Flood (15.8 and 15).

Thus the *City of God* sets out a comprehensive philosophy of history which at once subsumes and surpasses all its models. The Bible recounts the creation and the first sin, but neither of these expressly accounts for the forking destinies of the brothers who punctuate the line from Adam to

[14] Cf. MacCormack 1999: 205, citing *Civ. Dei* 2.21.

Jeroboam. In the *Clementine Recognitions* this antithetical scheme informs the whole economy of salvation, but there is more dramatic symmetry in the scheme than theological perspicuity. In Roman historiography the ills of the body politic are traced at times to the wrongdoing of a single man and at others to some collective indiscretion, but the vista is never long enough, and it is always falsely assumed that the corruption which is now inevitable could have been avoided by an exercise of freedom in the past.

Ancestral vices

But how is sin inherited if not by imputation? From the Decalogue we learn that God is a jealous God who visits the sins of the parents on the third and the fourth generation (Exod. 20:5); but neither this terrible dictum nor the Psalmist's ejaculation, 'I was shapen in iniquity and in sin did my mother conceive me' (Ps. 51:7), affords any purchase for scientific reasoning. Some Greek physicians believed in the transmission of acquired blemishes, and Aristotle, though he denies the rule, admits that exceptions have been proved. For Augustine, however, it is not enough to say that all children have in fact sinned because their parents sinned: the law must be not merely an empirical but a necessary one, not merely true in the main but universally binding. His own solution, adumbrated in the *City of God* and taken up with force in his writings against the Pelagians, is infamous. The cause of Adam's Fall was not the victory of the senses over the intellect but the capitulation of intellect to its own pride. Its consequence is that intellect, in Adam and in his descendants, no longer enjoys a perfect mastery over the sensual appetites, so that we not only sin through ignorance but do evil in the knowledge that it is evil. If this theory implied that sin (as we might say) is part of our genetic inheritance, some proviso would be required to explain the impeccability of Christ. Augustine has no doctrine of the immaculate conception of Mary;[15] he does, however, purport to draw a corollary from experience which will account for the universality of sin in every other case while making Christ a necessary exception. The body, he says, carries evidence of the Fall in the fact that, while we can move our other limbs as we please, the genital organ in men answers only to lust, and cannot be aroused by the rational prompting of the will (*Civ. Dei* 14.26). An erection may occur in defiance of a man's resolution; on the other hand, his desire to have children with his lawful wife may be frustrated by his physical

[15] *De nat. et gr.* 36.42 implies at most that she was preserved from actual sin.

impotence. It follows that even Christians duly married, for whom procreation is not merely no sin but a duty, will be unable to conceive without sin:

> We do not disparage the honourable union of spouses on account of the shame that bodily lust entails. For, had there been no antecedent commission of sin, there could have been such a union as would have brought no shame to the wedded couple; the lust, on the other hand, arose from the sin, and thus they were forced to cover it in their confusion. Thus it was the duty of all couples after them, even when they were making good and lawful use of this evil, to shun the gaze of other folk in this activity, and thus to admit its shamefulness, whereas no-one ought to be ashamed of a good activity. From this two inferences follow: first the goodness of marriage, which is laudable insofar as children are born from it; and secondly the evil of lust, which is shameful insofar as those who are born of it must be reborn lest they be damned. (*De nupt. et con.* 2.21.36)

Post-lapsarian modesty, says Augustine, forbids us to canvass the medical theory that the wife produces her own seed in addition to that of the husband (*De nupt. et con.* 2.13.26).[16] Since he assumes, however, that the woman also experiences lust in the act of conception, he seems to agree with those doctors who thought it impossible for a woman to conceive unless she had taken pleasure in sexual intercourse (Dean-Jones 1994: 157). Mary's case, however, is a miraculous anomaly: she felt no pleasure because there was no intercourse, and therefore the inexorable law that governs coition did not apply to the fruit of her womb. The argument that birth by insemination entails defilement, and that Jesus therefore owed his impeccability to his virgin birth, had already been stated in Origen's eighth Homily on Leviticus, and it is possible that Gregory of Nyssa anticipates Augustine's teaching more precisely, though his phrasing is obscure (*Orat. catech.* 16, in Mühlenberg 1996: 47.3–7): 'If birth itself is not an affliction, neither could anyone call life an affliction. But the affliction of pleasure precedes human birth, and the impulse of living creatures to vice (this is the infirmity of our nature).' Such reasoning will account for Christ's resistance to temptation – perhaps too well, as the most exacting of Augustine's critics urged, as they seem to entail that he was not even 'tempted as we are'. It is not clear that they can save Christ from the guilt of the primordial sin, which ought to fall upon anyone whose ancestors

[16] The opposing views of Aristotle (*De gen. animal.* 1.20) and Galen (*De sem.* 2.1) had been made familiar to Christian readers by Nemesius of Emesa *De nat. hom.* 2: see Sharples and van der Eijk 2008: 70–1. James Wilberding observes in his contribution to this volume (Chapter 9) that no Neoplatonist entertained the notion of female seed.

were in the loins of Adam.[17] The scope of the present chapter does not require us to vindicate the theology of Augustine or his Christian predecessors,[18] but it does require us to ask how far the science or lore of procreation in his day would support his belief that the character of a child will be affected by the mental state of the parents at the instant of conception. In the book of Genesis – for Augustine as scientific a resource as any other – Jacob causes Laban's ewes to bring forth lambs of the same colour as the rods that he sets before them (Gen. 30:34–43). Jerome, to preempt the incredulity of his readers, quotes a corroborative anecdote from Quintilian, in which the birth of a black son to a white queen is explained as a consequence of her seeing a picture of an Ethiopian at the time of intercourse.[19] The novelist Heliodorus, who may have been a contemporary of Jerome and Augustine, inverts the tale in an episode of his Ethiopian Story, in which the queen is African and the child is white (*Aethiopica* 4.8.5). Augustine, however, traces a similar anecdote to Hippocrates (*Loc. Hept.* 1.93), and appears have drawn another variant of it from Soranus at *Retractationes* 2.62.[20] He was exercised throughout his life by the question of the soul's origin (see O'Daly 1983), though he failed to discover a philosophic doctrine that was equally consistent with his metaphysics and his theory of original sin. His metaphysics told him that the soul is incorporeal, free of colour, shape or magnitude, and acting in the place where it is not. According to his hamartiology sin is transmitted by insemination, as physical characteristics are except where an Ethiopian or a Jacob crosses the mother's line of vision. Can incorporeal properties, or the properties of an incorporeal entity, be passed down by intercourse?

'Traducianist' was the ancient name for those who could answer this question in the affirmative; it is the name maliciously given to Augustine himself by Julian of Eclanum, with the purpose of singeing him by association.[21] Tertullian, the most celebrated Christian proponent of traducianism,[22] also held that the soul is a *corpus* or body, defending this position with agile casuistry and a dossier of quotations from the Stoics.

[17] *Pecc. mer.* 1.10.12, 2.25.39, appealing to Hebrews 7:5.

[18] For well-known critiques see Rist 1969; Wetzel 2000.

[19] Jerome 1995: 67. At 204 Hayward compares *Genesis Rabbah* 73.10. The original dissertation by Quintilian is lost.

[20] Brisson 2012 (see n. 25) cites Soranus, *On Female Illnesses* 1.39.

[21] Augustine *C. Julian.* 1.1, 1.6, 1.13 and *passim*. The first usage associates Augustine with his former teachers, the Manichaeans, whom he himself asserted to be incapable of grasping the notion of incorporeality.

[22] See Tertullian *De an.* 29.7. Osborn 1997: 166–7 points out that he is not a proponent of original sin in Augustine's sense.

To Augustine he might have been a priceless ally, as he maintains that a *vitium carnis*, a moral debility of the flesh, is necessarily inherited by all who are born of carnal intercourse.[23] Augustine, however, could not accept that a Christian can also be a materialist, and did not take note of the (cursory and unexplained) distinction between the corporeal and the material which Tertullian draws in *On the Soul* (*De an.* 11.2). Nor can he reply without satire to Vincentius Victor, who, if Augustine's paraphrase can be trusted, held that the soul is one of three *corpora* or bodies which make up the human person, but a body which remains unsullied before its incarceration in the womb (*De an. et orig.* 1.5.5). Having disposed of Victor to his own satisfaction, Augustine turns from philosophy to the scriptures, laying texts that seem to favour traducianism side by side with those that might be thought to imply a simultaneous creation of the soul and the tangible body. His verdict, as in his earlier work *On Genesis according to the Letter*, is that the scriptures offer no decisive evidence in favour of either view.

Although the mature Augustine canvasses only two opinions, he implies in letter 166, written to Jerome in 415, that the pre-existence of the soul was a notion worthy of Christian scrutiny.[24] Even when he had formally rejected it (or the Platonic variant of it), he was not ashamed to go on reading Porphyry. In his letter *To Gaurus*, the pagan Neoplatonist had denied that the child in the womb possesses a rational soul, which grows in maturity along with the physical members, since this doctrine seemed to him inconsistent with the Platonic teaching that the rational soul descends from the 'supercelestial place' to a body allotted to it in accordance with the merits and demerits of its past lives. He attributes the early signs of animation to the presence of an irrational soul, and, while he concedes that the embryo is capable of appetitive motions, he argues that these originate not in a rational choice of ends, but in affections of the phantastic or imaginative faculty of the mother, which at this stage are directly communicable to her offspring:[25]

> I happily grant to [my interlocutors] that the reason why embryos share in what the mother imagines and apprehends is that they have a share in the

[23] *De carn. Christ.* 16.3. M. C. Steenberg 2009: 79–87 attributes more to the lust of the parents than I can find in *De carn. Christ.* 4 or *De an.* 40, though he argues that Tertullian does not hold the same position as Augustine.

[24] For recent bibliography see Hürst 2011: 354–5 n. 90.

[25] Porphyry *AG* 5, in the edition edited by Brisson 2012: 160–2. For a critical analysis of *To Gaurus* see James Wilberding in this volume (Chapter 9); and for a more jejune assertion of the transmissibility of psychic traits see Plutarch *De sera numinis vindicta* 17.

> mother's imaginative and apprehensive soul. For it is agreed with regard to many living creatures, and indeed with regard to women during intercourse, that when forms of the same species become embedded in their imaginations, they bring forth offspring that resemble them. For this reason we place images fashioned into beauty of form before horses, dogs and doves, and on occasion before a woman, so that by contemplating these forms during intercourse, and committing them to memory, they may bring forth offspring of a similar character. We may readily admit that this would not happen if the seed did not partake of the imaginative soul.

Commenting on Jacob's deception of Laban, Augustine also grants that impressions received by the *phantasia* of the mother may be transmitted to the seed (*Trin.* 3.8.15). Nevertheless, he concurred with other Christians, against Porphyry, in holding that the rational soul is present as soon as the mother conceives, notwithstanding the changes that occur in its physical envelope during the ninth months of gestation. Any other opinion would have justified abortion and would have failed to account for John the Baptist's recognition of Jesus when both were in the womb.[26] Even if such considerations logically precluded the pre-existence of the rational soul (and perhaps they do not), Porphyry's belief that maternal images can produce appetitive motions in the embryo remained at least as credible as Jerome's story of the Ethiopian.

For Porphyry all such stirrings were perturbations of the lower, or irrational, soul.[27] Augustine too distinguishes the *anima irrationalis*, which we share with beasts, from the ensemble of higher functions which he styles the *mens* or 'mind'.[28] In the twelfth book of *On Genesis according to the Letter* Augustine distinguishes three forms of vision, the ocular, the spiritual and the mental (see further Finan 1992), of which the spiritual corresponds to the 'phantastic' in the vocabulary of Neoplatonism (*Gn. litt.* 12.7.16). He declines, however, to equate the spirit with the soul in contradistinction to the intellect (12.7.17), and finds authority for his own usage at 1 Corinthians 14:14, rather than in Plato's postulation of the *thumos* as a part or power of the soul that mediates between intellect and appetite (12.8.19). While his bifurcation of soul and mind may have lent some colour to the charge of Apollinarianism,[29] it does not play for him

[26] On Luke 1:41 see Origen *De princ.* 1.7.1 and Augustine, sermon 293.
[27] Which, according to Augustine *Civ. Dei* 10.9.2, cannot be saved by Porphyry's theurgic rites: see Clark 2007.
[28] O'Daly 1987: 7, citing *Div. qu.* 46.2, *Imm. an.* 25, *Civ. Dei* 5.11 and 11.3 and *Trin.* 14.26.
[29] See Keech 2012: 142–89. On the supposed traducianism of Apollinarius, who denied that Christ possessed a human intellect, see Nemesius of Emesa *De nat. hom.* 25 in Sharples and van der Eijk 2008: 155–6.

the same role that it plays for Porphyry in accounting for the inheritance of psychic properties. Just as Adam's fall was the result of an intellectual rebellion, rather than of incontinence in the lower appetites, so the transmission of lust by parent to child entails a radical vitiation of at least one of our hegemonic faculties, the deliberative will in contradistinction to mere desire.[30]

Afterword

To speak of Augustine's theory of the will without reference to its theological foundations is rather like transplanting the Elgin marbles from Athens to London.[31] On the one hand, it can be argued that the eviscerated structure has ceased to function as a temple, and that the contents are safer in their foreign tabernacle from the attrition of time and unbelief; on the other hand, confinement in a museum exposes an artefact to subtler forms of weathering and a more insidious process of neglect. Just as a full aetiology may require us to look for remote antecedents and latent conditions as well as the manifest causes of a phenomenon, so an examination of Augustine's thoughts on sin and volition may require us to take account of subliminal factors as well as the exegesis and logical reasoning which are all that he himself wishes us to see. The visceral sense that many Christians have of being both powerless and all-powerful in the strength of God does not yield to philosophical analysis; neither did the combination in Roman minds of indelible guilt with a mandate from the gods to rule all nations. Without an effort of historical sympathy we can make nothing of the credulous tales that attached themselves to childbirth in an age when only heaven could decide whether both the infant and the mother would survive. It has been the aim of this chapter to restore the shock of the old – that is, to recover some lost features of the milieu in which Augustine's thought developed – in the hope not so much of rendering it more palatable as of showing that it would not have ceased to be palatable if we were still free to entertain the other beliefs with which it was coupled in Augustine's mind.

[30] On the rationality of choice in Augustine and Aristotle see Chappell 1995: 154–62. Chappell goes on to argue that, if Augustine has a 'theory of the will' at all, its purpose is to account for our illogical possession of a bad will (199–200).

[31] Cf. Hanby 2003: 7–26 on readings of Augustine as a predecessor of Descartes.

References

Accattino, P. 2003. 'Processi Naturali e Comparsa dell'*Eidos* in Alessandro di Afrodisia', in Movia (ed.), 167–86

Adams, R. M. 1979. 'Primitive Thisness and Primitive Identity', *Journal of Philosophy* 76(1): 5–26

Adamson, P. 2007. '*Porphyrius Arabus* on Nature and Art: 463 Smith in Context', in Karamanolis and Sheppard (eds.), 141–63

2008. 'Plotinus on Astrology', *Oxford Studies in Ancient Philosophy* 35: 265–91

2015. 'Freedom, Providence, and Fate', in Remes and Slaveva-Griffin (eds.), 437–52

Aeneas of Gaza 1958. *Teofrasto*, ed M. Colonna. Naples: Iodice

Agaësse, P. and Solignac, A. 1972. *La Genèse au Sens Littéral*, 2 vols. Oeuvres de Saint Augustin, Bibliothèque Augustinienne 48–49. Paris: Desclée de Brouwer

Ahbel-Rappe, S. 2010. *Damascius' Problems and Solutions Concerning First Principles*. Oxford: Oxford University Press

Alexander of Alexandria 1857. *Epistolae de Ariana haeresi deque Arii depositione*, Patrologia Graeca 18. Paris: Imprimerie Catholique

Algra, K. et al. (eds.) 1999. *The Cambridge History of Hellenistic Philosophy*. Cambridge: Cambridge University Press

Alt, K. 2005. 'Man and Daimones: Do the Daimones Influence Man's Life?', in Smith and Alt (eds.), 73–90

Amand, D. 1973. *Fatalisme et liberté dans l'antiquité grecque*. Amsterdam: Hakkert

Ambrose 1896. *Hexaemeron*, Corpus Scriptorum Ecclesiasticorum Latinorum 32, part 1. Vienna: Verlag der Österreichischen Akademie der Wissenschaften

Ammonius 1897. *Commentary on Aristotle On Interpretation*, Commentaria in Aristotelem Graeca 4, part 5. Berlin: Reimer

Anghelina, C. 2010. 'The Homeric Gates of Horn and Ivory', *Museum Helveticum* 67: 65–72

Aristotle 1984. *Complete Works of Aristotle: The Revised Oxford Translation*, ed. and trans. J. Barnes. Princeton: Princeton University Press

Armstrong, A. H. 1940. *The Architecture of the Intelligible Universe in the Philosophy of Plotinus*. Cambridge: Cambridge University Press

1962 'The Theory of the Non-Existence of Matter in Plotinus and the Cappadocians', *Studia Patristica* 5: 427–9

1966–88. *Plotinus. Enneads. Text and Translation with Notes*, 7 vols. Cambridge, MA: Harvard University Press
1982. 'Two Views of Freedom: A Christian Objection to Plotinus VI 8 (39) 7, 11–15', in Livingstone (ed.), 397–406
Athanasius 1887. *Contra Arianos, Oratio 1*, Patrologia Graeca 26. Paris: Imprimerie Catholique
Atkinson, M. 1983. *Plotinus. Ennead V.1*. Oxford: Oxford University Press
Aubry, G. 2007. 'Conscience, pensée et connaissance de soi selon Plotin: le double héritage de l'*Alcibiade* et du *Charmide*', *Études Platoniciennes* 4: 163–81
Aujoulat, N. 1986. 'Le Théophraste d' Énée de Gaza: problèmes de chronologie', *Koinonia* 10: 67–80
Ayres, L. (ed.) 1995. *The Passionate Intellect: Essays on the Transformation of Classical Traditions, Presented to Professor I. G. Kidd*. New Brunswick, NJ: Transaction
Babcock, W. 2012. *Saint Augustine: The City of God (De civitate Dei) Books 1–10*. New York: New City Press
Baltes, M. 1976–8. *Die Weltentstehung des Platonischen Timaios nach den Antiken Interpreten*. 2 vols. Leiden: Brill
1983. 'Zur Philosophie des Platonikers Attikos', in Blume and Mann (eds.), 38–57; repr. in Baltes 1999, 81–111
1999. *Dianoêmata: Kleine Schriften zu Platon und zum Platonismus*. Stuttgart: Teubner
Barney, R. ms. 'On Hippo of (Possibly) Rhegium' [DOI 10.2139/ssrn.1607792] at http://apaclassics.org/sites/default/files/documents/abstracts/Barney.pdf
Baltussen, H. 2002. 'Wehrli's Edition of Eudemus of Rhodes: The Physical Fragments from Simplicius, *On Aristotle's Physics*', in Fortenbaugh and Bodnàr (eds.), 127–56
2003. 'Early Reactions to Plato's Timaeus: Polemic and Exegesis in Theophrastus and Epicurus', in Sharples and Sheppard (eds.), 49–71
2006. 'Addenda Eudemea', *Leeds International Classical Studies* 5:01: 1–28, available at www.leeds.ac.uk/classics/lics/2006/200601.pdf
2008. *Philosophy and Exegesis in Simplicius: The Methodology of a Commentator*. London: Duckworth
2012. 'Translation of Simpl. in Phys. 1.5', in Baltussen, Atkinson, Share and Mueller (eds.), 17–29
Baltussen, H., Atkinson, M., Share, M. and Mueller, I. (eds.) 2012. *Simplicius on Aristotle Physics 1.5–9*. London: Bristol Classical Press
Baltzly, D., Tarrant, H., Runia, D. T. and Share, M. J. (eds.) 2007. *Commentary on Plato's Timaeus*. Cambridge: Cambridge University Press
Barbanti, M. and Iozzia, D. (eds.) 2009. *Anima e Libertà in Plotino*. Catania: CUECM
Barnes, J., Brunschwig, J., Burnyeat, M. and Schofield, M. (eds.) 1982. *Science and Speculation: Studies in Hellenistic Theory and Practice*. Cambridge: Cambridge University Press
Basil of Caesarea 1857–66. *De Spiritu Sancto*, Patrologia Graeca 32. Paris: Imprimerie Catholique

Beatrice, P. F. 2013. *The Transmisison of Sin: Augustine and Pre-Augustinian Sources*. Oxford: Oxford University Press

Beauchamp, T. L. and Frey, R. G. (eds.) 2011. *The Oxford Handbook of Animal Ethics*. Oxford: Oxford University Press

Beierwaltes, W. 1991. *Selbsterkenntnis und Erfahrung der Einheit*. Frankfurt am Main: Vittorio Klostermann

2001a. '*Causa sui*: Plotins Begriff des Einen als Ursprung des Gedankens der Selbstursächlichkeit', in Beierwaltes (ed.), 123–59

(ed.) 2001b. *Das wahre Selbst: Studien zu Plotins Begriff des Geistes und des Einen*. Frankfurt am Main: Vittorio Klostermann

Benz, E. 1932. *Marius Victorinus und die Entwicklung der Abendländischen Willensmetaphysik*. Stuttgart: W. Kohlhammer

Bettenson, H. 1972. *Augustine: City of God*. Harmondsworth: Penguin

Bidez, J. 1913. *Vie de Porphyre*. Gand: Van Goethem

Blowers, P. 2012. *Drama of the Divine Economy: Creator and Creation in Early Christian Theology and Piety*. Oxford: Oxford University Press

Blume, H. D. and Mann, F. (eds.) 1983. *Platonismus und Christentum: Festschrift für Heinrich Dörrie*. Münster: Aschendorff

Blumenthal, H. J. 2000. *'Simplicius' on Aristotle On the Soul. 3.1–5*. London: Duckworth

Bobzien, S. 1998. 'The Inadvertent Conception and Late Birth of the Free-Will Problem', *Phronesis* 43: 133–75

1999. 'Chrysippus's Theory of Causes', in Ierodiakonou (ed.), 197–242

2001. *Determinism and Freedom in Stoic Philosophy*. Oxford: Oxford University Press

Boys-Stones, G. 2007. 'Middle Platonists on Fate and Human Autonomy', in Sharples and Sorabji (eds.), vol. II: 431–47

Bréhier, É. 1924–38. *Plotin. Ennéades. Texte grec et traduction française*, 7 vols. Paris: Les Belles Lettres

Brisson, L. (ed.) 2012. *Porphyrye: Sur la Manière dont l'Embryon Reçoit l'Âme*. Paris: Vrin

Brisson, L. et al. 2005. *Porphyre: Sentences*, 2 vols. Paris: Vrin

Brisson, L. and Pradeau, J. F. (eds.) 2003–10. *Plotin Ennéades, Traités*, 9 vols.: *1–6, 7–21, 22–26, 27–29, 30–37, 38–41, 42–44, 45–50, 51–54*. Paris: Flammarion

Burnyeat, M. 1982. 'Idealism and Greek Philosophy: What Descartes saw and Berkeley Missed', *Philosophical Review* 91:3–40

2000. 'Plato on Why Mathematics is Good for the Soul', in Smiley (ed.), 1–81

Busse, A. 1891. *Ammonius In Porphyrii isagogen sive v voces*. Berlin: Reimer

1900. *Eliae In Porphyrii isagogen et Aristotelis Categorias commentaria*. Berlin: Reimer

Butler, E. P. 2005. 'Polytheism and Individuality in the Henadic Manifold', *Dionysius* 23: 83–104

Cain, A. and Lenski, N. (eds.) 2009. *The Power of Religion in Late Antiquity*. Farnham: Ashgate

Cameron, A. and Hoyland, R. G. (eds.) 2011. *Doctrine and Debate in the East Christian World, 300–1500*. Farnham: Ashgate

Champion, M. 2010. 'Cultures and Creation in Late Antique Gaza: Christian and Neoplatonic Interactions', Ph.D. thesis, King's College London

Chappell, T. D. J. 1995. *Aristotle and Augustine on Freedom*. Basingstoke: Macmillan

Charbonnier, L. (ed.) 2012, *Individualität: Genese und Konzeption einer Leitkategorie humaner Selbstdeutung*. Berlin: Berlin University Press

Charles, D. (ed.) 2010. *Definition in Ancient Philosophy*. Oxford: Clarendon Press

Charlton, W. 1994. 'Aristotle on Identity', in Scaltsas, Charles and Gill (eds.), 41–53

Charrue, J. M. 2013. 'Plotin et la liberté humaine', paper presented at the meeting of the International Society for Neoplatonic Studies, Cardiff, Wales, June 2013

Chase, M. 2003. *Simplicius: On Aristotle's Categories 1–4*. London: Duckworth

Cherniss, H. 1935. *Aristotle's Criticism of Presocratic Philosophy*. Baltimore: Johns Hopkins Press

1944. *Aristotle's Criticism of Plato and the Academy*. Baltimore: Johns Hopkins University Press

1962. *Aristotle's Criticism of Plato and the Academy*. New York: Russel & Russel

Chiaradonna, R. (ed.) 2005. *Studi sull'Anima in Plotino*. Naples: Bibliopolis

2008. 'Hylémorphisme et causalité des intelligibles: Plotin et Alexandre d'Aphrodise', *Les Études Philosophiques* 3: 379–97

2012. 'Plotinus' Account of the Cognitive Powers of the Soul: Sense Perception and Discursive Thought', *Topoi* 31: 191–207

2014a. 'Intelligibles as Causes in Plotinus' Metaphysics: *Enn.* 6.7 (38)', in Natali and Viano (eds.), 185–213

2014b. 'Plotinus' Metaphorical Reading of the *Timaeus*: Soul, Mathematics, Providence', in d'Hoine and van Riel (eds.), 187–210

Chiaradonna, R. and Trabattoni, F. (eds.) 2006. *Physics and Philosophy of Nature in Greek Neoplatonism*. Leiden: Brill

Chlup, R. 2012. *Proclus: An Introduction*. Cambridge: Cambridge University Press

Cilento, V. 1947–9. *Enneadi*, 3 vols. Bari: G. Laterza & Figli

1971. *Paideia Antignostica (Enn. III.8, V.8, V.5, II.9)*. Florence: Felice le Monnier

1973. *Saggi su Plotino*. Milan: U. Mursia

Claghorn, G. S. 1954. *Aristotle's Criticism of Plato's Timaeus*. The Hague: Nijjoff

Clark, G. 2000. *Porphyry of Tyre: On Abstinence from Killing Animals*. Ithaca: Cornell University Press

2004. 'Pilgrims and Foreigners: Augustine on Going Home', in Ellis and Kidner (eds.), 149–58

2007. 'Augustine's Porphyry and the Universal Way of Salvation', in Karamanolis and Sheppard (eds.), 126–40

2009. 'The Truth Shall Make You Free: Augustine on the Power of Religion', in Cain and Lenski (eds.), 189–200

Clark, S. 2011. 'Animals in Classical and Late Antique Philosophy', in Beauchamp and Frey (eds.), 35–60
Clarke, E. C., Dillon, J. M. and Hershbell, J. P. 2003. *Iamblichus, De Mysteriis*. Atlanta: Society of Biblical Literature
Cleary, J. J. (ed.) 1999. *Traditions of Platonism: Essays in Honour of John Dillon*. Aldershot: Ashgate
Cochrane, C. N. 1957. *Christianity and Classical Culture*. New York: Oxford University Press
Combès, G. 1959. *La Cité de Dieu, livres VI–X*. Oeuvres de Saint Augustin, Bibliothèque Augustinienne 34. Paris: Institut des Études Augustiniennes
1994. *La Cité de Dieu, livres XI–XVIII*. Oeuvres de Saint Augustin, Nouvelle Bibliothèque Augustinienne 4/1. Paris: Institut des Études Augustiniennes
Cooper, J. M. 2009. 'Chrysippus on Physical Elements', in Salles (ed.), 93–117
2012. *Pursuits of Wisdom: Six Ways of Life in Ancient Philosophy*. Princeton: Princeton University Press
Cornford, F. M. 1956 [1937]. *Plato's Cosmology: The Timaeus of Plato*. London: Routledge
Corrigan, K. 2000. 'Platonism and Gnosticism: The Anonymous Commentary on the Parmenides: Middle or Neoplatonic?', in Majercik and Turner (eds.), 141–77
2004. *Reading Plotinus: A Practical Introduction to Neoplatonism*. West Lafayette, IN: Purdue University Press
2008. '*Ousia* and *Hypostasis* in the Trinitarian Theology of the Cappadocian Fathers: Basil and Gregory of Nyssa', *Zeitschrift für Antikes Christentum* 12: 113–34
Cotnoir, A. and Baxter, D. (eds.) 2014. *Composition as Identity*. Oxford: Oxford University Press
Courcelle, P. 1943. *Les Lettres grecques en occident: De Macrobe à Cassiodore*. Paris: de Boccard
Crocker, J. B. 1956. 'The Freedom of Man in Plotinus', *Modern Schoolman* 34: 23–35
Croke, B. and Harries, J. 1982. *Religious Conflict in Fourth-Century Rome: A Documentary Study*. Sydney: Sydney University Press
D'Ancona, C. 1992. 'ΑΜΟΡΦΟΝ ΚΑΙ ΑΝΕΙΔΕΟΝ: causalité des Formes et causalité de l'Un chez Plotin', *Revue de Philosophie Ancienne* 10: 71–113
2009. 'Modèles de causalité chez Plotin', *Les Études Philosophiques* 3: 361–85
2012. 'Plotin', in Goulet (ed.), 885–1070
Daniélou, J. 1944. *Platonisme et théologie mystique*. Paris: Aubier
Dean-Jones, L. 1994. *Women's Bodies in Classical Greek Science*. Oxford: Clarendon Press
Deck, J. 1967. *Nature, Contemplation and the One: A Study in the Philosophy of Plotinus*. Toronto: University of Toronto Press
De Falco, V. and Klein, U. 1975. *[Iamblichi] Theologumena Arithmeticae*. Stuttgart: Teubner
De Lacy, P. 1992. *Galeni De Semine*. Berlin: Akademie

des Places, É. 1973. *Numénius: Fragments*. Paris: Les Belles Lettres
1977. *Atticus: Fragments*. Paris: Les Belles Lettres
Deuse, W. 1983. *Untersuchungen zur Mittelplatonischen und Neuplatonischen Seelenlehre*. Wiesbaden: Franz Steiner Verlag
Devereux, D. and Pellegrin, P. (eds.) 1990. *Biologie, logique et métaphysique chez Aristote*. Paris: CNRS Éditions
d'Hoine, P. 2006. 'Platonic Problems and Neoplatonic Ideas: A Critical Edition of Proclus *In Parmenidem* III, with Interpretative Essays', Ph.D. thesis, Katholieke Universiteit Leuven
d'Hoine, P. and van Riel, G. (eds.) 2014. *Fate, Providence, and Moral Responsibility in Ancient, Medieval and Early Modern Thought: Collected Studies in Honour of Carlos Steel*. Leuven: Leuven University Press
Diehl, E. 1903–6. *Procli Diadochi in Platonis Timaeum Commentaria*, 3 vols. Leipzig: Teubner
Diels, H. 1882–95. *Simplicii in Aristotelis Physica*. Berolini: Reimer
Dihle, A. 1982. *The Theory of Will in Classical Antiquity*. Berkeley: University of California Press
Dillon, J. 1989. 'Tampering with the *Timaeus*: Ideological Emendations in Plato, with Special Reference to the *Timaeus*', *American Journal of Philology* 110: 50–72
1996a. 'An Ethic for the Late Antique Sage', in Gerson (ed.), 315–35
1996b. *The Middle Platonists*. Ithaca: Cornell University Press
Dillon, J. M. and Gerson, L. 2004. *Neoplatonic Philosophy: Introductory Readings*. Indianapolis: Hackett
Dillon, J. and Long, A. A. (eds.) 1988. *The Question of 'Eclecticism': Studies in Later Greek Philosophy*. Berkeley: University of California Press
Dionysius of Alexandria *apud* Athanasius 1857–66. *De sententia Dionysii*, Patrologia Graeca 25. Paris: Imprimerie Catholique
Dodaro, R. and Lawless, G. (eds.) 2000. *Augustine and his Critics*. London: Routledge
Dodds, E. R. 1963 [1933]. *The Elements of Theology*, 2nd edn. Oxford: Oxford University Press
Donini, P. 1996. 'θεῖα δύναμις in Alessandro di Afrodisia', in Romano and Cardullo (eds.), 11–30; repr. in Donini 2011, 125–38
2011. *Commentary and Tradition: Aristotelianism, Platonism and Post-Hellenistic Philosophy*, ed. M. Bonazzi. Berlin: de Gruyter
Dörrie, H. (†) and Baltes, M. 1998. *Der Platonismus in der Antike*, vol. V: *Die philosophische Lehre des Platonismus (2)*. Stuttgart: Frommann-Holzboog
Dragona-Monachou, M. 1994. 'Divine Providence in the Philosophy of the Early Empire', in Haase and Temporini (eds.), vol. II, 36.7: 4417–4490
Drobner H. R. 2000. 'Gregory of Nyssa as philosopher: De Anima et Resurrectione and De Hominis Opificio', *Dionysius* 18: 69–101
Dyson, R. 1998. *Augustine: The City of God Against the Pagans*. Cambridge Texts in the History of Political Thought. Cambridge: Cambridge University Press

Edwards, M. 2000. *Neoplatonic Saints: The Lives of Plotinus and Proclus by their Students*. Liverpool: Liverpool University Press

2012. 'Astrologie und Freiheit: der fall des Firmicus Maternus', in Charbonnier (ed.), 103–31

2013. 'Christians against Matter: A Bouquet for Bishop Berkeley', in Turner, Corrigan and Rasimus (eds.), 569–80

Eliasson, E. 2008. *The Notion of that which Depends on us in Plotinus, and its Background*. Leiden: Brill

Elingsen, M. 2005. *The Richness of Augustine: His Contextual and Pastoral Theology*. Louisville, KY: Westminster John Knox Press

Ellis, L. and Kidner, F. (eds.) 2004. *Travel, Communication and Geography in Late Antiquity: Sacred and Profane*. Farnham: Ashgate

Emerson, R. W. 1979. *The Collected Works of Ralph Waldo Emerson*, vol. II, ed. A. R. Ferguson, J. Slater and J. F. Carr. Cambridge, MA: Belknap Press

Emilsson, E. K. 1988. 'Critical Comments on John Dillon, "The Mind of Plotinus"', *Proceedings of the Boston Area Colloquium in Ancient Philosophy* 3: 359–63

2007. *Plotinus on Intellect*. Oxford: Clarendon Press

Everson, S. (ed.) 1991. *Companions to Ancient Thought II: Psychology*. Cambridge: Cambridge University Press

Falcon, A. 2012. *Aristotelianism in the First Century BCE: Xenarchus of Seleucia*. Cambridge: Cambridge University Press

Fazzo, S. and Zonta, M. 1999. *Alessandro di Afrodisia. La Provvidenza. Questioni sulla Provvidenza*. Milan: Rizzoli

Ferguson, J. 1969. 'Aspects of Early Christianity in North Africa', in Thomson and Ferguson (eds.), 182–91

Festugière, A. J. 1932. *L'ideal religieux des Grecs et l'Évangile*. Paris: J. Gabalda et cie

Fiedrowicz, M. 2002. 'General Introduction', in Hill 2002, 13–22

Finamore, J. F. and Dillon, J. M. 2002. *Iamblichus De Anima*. Leiden: Brill

Finan, T. 1992. 'Modes of Vision in St Augustine: De Genesi ad Litteram XII', in Finan and Twomey (eds.), 141–54

Finan, T. and Twomey, V. (eds.) 1992. *The Relationship between Neoplatonism and Christianity*. Dublin: Four Courts Press

Fine, K. 1994. 'A Puzzle Concerning Matter and Form', in Scaltsas, Charles and Gill (eds.), 13–40

2002. *The Limits of Abstraction*. Oxford: Clarendon Press

Fitzgerald, A. (ed.) 1999. *Augustine through the Ages: An Encyclopedia*. Grand Rapids, MI: Eerdmans

Fortenbaugh, W. W. and Bodnàr, I. (eds.) 2002. *Eudemus of Rhodes*. London: Transaction Books

Fortenbaugh, W. W., Huby, P., Sharples, R. W. and Gutas, D. (eds.) 1992. *Theophrastus of Eresus: Sources for his Life, Writings, Thought and Influence*, 2 vols. Leiden: Brill (in text FHSG)

Fortenbaugh, W. W. and Steinmetz, P. (eds.) 1989. *Cicero's Knowledge of the Peripatos*. New Brunswick, NJ: Transaction
Frankfurt, H. 1971. 'Freedom of the Will and the Concept of a Person', *Journal of Philosophy* 68(1): 5–20
Frede, M. 1990. 'The Definition of Sensible Substances in *Metaphysics* Z', in Devereux and Pellegrin (eds.), 113–29
2007. 'The *EΦ' HMIN* in Ancient Philosophy', *Philosophia* 37: 110–23
2011. *A Free Will: Origins of the Notion in Ancient Thought*. Berkeley: University of California Press
Fronterotta, F. 2007. *Plotin: Traité 38 (VI, 7)*, in Brisson and Pradeau (eds.), *Plotin Ennéades, traités 38–41*, 15–171
Furley, D. 1999. 'Cosmology', in Algra et al. (eds.), 412–51
Fürst, A., 2011. 'Origen: Exegesis and Philosophy in Early Christian Alexandria', in Lössl and Watt (eds.), 13–32
Gassi, V. 1969. 'Il Battesimo e la Polemica Pelagiana negli Anni 411/13', *Augustinianum* 9: 30–61
Gauthier, R. A. and Jolif, J. Y. 1970. *L'Éthique à Nicomaque*, 4 vols. Paris: Béatrice-Nauwelaerts
Genequand, C. 1984. 'Quelques aspects de l'idée de nature, d'Aristote à al-Ghazālī', *Revue de Théologie et de Philosophie* 116: 105–29
Gerson, L. P. 1993. 'Plotinus's Metaphysics: Emanation or Creation?' *Review of Metaphysics* 46(3): 559–74
1994. *Plotinus*. New York: Routledge
(ed.) 1996. *The Cambridge Companion to Plotinus*. Cambridge: Cambridge University Press
2003. *Knowing Persons: A Study in Plato*. Oxford: Oxford University Press
(ed.) 2010. *The Cambridge History of Philosophy in Late Antiquity*. Cambridge: Cambridge University Press
2012. 'Plotinus on Logos', in Horn and Wilberding (eds.), 17–29
Gill, C. 2006. *The Structured Self in Hellenistic and Roman Thought*. Oxford: Oxford University Press
Ginzberg, L. 1998. *Legends of the Jews*, vol. I. Baltimore: Johns Hopkins University Press
Goulet, R. (ed.) 2012 [1989]. *Dictionnaire des philosophes antiques*, vol. Va. Paris: CNRS Éditions
Gourinat, J. B. 2002. 'Éternel retour et temps périodique dans la philosophie stoïcienne', *Revue Philosophique de la France et de l'Étranger* 127: 213–27
Gregory of Nyssa 1857–66a. *On the Creation of Man*, Patrologia Graeca 44. Paris: Imprimerie Catholique
1857–66b. *De fide*, Patrologia Graeca 45. Paris: Imprimerie Catholique
1857–66c. *On the Soul and Resurrection*, Patrologia Graeca 46. Paris: Imprimerie Catholique
Griffin, M. J. 2013. 'What Has Aristotelian Dialectic to Offer a Neoplatonist? A Possible Sample of Iamblichus at Simplicius *On the Categories*, 12, 10–13, 12', *International Journal of the Platonic Tradition* 6(2): 173–85

2014. 'Hypostasizing Socrates', in Layne and Tarrant (eds.)
Gurtler, G. 2002. 'Providence: The Platonic Demiurge and Hellenistic Causality', in Wagner (ed.), 99–124
Haase, W. and Temporini, H. (eds.) 1972–. *Aufstieg und Niedergang der Römischen Welt*. Berlin: De Gruyter
Hadot, I. 1987a. 'Les Introductions aux commentaires exégètiques chez les auteurs néoplatoniciens et les auteurs chrétiens', in Tardieu (ed.), 99–122
(ed.) 1987b. *Simplicius: sa vie et son oeuvre, sa survie*. Berlin: De Gruyter
1996. *Commentaire sur le Manuel d'Épictète: introduction et edition critique du texte grec*. Cologne: Brill
Hadot, P. 1960. 'Citations de Porphyre chez Augustin', *Revue des Études Augustiniennes* 6: 205–44
1971. 'Causa Sui', in Ritter (ed.), vol. 1, 976–7
1988. *Plotin. Traité 38 (VI,7). Introduction, traduction, commentaire et notes*. Paris: Editions du Cerf
1989. *Plotin ou la Simplicité du Regard*. Paris: Études Augustiniennes
Hahm, D. E. 1977. *The Origins of Stoic Cosmology*. Columbus: Ohio State University Press
Hammerstaedt, J. and Smith, M. F. 2008. 'Diogenes of Oinoanda: The Discoveries of 2008 (NF 142–167)', *Epigraphica Anatolica* 41: 1–37
Hanby, M. 2003. *Augustine and Modernity*. London: Routledge
Hankinson, R. J. 2002. *Simplicius on Aristotle On the Heavens 1.1–4*. London: Duckworth
2004. *Simplicius on Aristotle On the Heavens 1.5–9*. London: Duckworth
Hansberger, R. E., al-Akiti, M. A. and Burnett, C. (eds.) 2012. *Medieval Arabic Thought: Essays in Honour of Fritz Zimmermann*. London: Warburg Institute
Harder, R. 1936. 'Eine neue Schrift Plotins', *Hermes* 71: 1–10
Harder, R., Beutler, R. and Theiler, W. 1960–7. *Plotins Schriften*. Hamburg: F. Meiner
Harrison, S. 2006. *Augustine's Way into the Will*. Oxford: Oxford University Press
Hawley, K. 2014. 'Ontological Innocence', in Cotnoir and Baxter (eds.), 70–89
Hayduck, M. 1882. *Simplicii In Libros Aristotelis De Anima Commentaria*. Berlin: Reimer
(ed.) 1888. *Asclepii in Aristotelis Metaphysicorum Libros A–Z Commentaria*. Berlin: Reimer
1897. *Ioannis Philoponi in Aristotelis De Anima Libros Commentaria*. Berlin: Reimer
Heiberg, I. L. 1894. *Simplicii De Caelo*. Berlin: Reimer
Helmig, C. and Steel, C. 2012. 'Proclus', in *The Stanford Encyclopedia of Philosophy*, ed. Edward N. Zalta, available at plato.stanford.edu/archives/sum2012/entries/proclus/
Henry, P. 1931. 'Le Problême de la liberté chez Plotin', *Revue Néoscolastique* 33: 50–79, 180–215, 318–39
Hibbs, D. 2005. 'Was Gregory of Nyssa a Berkeleyan Idealist?' *British Journal for the History of Philosophy* 13(3): 425–35

Hill, E. 2002. *On Genesis*. The Works of Saint Augustine, a Translation for the 21st Century I/13. New York: New City Press

Hill, J. 2009. 'Gregory of Nyssa, Material Substance and Berkeleyan Idealism', *British Journal for the History of Philosophy* 17(4): 653–83

Hirsch-Luipold, R. (ed.) 2005. *Gott und die Götter bei Plutarch*. Berlin: De Gruyter

Hoffmann, P. 1987a. 'Aspects de la polémique de Simplicius contre Philopon', in Hadot (ed.) 1987b, 183–221

1987b. 'Catégories et langage selon Simplicius – la question du "Skopos" du traité Aristotélicien des Catégories', in Hadot (ed.) 1987b, 61–90

Horn, C. 2007. 'The Concept of Will in Plotinus', in Stern-Gillet and Corrigan (eds.), vol. II, 153–78

2012. 'Aspects of Biology in Plotinus', in Horn and Wilberding (eds.), 214–28

Horn, C. and Wilberding, J. (eds.) 2012. *Neoplatonism and the Philosophy of Nature*. Oxford: Oxford University Press

Hürst, A. 2011. *Von Origenes und Hieronymus zu Augustinus*. Berlin: De Gruyter

Hutchinson, D. M. 2011. 'Apprehension of Thought in *Ennead* 4.3.30', *International Journal of the Platonic Tradition* 5: 262–82

2012 'Sympathy, Awareness, and Belonging to Oneself in Plotinus', in Patterson, Karasmanis and Hermann (eds.), 491–510

Ierodiakonou, K. (ed.) 1999. *Topics in Stoic Philosophy*. Oxford: Clarendon Press

Irwin, T. H. 1988. *Aristotle's First Principles*. Oxford: Clarendon Press

1992. 'Who Discovered the Will?', *Philosophical Perspectives* 6: 453–73

Isaac, D. 1977. *Proclus. Trois études sur la providence I: dix problèmes concernant la providence*. Paris: Les Belles Lettres

Jackson, B. D. 1966. 'Sources of Origen's Doctrine of Freedom', *Church History* 35: 13–23

Jerome. 1995. *Saint Jerome's Hebrew Questions on Genesis*, trans. C. T. R. Hayward. Oxford: Clarendon Press

Joachim, H. H. 1970. *Aristotle: The Nicomachean Ethics*. Oxford: Clarendon Press

Johnson, A. P. 2012. 'Philosophy, Hellenicity, Law: Porphyry on Origen, Again', *Journal of Hellenic Studies* 132: 55–69

2013a. *Religion and Identity in Porphyry of Tyre: The Limits of Hellenism in Late Antiquity*. Cambridge: Cambridge University Press

2013b. 'The Author of the *Against Hierocles*: A Response to Borzì and Jones', *Journal of Theological Studies* 64(2): 574–94

Joly, R. 1970. *Hippocrate: Tome XI: De la Génération; De la Nature de l'Enfant; De Maladies IV*. Paris: Les Belles Lettres

Judson, L. 2010. 'God or Nature? Philoponus on Generability and Perishability', in Sorabji (ed.), 221–37

Kahn, C. 1988. 'Discovering the Will: From Aristotle to Augustine', in Dillon and Long (eds.), 234–59

Kalbfleisch, K. 1895. *Die Neuplatonische fälschlich dem Galen zugeschriebene Schrift πρὸς Γαῦρον περὶ τοῦ πῶς ἐμψυχοῦται τὰ ἔμβρυα*. Berlin: Verlag der königlichen Akademie der Wissenschaften

1907. *Simplicii In Aristotelis Categorias Commentarium*. Berlin: Reimer

Kalligas, P. 1997. 'Forms of Individuals in Plotinus: A Re-Examination', *Phronesis* 42: 207–27

2000. 'Living Body, Soul, and Virtue in the Philosophy of Plotinus', *Dionysius* 18: 25–38

2011. 'The Structure of Appearances: Plotinus on the Constitution of Sensible Objects', *Philosophical Quarterly* 61: 762–82

2014. *Plotinus: A Commentary to the Enneads*, vol. 1. Princeton: Princeton University Press

Karamanolis, G. 2006. *Plato and Aristotle in Agreement? Platonists on Aristotle from Antiochus to Porphyry*. Oxford: Clarendon Press

Karamanolis, G. and Sheppard, A. (eds.), 2007. *Studies on Porphyry*. London: Institute of Classical Studies

Karfík, F. 2011–12. 'L'Âme *logos* de l'intellect et le *logismos* de l'âme: à propos des *Ennéades* v, 1 [10] et iv, 3 [27]'. *Chôra* 9–10: 67–80

2012. 'Le temps et l'âme chez Plotin: à propos des *Ennéades* vi, 5 [23], 11; iv, 4 [28], 15–16; iii, 7 [45], 11', *Elenchos* 33: 227–58

Keech, D. 2012. *The Anti-Pelagian Christology of Augustine of Hippo 396–430*. Oxford: Oxford University Press

Keyser, P. T. 1995. 'The *Onkoi* and *Dunameis* of Plato, *Tim.* 31c–32b', *Echos du Monde Classique/Classical Views* 39, ns 14: 251–66

King, D. 2011. 'The Genesis and Development of a Logical Lexicon in the Syriac Tradition', in Lössl and Watt (eds.), 225–38

Konstan, D. 1989. *Simplicius, On Aristotle Physics 6*. London: Duckworth

Korsgaard, C. 1996. *Creating the Kingdom of Ends*. Cambridge: Cambridge University Press

2009. *Self-Constitution: Agency, Identity, and Integrity*. Oxford: Oxford University Press

Koslicki, K. 2010. *The Structure of Objects*. Oxford: Oxford University Press

Kremer, K. 1987. '*Bonum est Diffusivum Sui*: Ein Beitrag zum Verhältnis von Neuplatonismus und Christentum', in Haase and Temporini (eds.), vol. ii, 36.2: 994–1032

Kroll, W. (ed.), 1899–1901. *Procli Diadochi in Platonis Rem Publicam Commentarii*, 2 vols. Leipzig: Teubner

1902. *Syriani in Metaphysica Commentaria*, CAG 6.1. Berlin: Reimer

Kühner, R. and Gerth, B. 1904. *Ausführliche Grammatik der Griechischen Sprache. Zweiter Teil: Satzlehre. Zweiter Band.* Hannover: Hahn

Laks, A. and Most, G. forthcoming. *Early Greek Philosophy*. Cambridge, MA: Harvard University Press

Lampe, G. W. H. 1961. *A Patristic Greek Lexicon*. Oxford: Clarendon Press

Larsen, B. D. and Dörrie, H. (eds.) 1975. *De Jamblique à Proclus*. Geneva: Entretiens Fondation Hardt

Lavaud, L. 2007. *Traité 39 (VI 8): 'Présentation, traduction et notes'*, in Brisson and Pradeau (eds.), 175–326

2012. 'La métaphore de la liberté. Liberté humaine et liberté divine chez Plotin', *Archives de Philosophie* 75(1): 11–25

Layne, D. and Tarrant, H. (eds.) 2014. *The Neoplatonic Socrates*. Philadelphia: Pennsylvania University Press

Lear, J. 1988. *Aristotle: The Desire to Understand*. Cambridge: Cambridge University Press

Leinkauf, T. and Steel, C. (eds.) 2005. *Plato's Timaeus and the Foundations of Cosmology in Late Antiquity, the Middle Ages and Renaissance*. Leuven: Leuven University Press

Leone, G. 1984. 'Epicuro, *Della Natura*, Libro xiv', *Cronache Ercolanesi* 14: 17–107

Leroux, G. 1990. *Plotin: traité sur la liberté et la volonté de l'Un [Ennéade VI, 8 (39)]*. Paris: J. Vrin

1996. 'Human Freedom in the Thought of Plotinus', in Gerson (ed.), 292–314

Linguiti, A. 2009. 'Plotino e Platonici Sull' Autodeterminazione e la Possibilita di Scelta', in Barbanti and Iozzia (eds.), 213–26

Livingstone, E. (ed.) 1982. *Studia Patristica XVIII*. Oxford: Pergamon Press

Lloyd, A. C. 1990. *The Anatomy of Neoplatonism*. Oxford: Clarendon Press

Long, A. A. 1982. 'Astrology: Arguments Pro and Contra', in Barnes, Brunschwig, Burnyeat and Schofield (eds.), 165–92

1985. 'The Stoics on World-Conflagration and Everlasting Recurrence', *Southern Journal of Philosophy* 23(1): 13–37

1991. 'Representation and the Self in Stoicism', in Everson (ed.), 101–20

Long, A. A. and Sedley, D. N. 1987. *The Hellenistic Philosophers*. Cambridge: Cambridge University Press

Lonie, I. M. 1981. *The Hippocratic Treatises 'On Generation' 'On the Nature of the Child' 'Diseases IV'*. Berlin: De Gruyter

Lössl, J. and Watt, J. W. (eds.) 2011. *Interpreting the Bible and Aristotle in Late Antiquity: The Alexandrian Commentary Tradition between Rome and Baghdad*, Farnham: Ashgate

Luna, C. and Segonds, A. P. 2007–11. *Proclus: commentaire sur le Parménide de Platon*, 3 vols. (to date). Paris: Les Belles Lettres

Lunn-Rockliffe, S. 2013. 'The Diabolical Problem of Satan's First Sin: Self-Moved Pride or a Response to the Goads of Envy?', in Vinzent (ed.) (2013b), 121–40

MacCormack, S. 1999 'Classical Authors', in Fitzgerald (ed.), 202–6

Macdonald, C. and Laurence, S. (eds.) 1998. *Contemporary Readings in the Foundations of Metaphysics*. Oxford: Blackwell

Magrin, S. 2010. 'Sensation and Scepticism in Plotinus', *Oxford Studies in Ancient Philosophy* 39: 249–97

Mahé, J. P. and Poirier, P. H. 2007. *Écrits gnostique, la bibliothèque de Nag Hammadi*. Paris: Gallimard

Majercik, R. 1989. *The Chaldean Oracles*. Leiden: Brill

Majercik, R. and Turner, J. (eds.) 2000. *Gnosticism and Later Platonism: Themes, Figures, and Texts*. Atlanta: Society of Biblical Literature

Mann, W. R. 2000. *The Discovery of Things: Aristotle's Categories and their Context.* Princeton: Princeton University Press

2011. 'Elements, Causes, and Principles: A Context for *Metaphysics* Z 7', *Oxford Studies in Ancient Philosophy* 40: 29–61

Mansfeld, J. 1979. 'Providence and the Destruction of the Universe in Early Stoic Thought: With Some Remarks on the Mysteries of Philosophy', in Vermaseren (ed.), 129–88

1981. 'Bad World and Demiurge: A "Gnostic" Motif from Parmenides and Empedocles to Lucretius and Philo', in van den Broek and Vermaseren (eds.), 261–314

1989. 'Gibt es Spuren von Theophrasts *Phys. Op.* bei Cicero?', in Fortenbaugh and Steinmetz (eds.), 133–58

1991. 'The Idea of the Will in Chrysippus, Posidonius, and Galen', *Boston Area Colloquium in Ancient Philosophy* 7: 107–45

1994. *Prolegomena: Questions to be Settled before the Study of an Author, or a Text.* New York: A. J. Brill

1999. 'Alcinous on Fate and Providence', in Cleary (ed.), 139–50

Mansfeld, J. and Primavesi, O. 2012. *Die Vorsokratiker Griechisch-Deutsch.* Stuttgart: Reclam

Mansfeld, J. and van der Horst, P. W. 1974. *An Alexandrian Platonist against Dualism: Alexander of Lycopolis' Treatise 'Critique of the Doctrines of Manichaeus'.* Leiden: Brill

Markus, R. 1990. '*De Civitate Dei*: Pride and the Common Good', in Schnaubelt and van Fleteren (eds.), 1–16; repr. in Markus (ed.) 1994, 245–59

(ed.) 1994. *Sacred and Secular: Studies on Augustine and Latin Christianity.* Aldershot: Ashgate

1996. *Signs and Meanings: World and Text in Ancient Christianity.* Liverpool: Liverpool University Press

Martin, A. and Primavesi, O. 1999. *L'Empédocle de Strasbourg (P. Strasb. gr. Inv. 1665–1666).* Berlin: De Gruyter

McDiarmid, J. B. 1953. 'Theophrastus on the Presocratic Causes', *Harvard Studies in Classical Philology* 61: 85–156

McLaughlin, B., Beckermann, A. and Walter, S. (eds.) 2009. *The Oxford Handbook of Philosophy of Mind.* Oxford: Clarendon Press

Melinkoff, R. 1979 and 1980. 'Cain's Monstrous Progeny in Beowulf', *Anglo-Saxon England* 8: 142–62 and 9: 183–97

Menn, S. 2001. 'Plotinus on the Identity of Knowledge with its Object', *Apeiron* 34:3: 233–46

Meredith, A. 1999. *Gregory of Nyssa.* London: Routledge

Mesch, W. 2005. 'Plotins Deutung der Platonischen Weltseele', in Leinkauf and Steel (eds.), 41–66

Meyer, M. 2007. *The Nag Hammadi Scriptures.* New York: Harper Collins

Meyer, S. 2011 [1993]. *Aristotle on Moral Responsibility.* Oxford: Oxford University Press

Migne, J. P. 1857–66. *Patrologiae Cursus Completus (Series Graeca) Vol. 44*. Paris: Migne
Moore, G. E. 1912. *Ethics*. London: Oxford University Press
Moraux, P. 1967. 'Alexander von Aphrodisias *Quaest.* 2, 3', *Hermes* 95: 159–69
1984. *Der Aristotelismus bei den Griechen von Andronikos bis Alexander von Aphrodisias, vol. 2: Der Aristotelismus im I. und II. Jh. n. Chr.* Berlin: De Gruyter
Morel, P. M. 2009. 'Comment parler de la nature? Sur le traité 30 de Plotin', *Les Études Philosophiques* 3: 387–403
Morrow, G. 1968. 'Plato's Theory of the Primary Bodies in the *Timaeus* and the Later Doctrine of Forms', *Archiv für Geschichte der Philosophie* 50: 12–28
Mouraviev, S. 2005. 'Zeno's Cosmology and the Presumption of Innocence: Interpretations and Vindications', *Phronesis* 50: 232–49
Movia, G. (ed.) 2003. *Alessandro di Afrodisia e la Metafisica di Aristotele*. Milan: Vita e Pensiero
Mueller, I. 2011. *Simplicius: On Aristotle On the Heavens 1.2–3*. London: Bristol Classical Press
2012. 'Translation of Simplicius, *In phys.* 1.7–9', in Baltussen, Atkinson, Share and Mueller (eds.), 87–156
Mühlenberg, E. 1996. *Gregorii Nysseni Oratio Catechetica: Opera Dogmatica Minora, Pars 4*. Gregorii Nysseni Opera. Leiden: Brill
Narbonne, J. M. 1993. 'Plotin, Descartes et la notion de *Causa Sui*', *Archives de Philosophie* 56: 177–95
2007. 'Divine Freedom in Plotinus and Iamblichus (Tractate VI.8 (39) and De Mysteriis III, 17–20)', in Stern-Gillet and Corrigan (eds.), vol. II, 179–97
2008. 'L'Énigme de la non-descente partielle de l'âme chez Plotin: la piste gnostique/hermétique de l'ὁμοούσιος', *Laval Théologique et Philosophique* 64 (3): 691–708
2011. *Plotinus in Dialogue with the Gnostics*. Leiden: Brill
Natali, C. and Viano, C. (eds.) 2014. *Aitia II. Avec où sans Aristote. Le Débat sur les causes à l'age hellénistique et impérial*. Leuven: Peeters
Nauck, A. 1886. *Porphyrii Philosophi Platonici Opusula Selecta*. Leipzig: Teubner
Nickel, D. 1989. *Untersuchungen zur Embryologie Galens*. Berlin: Akademie
1991. *Galen: Über die Ausformung der Keimlinge (de Foetuum Formatione)*, Corpus Medicorum Graecorum V 3,3. Berlin: Akademie
Noble, C. 2013. 'How Plotinus' Soul Animates his Body: The Argument for the Soul-Trace at *Ennead* IV.4.18.1–9', *Phronesis* 58: 249–79
Nutton, V. 2004. *Ancient Medicine*. New York: Routledge
O'Brien, D. 1977. 'Le Volontaire et la necessité: reflexions sur la descente de l'âme dans la philosophie de Plotin', *Revue Philosophique de la France et de l'Étranger* 167: 401–22
1993. *Theodicée plotinienne, theodicée gnostique*. Leiden: Brill
2012. 'Plotinus on the Making of Matter. Part III: The Essential Background', *International Journal of the Platonic Tradition* 6: 27–80
O'Daly, G. 1973. *Plotinus' Philosophy of the Self*. Shannon: Irish University Press

1983. 'Augustine on the Origin of Souls', in Blume and Mann (eds.), 184–91
1987. *Augustine's Philosophy of Mind*. Berkeley: University of California Press
1999. *Augustine's City of God: A Reader's Guide*. Oxford: Oxford University Press
O'Donnell, J. 1992. *Augustine: Confessions*, 3 vols. Oxford: Oxford University Press
O'Meara, D. J. 1989. *Michaelis Pselli Philosophica Minora. Vol. II: Opuscula Psychologica, Theologica, Daemonologica*. Leipzig: Teubner
1992. 'The Freedom of the One', *Phronesis* 37: 343–9
2003. *Platonopolis: Platonic Political Philosophy in Late Antiquity*. Oxford: Oxford University Press
O'Neill, W. 1965. *Proclus: Alcibiades I: A Translation and Commentary*. The Hague: Nijhoff
Opsomer, J. 2000. 'Proclus on Demiurgy and Procession', in Wright (ed.), 113–43
2005a. 'A Craftsman and his Handmaiden: Demiurgy According to Plotinus', in Leinkauf and Steel (eds.), 67–102
2005b. 'Demiurges in Early Imperial Platonism', in Hirsch-Luipold (ed.), 51–99
Opsomer, J. and Steel, C. 2012. *Proclus: Ten Problems Concerning Providence*. London: Duckworth
Origen. 1831–48. *Origenis Opera Omnia*, ed. C. H. E. Lommatzsch. Berlin: Haude & Spener
Origen, Koetschau, P., Butterworth, G. W., Rufinus and de Lubac, H. 1973. *Origen on First Principles: Being Koetschau's Text of the De Principiis*. Gloucester, MA: Peter Smith
Osborne, C. 1987. *Rethinking Early Greek Philosophy*. London: Duckworth
2006. *On Aristotle Physics 1.1–3*. London: Duckworth
2009. *On Aristotle Physics 1.4–9*. London: Duckworth
Osborn, E. F. 1997. *Tertullian, First Theologian of the West*. Cambridge: Cambridge University Press
Ousager, A. 2004. *Plotinus, on Selfhood, Freedom, and Politics*. Aarhus: Aarhus University Press
Parfit, D. 1984. *Reasons and Persons*. Oxford: Oxford University Press
Patterson, R., Karasmanis, V. and Hermann, A. 2012. *The Presocratics and Plato: A Festschrift at Delphi in Honor of Professor Charles H. Kahn*. Zurich and Las Vegas: Parmenides Publishing
Peramatzis, M. 2011. *Priority in Aristotle's Metaphysics*. Oxford: Oxford University Press
Petty, R. 2012. *Fragments of Numenius of Apamea*. Westbury: Prometheus Trust
Phillips, J. 2002. 'Plato's *Psychogonia* in Later Platonism', *Classical Quarterly* 52: 231–47
Philoponus. 1899. *De Aeternitate Contra Proclum*, ed. H. Rabe. Leipzig: Teubner
2004. *Against Proclus on the Eternity of the World 1–5.*, trans. M. Share. London: Duckworth
2005. *Against Proclus on the Eternity of the World 6–8*, trans. M. Share. London: Duckworth

2006. *Against Proclus on the Eternity of the World 12–18*, trans. J. Wilberding. London: Duckworth

2010. *Against Proclus on the Eternity of the World 9–11*, trans. M. Share. London: Duckworth

Pink, T. and Stone, M. W. F. (eds.) 2004. *The Will and Human Action*. London: Routledge

Pistelli, H. and Klein, U. 1975. *Iamblichi In Nicomachi Arithmeticam Introductionem Liber*. Stuttgart: Teubner

Plotinus. 1951–73. *Plotini Opera*, ed. P. Henry and H. R. Schwyzer, 3 vols. Paris: Desclée de Brouwer

1964–82. *Plotini Opera*, ed. P. Henry and H. R. Schwyzer, 3 vols. Oxford: Clarendon Press

Pollmann, K. 2007. 'Augustine, Genesis, and Controversy', *Augustinian Studies* 38 (1): 203–16

Powers, N. 2013. 'Plato's Demiurge as Precursor to the Stoic Providential God', *Classical Quarterly* 63(2), 713–22

Proclus. 1903. *Procli Diadochi in Platonis Timaeum Commentaria*, ed. E. Diehl. Leipzig: Teubner

Procopius of Gaza. 1857–66. *Commentary on Genesis*, Patrologia Graeca 87. Paris: Imprimerie Catholique

Quandt, K. 1983. 'Αἱ γὰρ τῶν ἐναντίων ἀποδείξεις ἀπορίαι περὶ τῶν ἐναντίων εἰσίν. Philosophical Program and Expository Practice in Aristotle', *Classical Antiquity* 2: 279–98

Rappe, S. 2002. 'Explanation and Nature in Enneads VI.7.1–15', in Wagner (ed.), 71–98

Rashed, M. 2007. *Essentialisme: Alexandre d'Aphrodise entre logique, physique et cosmologie*. Berlin: de Gruyter

2011. *Alexander d'Aphrodise, commentaire perdu à la Physique d'Aristote (livres 4–8): les scholies byzantines*. Berlin: de Gruyter

Remes, P. 2007. *Plotinus on Self, the Philosophy of the We*. Cambridge: Cambridge University Press

2008. 'Inwardness and Infinity of Selfhood: From Plotinus to Augustine', in Remes and Sihvola (eds.), 155–76

Remes, P. and Sihvola, J. (eds.) 2008. *Ancient Philosophy of the Self.* Dordrecht: Springer

Remes, P. and Slaveva-Griffin, S. 2015. *The Routledge Handbook of Neoplatonism*. Stockfield: Acumen

Richard, M. 2011. 'Dyophysite Florilegia of the Fifth and Sixth Centuries CE', in Cameron and Hoyland (eds.), 321–46

Riedweg, C. and Gritti, E. 2010. 'Echi dal *Timeo* nelle Aporie sull'Impassibilità dell'Anima in *Enneadi* III.6, 1–5, Frutti di una *Synousia* Plotiniana', *Elenchos* 31: 123–50

Rist, J. M. 1967. *Plotinus: The Road to Reality*. Cambridge: Cambridge University Press

1969. 'Augustine on Freewill and Predestination', *Journal of Theological Studies* 20: 20–47

1975. 'Prohairesis, Plotinus, Philo and Origen', in Larsen and Dörrie (eds.), 103–22

1994. *Augustine: Ancient Thought Baptized.* Cambridge: Cambridge University Press

2000. 'On the Platonism of Gregory of Nyssa', *Hermathena* 169: 129–51

Ritter, J. (ed.) 1971 *Historisches Wörterbuch der Philosophie.* Basel: Schwabe

Romano, F. 1979. *Porfirio di Tiro: Filosofia e cultura nel III secolo D.C.* Catania: Università di Catania

1999. 'Azione Morale e Libero Arbitrio in Plotino – La Virtu non ha Padrone (arête adespoton)', in Vegetti and Abbate (eds.), 151–91

Romano, F. and Cardullo, R. L. (eds.) 1996. *Dunamis nel Neoplatonismo.* Florence: La Nuova Italia

Rudder Baker, L. 2009. 'Non-Reductive Materialism', in McLaughlin, Beckermann and Walter (eds.), 109–27

Runia, D. 1989. 'Festugière Revisited: Aristotle in the Greek Patres', *Vigiliae Christianae* 43: 1–34

Runia, D. T. and Share, M. 2008. *Proclus. Commentary on Plato's Timaeus, vol. II, Book 2: Proclus on the Causes of the Cosmos and its Creation.* Cambridge: Cambridge University Press

Rutten, C. 1956. 'La Doctrine des deux actes dans la philosophie de Plotin', *Revue Philosophique* 146: 100–6

Salles, R. 2005. 'Ekpúrosis and the Goodness of God in Cleanthes', *Phronesis* 50: 56–78

(ed.) 2009a. *God and Cosmos in Stoicism.* Oxford: Oxford University Press

2009b. 'Chrysippus on Conflagration and the Indestructibility of the Cosmos', in Salles (ed.) 2009a, 118–34

in progress. *Conflagration and Everlasting Recurrence in Stoicism and Greek Cosmology*

Sallustius 1926. *Concerning the Gods and the Universe*, ed. and trans. A. D. Nock. Cambridge: Cambridge University Press

Scaltsas, T., Charles, D. and Gill, M. L. (eds.) 1994. *Unity, Identity, and Explanation in Aristotle's Metaphysics.* Oxford: Clarendon Press

Schaff, P. and Wace, H. 1892. *A Select Library of Nicene and Post-Nicene Fathers of the Christian Church: Second Series. Volume 5: Gregory of Nyssa.* Grand Rapids, MI: W. B. Eerdmans 1954

Schiaparelli, A. 2010. 'Essence and Cause in Plotinus' *Ennead* VI.7 [38] 2: An Outline of Some Problems', in Charles (ed.), 467–92

Schibli, H. S. 2002. *Hierocles of Alexandria.* Oxford: Oxford University Press.

Schmid, W. 1936. *Epikurs Kritik der platonischen Elementenlehre.* Leipzig: Kommissions-Verlag Otto Harrassowitz

Schnaubelt, J. and van Fleteren, F. (eds.) 1990. *Augustine: 'Second Founder of the Faith'*, Collecteana Augustiniana. New York: Peter Lang

Schniewind, A. 2003. *L'Ethique du sage selon Plotin: le paradigme du spoudaios*. Paris: Vrin
Scholten, C. 1996. *Antike Naturphilosophie*. Berlin: De Gruyter
Schroeder, F. M. 1992. *Form and Transformation: A Study in the Philosophy of Plotinus*. Montreal: McGill
Schwyzer, H. R. 1935. 'Plotins Interpretation von *Timaios* 35 a', *Rheinisches Museum* 74: 360–8
1974. 'Plotinisches und Unplotinisches in den Aphormai des Porphyrios', in *Plotino e il Neoplatonismo in Oriente e in Occidente*, Rome: Accademia Nazionale dei Lincei, 221–52
Scott, A. 1991. *Origen and the Life of the Stars*. Oxford: Clarendon Press
Sedley, D. 1998. *Lucretius and the Transformation of Greek Wisdom*. Cambridge: Cambridge University Press
2007. *Creationism and its Critics in Antiquity*. Berkeley: University of California Press
Segonds, A, 1989. 'Ainéas de Gaza', in Goulet (ed.), 82–7
Share, M. and Atkinson, M. 2012. 'Translation of Simplicius, *In phys.* 1.6', in Baltussen, Atkinson, Share, Mueller (eds.), 30–49
Sharples, R. W. 1982. 'Alexander of Aphrodisias on Divine Providence: Two Problems'. *Classical Quarterly* 32: 198–211
1983. *Alexander of Aphrodisias on Fate*. London: Duckworth
1990. 'The School of Alexander', in Sorabji (ed.), 83–111
1992. *Alexander of Aphrodisias: Quaestiones 1.1–2.15*. London: Duckworth
1995. 'Counting Plato's Principles', in Ayres (ed.), 67–82
1998. *Theophrastus of Eresus: Sources for his Life, Writings, Thought and Influence*, Commentary vol. III.1, Sources on Physics, Leiden: Brill
2002. 'The *Physics* of Eudemus of Rhodes', in Fortenbaugh and Bodnár (eds.), 107–26
2003. 'Threefold Providence: The History and Background of a Doctrine', in Sharples and Sheppard (eds.), 107–27
2010. *Peripatetic Philosophy 100 BC to AD 200: An Introduction and Collection of Sources in Translation*. Cambridge: Cambridge University Press
Sharples, R. W. and Sheppard, A. (eds.) 2003. *Ancient Approaches to the 'Timaeus'*. London: Institute of Classical Studies
Sharples, R. W. and Sorabji, R. (eds.) 2007. *Greek and Roman Philosophy: 100 BC–200 AD*. London: Institute for Classical Studies
Sharples, R. W. and van der Eijk, P.W. (eds.) 2008. *Nemesius of Emesa, On Human Nature*. Liverpool: Liverpool University Press
Sheppard, A. 2007. 'Porphyry's Views on Phantasia', in Karamanolis and Sheppard (eds.), 71–6
Silverman, A. 1992. 'Timaean Particulars', *Classical Quarterly* 42(1): 87–113
Simplicius 1992. *Simplicius: Corrolaries on Place and Time*, trans. J. O. Urmson, annotated by Lucas Siorvanes. London: Duckworth
Siorvanes, L. 1997. *Proclus: Neo-Platonic Philosophy and Science*. New Haven: Yale University Press

Smiley, T. (ed.) 2000. *Mathematics and Necessity: Essays in the History of Philosophy*. Oxford: Oxford University Press

Smith, A. 1974. *Porphyry's Place in the Neoplatonic Tradition*. The Hague: Martinus Nijhoff

1984. 'Did Porphyry Reject the Transmigration of Human Souls into Animals?', *Rheinisches Museum für Philologie* 127: 276–84

1987. 'Porphyrian Studies since 1913', in Haase and Temporini (eds.), vol. II 36.2: 717–73

1993. *Porphyrius: Fragmenta*. Leipzig: Teubner

Smith, A. and Alt, K. (eds.) 2005. *The Philosopher and Society in Late Antiquity: Essays in Honour of Peter Brown*. Swansea: Classical Press of Wales

Solmsen, F. 1960. *Aristotle's System of the Physical World: A Comparison with his Predecessors*. Ithaca: Cornell University Press

Sorabji, R. 1983. *Time, Creation, and the Continuum*. London: Duckworth

(ed.) 1987. *Philoponus and the Rejection of Aristotelian Science*. London: Duckworth

1988. *Matter, Space and Motion*. London: Duckworth

(ed.) 1990. *Aristotle Transformed: The Ancient Commentators and their Influence*. London: Duckworth

1993. *Animal Minds and Human Morals*. London: Duckworth

2004a. 'The Concept of the Will from Plato to Maximus the Confessor', in Pink and Stone (eds.), 6–28

2004b. *The Philosophy of the Commentators, 200–600 AD*, vol. 2: *Physics*. London: Duckworth

2006. *Self: Ancient and Modern Theories about Individuality, Life, and Death*. Oxford: Clarendon Press

(ed.) 2010. *Philoponus and the Rejection of Aristotelian Science*. London: Institute of Classical Studies, School of Advanced Study, University of London

2012. 'Introduction', in Baltussen, Atkinson, Share and Mueller (eds.), 1–12

Spataro, L. 1994. 'Il Linguaggio della Libertà in *Enneadi* VI, 8', *Verifiche* 23(3–4): 329–73

Steel, C. 1997. *'Simplicius' on Aristotle On the Soul 2.5–12*. London: Duckworth

2007. *Proclus: On Providence*. London: Duckworth

2007–9. *Procli in Platonis Parmenidem Commentaria*, 3 vols. Oxford: Oxford University Press

Steel, C. and Ritups, A. 2013. *'Simplicius': On Aristotle On the Soul 3.6–13*. Bristol: Bristol Classical Press

Steenberg, M. C. 2009. *Of God and Man*. London: T. & T. Clark

Stern-Gillet, S. and Corrigan, K. (eds.) 2007. *Reading Ancient Texts: Essays in Honour of D. O'Brien*. Leiden: Brill

Stratton, G. M. (ed.) 1964. *Theophrastus and the Greek Physiological Psychology before Aristotle*. Amsterdam: E. J. Bonset

Strobel, B. (ed.) forthcoming. *The Art of Philosophical Exegesis*. Stuttgart: Veröffentlichungen der Karl und Gertrud Abel-Stiftung

Tardieu, M. (ed.) 1987. *Les Règles de l'interpretation*. Paris: Éditions du Cerf

Taylor, A. E. 1962 [1928]. *A Commentary on Plato's Timaeus*. Oxford: Oxford University Press
Taylor, J. V. 1963. *The Primal Vision*. London: SCM Press
Thaler, N. 2011. 'Traces of Good in Plotinus's Philosophy of Nature: *Ennead* VI.7.1–14', *Journal of the History of Philosophy* 49(2): 161–80
Theognostus *apud* Athanasius. 1857–66. *De Decretis Nicaenae Synodi*, Patrologia Graeca 25. Paris: Imprimerie Catholique
Thillet, P. 2003. *Alexandre d'Aphrodise: traité de la Providence*. Paris: Verdier
Thomassen, E. and Painchaud, L. 1989. *Le Traité Tripartite (NH I, 5)*. Quebec: Presses de L'Université Laval
Thomson, L. and Ferguson, J. (eds.) 1969. *Africa in Classical Antiquity*. Ibadan: Ibadan University Press
Tornau, C. 2001. *Plotin: Ausgewählte Schriften*. Stuttgart: Reclam
2005. 'Plotinus' Criticism of Aristotelian Entelechism in *Enn.* IV 7 [2], 85.25–50', in Chiaradonna (ed.), 149–78
Tracy, T. T. 1994. *The God Who Acts: Philosophical and Theological Explorations*. Philadelphia: Pennsylvania University Press
Trouillard, J. 1949. 'La Liberté selon Plotin', *Actes du IVè Congrès de Philosophie de Langue Française*. Neuchatel: LaBaconnière, 353–7
1955. *La Procession Plotinienne*. Paris: Presses Universitaires de France
Turner, J. D. 2001. *Sethian Gnosticism and the Platonic Tradition*. Quebec: Presses de L'Université Laval
Turner, J. D., Corrigan, K. and Rasimus, T. (eds.) 2013. *Gnosticism, Platonism and the Late Ancient World: Essays in Honour of John D Turner*. Leiden: Brill
Unger, P. 1979. 'There Are No Ordinary Things', *Synthese* 41(2): 117–54
Urmson, J. O. and Lautner, P. 1995. *Simplicius on Aristotle On the Soul 1.1–2.4*. London: Duckworth
Van den Broek, R. and Vermaseren, M. J. (eds.) 1981. *Studies in Gnosticism and Hellenistic Religions*. Leiden: Brill
Van Inwagen, P. 1995. *Material Beings*. Ithaca: Cornell University Press
van Roey, A. 1984. 'Un Traité Cononite contre Jean Philopon sur la Résurrection', in *ANTIΔΩPON: Hommage à Maurits Geerard*, Wetteren: Cultura, 123–39
Vegetti, M. and Abbate, M. (eds.) 1999. *La Republica di Platone nella Tradizione Antica*. Naples: Bibliopolis
Verde, F. 2013. *Elachista: La Dottrina dei Minimi nell'Epicureismo*. Leuven: Leuven University Press
Vermaseren, M. J. (ed.) 1979. *Studies in Hellenistic Religions*. Leiden: Brill
Vessey, M. 2012a. 'The History of the Book: Augustine's *City of God* and Post-Roman Cultural Memory', in Wetzel (ed.) 2012a, 14–32
(ed.) 2012b. *A Companion to Augustine*. Oxford: Blackwell
Vinzent, M. (ed.) 2013a. *Studia Patristica 58*. Leuven: Peeters
(ed.) 2013b. *Studia Patristica 63*. Leuven: Peeters
Vitelli, H. 1887. *Ioannis Philoponi in Aristotelis Physicorum Libros Tres Priores Commentaria*, CAG 16. Berlin: Reimer

1897. *Ioannis Philoponi in Aristotelis Libros De Generatione et Corruptione Commentaria*. CAG 14.2. Berlin: Reimer

Vlastos, G. 1939–40. 'The Disorderly Motion in the *Timaeus*', *Classical Quarterly* 33–4: 71–83

1967. 'Plato's Supposed Theory of Irregular Atomic Figures', *Isis* 57: 204–9; repr. in Vlastos (ed.) 1973, 387–91

(ed.) 1973. *Platonic Studies*. Princeton: Princeton University Press

Wagner, M. F. (ed.) 2002. *Neoplatonism and Nature: Studies in Plotinus' Enneads*. Albany: State University of New York Press

Wallis, R. T. 1995. *Neoplatonism*. London: Duckworth

Watts, E. 2005. 'An Alexandrian Christian Response to Fifth-century Neoplatonic Influence', in Smith and Alt (eds.), 215–29

2006. *City and School in Late Antique Athens and Alexandria*. Berkeley: University of California Press

2009. 'The Enduring Legacy of the Iatrosophist Gessius', *Greek, Roman and Byzantine Studies* 49: 113–33

2010. *Riot in Alexandria: Tradition and Group Dynamics in Late Antique Pagan and Christian Communities*. Berkeley: University of California Press

Weismann, F. J. 1997. 'Plotino e il Tema de la Libertad', *Revista Agustiniana* 38: 1163–97

Westerink, L. G. 1956. *Olympiodorus: Commentary on the First Alcibiades of Plato*. Amsterdam: North-Holland

1975. *The Greek Commentaries on Plato's Phaedo, Vol. 1 – Olympiodorus*. Amsterdam: North-Holland

2010 [1962]. *Anonymous Prolegomena to Platonic Philosophy*. Westbury: Prometheus Trust

Westerink, L. G. and Combès, J. 1986–1991. *Damascius: traité des Premiers Principes*, 3 vols. Paris: Les Belles Lettres

2002–3. *Damascius: commentaire du Parménide de Platon*, 4 vols. Paris: Les Belles Lettres

Westerink, L. G. and Saffrey, H. D. 1968. *Proclus, théologie platonicienne*. Paris: Les Belles Lettres

Westra, L. 1990. *Plotinus on Freedom, a Meditation on Ennead VI, 8*. Lewiston, NY: E. Mellen Press

2002. 'Freedom and Providence in Plotinus', in Wagner (ed.), 125–47

Wetzel, J. 2000. 'Augustine on Free Will and Predestination', in Dodaro and Lawless (eds.), 124–41

(ed.) 2012a. *Augustine's City of God: A Critical Guide*. Cambridge: Cambridge University Press

2012b. 'Augustine on the Will', in Vessey 2012b (ed.), 339–52

2012c. 'Augustine on the Origin of Evil: Myth and Metaphysics', in Wetzel 2012a, 167–85

Whittaker, J. 1990. *Alcinoos: enseignement des doctrine de Platon*. Paris: Les Belles Lettres

Wilberding, J. 2006. *Plotinus' Cosmology: A Study of Ennead II.1 (40)*. Oxford: Oxford University Press

2008a. 'Automatic Action in Plotinus', *Oxford Studies in Ancient Philosophy* 34: 373–407

2008b. 'Porphyry and Plotinus on the Seed', *Phronesis* 53: 406–32

2011a. *Porphyry: To Gaurus on How Embryos are Ensouled and On What is in our Power*. Bristol: Bristol Classical Press

2011b. 'Intelligible Kinds and Natural Kinds in Plotinus', *Études Platoniciennes* 8: 53–73

2012. 'Neoplatonists on "Spontaneous" Generation', in Wilberding and Horn (eds.), 197–213

Wilberding, J. and Horn, C. (eds.) 2012. *Neoplatonism and the Philosophy of Nature*. Oxford: Oxford University Press

Wildberg, C. 1987. *Philoponus Against Aristotle on the Eternity of the World*. Ithaca: Cornell University Press

2006. 'A World of Thoughts: Plotinus on Nature and Contemplation (ENN. III.8 [30] 1–6)', in Chiaradonna and Trabattoni (eds.), 121–43

Williams, N. P. 1927. *The Ideas of the Fall and Original Sin*. London: Longmans, Green & Company

Williams, R. 2000. 'Insubstantial Evil', in Dodaro and Lawless (eds.), 105–23

1999. 'Creation', in Fitzgerald (ed.), 251–4

Wolfson, H. 1970. 'The Identification of *Ex Nihilo* with Emanation in Gregory of Nyssa', *Harvard Theological Review* 63: 53–60

Wright, M. R. (ed.) 2000. *Reason and Necessity: Essays on Plato's Timaeus*. London: Duckworth

Yudin, V. 2013. 'A Stoic Conversion: Porphyry by Plato. Augustine's Reading of the Timaeus 41a7–b6', in Vinzent (ed.) 2013a, 127–79

Zacharias. 1973. *Ammonio*, ed. M. Colonna. Naples: Tipolitografia "La Buona Stampa"

2008. *The Life of Severus*, trans. L. Ambjörn. Piscataway, NJ: Gorgias Press

Zacharias and M. A. Kugener (ed.) 1903. *Vie de Sévère*. Paris: Firmin-Didot et Cie

Zachhuber, J. 2000. *Human Nature in Gregory of Nyssa*. Leiden: Brill

2005. 'Once Again: Gregory of Nyssa on Universals', *Journal of Theological Studies* 56(1): 30–74

n.d. 'Gregory of Nyssa, Contra Eunomium III/4', unpublished manuscript.

Zalta, E. 2006. 'Essence and Modality', *Mind* 115(459): 659–93

Zambon, M. 2002. *Porphyre et le Moyen-Platonisme*. Paris: Vrin

Zekl, H. G. 2010. *Proklos Diadochos: Kommentar zum Platonischen Parmenides*. Würzburg: Köningshausen und Neumann

Zimmerman, B. 2013. 'Does Plotinus Present a Philosophical Account of Creation?', *Review of Metaphysics* 67(1): 55–105

Index locorum

General index